Peter Deuflhard
Andreas Hohmann

Numerical Analysis

A First Course in Scientific Computation

Translated from the German by F. A. Potra and F. Schulz

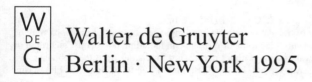
Walter de Gruyter
Berlin · New York 1995

Authors

Peter Deuflhard
Andreas Hohmann

Konrad-Zuse-Zentrum für   and   Freie Universität Berlin
Informationstechnik Berlin   Institut für Mathematik
Heilbronner Str. 10   Arnimallee 2−6
D-10711 Berlin   D-14195 Berlin
Germany   Germany

1991 Mathematics Subject Classification: Primary: 65−01
Secondary: 65 Bxx, 65 Cxx, 65 Dxx, 65 Fxx, 65 Gxx

Title of the German original edition: Numerische Mathematik I, Eine algorithmisch
orientierte Einführung, 2. Auflage, Walter de Gruyter · Berlin · New York, 1993

With 62 figures and 14 tables.

⊗ Printed on acid-free paper which falls within the guidelines of the ANSI to ensure permanence and durability.

Library of Congress Cataloging-in-Publication-Data

Deuflhard, P. (Peter)
    [Numerische Mathematik I. English]
    Numerical analysis : a first course in scientific computation / Peter
Deuflhard, Andreas Hohmann ; translated by F. A. Potra and F. Schulz.
        p.   cm.
    Includes bibliographical references (p.   −   ) and index.
    ISBN 3-11-014031-4 (cloth : acid-free). −
    ISBN 3-11-013882-4 (pbk. : acid-free)
        1. Numerical analysis − Data processing.   I. Hohmann, Andreas,
    1964−   . II. Title.
    QA297.D4713   1995
    519.4−dc20   94-46993
                                                                            CIP

Die Deutsche Bibliothek − Cataloging-in-Publication-Data

Numerical analysis / Peter Deuflhard ; Andreas Hohmann. Transl. by
F. A. Potra and F. Schulz. − Berlin ; New York : de Gruyter.
    (De Gruyter textbook)
    Bd. 2 verf. von Peter Deuflhard und Folkmar Bornemann
NE: Deuflhard, Peter; Hohmann, Andreas; Bornemann, Folkmar
1. A first course in scientific computation. − 1995
    ISBN 3-11-013882-4 kart.
    ISBN 3-11-014031-4 Pp.

Preface

The topic of Numerical Analysis is the development and the understanding of computational methods for the numerical solution of mathematical problems. Such problems typically arise from areas outside of Mathematics — such as Science and Engineering. Therefore Numerical Analysis is directly situated at the confluence of Mathematics, Computer Science, Natural Sciences, and Engineering. A new interdisciplinary field has been evolving rapidly called *Scientific Computing*. Driving force of this evolution is the recent vigorous development of both computers and algorithms, which encouraged the refinement of mathematical models for physical, chemical, technical or biological phenomena to such an extent that their computer simulations are now describing reality to sufficient accuracy. In this process, the complexity of solvable problems has been expanding continuously. New areas of the natural sciences and engineering, which had been considered rather closed until recently, thus opened up. Today, Scientific Computing is contributing to numerous areas of industry (chemistry, electronics, robotics, automotive industry, air and space technology, etc.) as well as to important problems of society (balance of economy and ecology in the use of primary energy, global climate models, spread of epidemics).

The movement of the entire interdisciplinary net of Scientific Computing seizes each of its knots, including Numerical Analysis, of course. Consequently, fundamental changes in the selection of the material and the presentation in lectures and seminars have occurred — with an impact even to introductory courses. The present book takes this development into account. It is understood as an introductory course for students of mathematics and natural sciences, as well as mathematicians and natural scientists working in research and development in industry and universities. Possible readers are assumed to have basic knowledge of undergraduate Linear Algebra and Calculus. More advanced mathematical knowledge, say about differential equations, is not a required prerequisite, since this elementary textbook is intentionally excluding the numerics of differential equations. As a further deliberate restriction, the presentation of topics like interpolation or integration is limited to the one–dimensional case. Nevertheless, many essential concepts of modern Numerical Analysis, which play an important role in numerical differential equation solving, are treated on the simplest possible model problems.

The aim of the book is to develop algorithmic feeling and thinking. After all, the algorithmic approach is historically one of the roots of todays Mathematics. That is why historical names like Gauss, Newton and Chebyshev are found in numerous places of the subsequent text together with contemporary names. The orientation towards algorithms should by no means be misunderstood. In fact, the most efficient algorithms do require a substantial amount of mathematical theory, which will be developed in the text. As a rule, elementary mathematical arguments are preferred. Wherever meaningful, the reasoning appeals to geometric intuition — which also explains the quite large number of graphical representations. Notions like scalar product and orthogonality are used throughout — in the finite dimensional case as well as in function spaces. In spite of the elementary presentation, the book does contain a significant number of rather recent results, some of which have not been published elsewhere. In addition, some of the more classical results are derived in a way, which significantly differs from more standard derivations.

Last, but not least, the selection of the material expresses the scientific taste of the authors. The first author has taught Numerical Analysis courses since 1978 at different German institutions such as the University of Technology in Munich, the University of Heidelberg, and now the Free University of Berlin. Over the years he has co–influenced the dynamic development of Scientific Computing by his research activities. Needless to say, he has presented his research results in numerous invited talks at international conferences and seminars at renowned university and industry places all over the world. The second author rather recently entered the field of Numerical Analysis, after having graduated in pure mathematics from the University of Bonn. Both authors hope that this combination of a senior and a junior has had a stimulating effect on the presentation in this book. Moreover, it is certainly a clear indication of the old dream of unity of pure and applied mathematics.

Of course, the authors stand on the shoulders of others. In this respect, the first author remembers with gratitude the time, when he was a graduate student of Roland Bulirsch. Numerous ideas of the colleagues Ernst Hairer and Gerhard Wanner (University of Geneva) and intensive discussions with Wolfgang Dahmen (Technical University of Aachen) have influenced our presentation. Cordial thanks go to Folkmar Bornemann for his many stimulating ideas and discussions especially on the formulation of the error analysis in Chapter 2. We also want to thank our colleagues at the Konrad Zuse Center Berlin, in particular Michael Wulkow, Ralf Kornhuber, Ulli Nowak and Karin Gatermann for many suggestions and a constructive atmosphere.

This book is a translation of our German textbook "Numerische Mathematik I (Eine algorithmisch orientierte Einführung)", second edition. Many thanks to our translators, Florian Potra and Friedmar Schulz, and to Erlinda Cadano–Körnig for her excellent work in the final polishing of the English version. May this version be accepted by the Numerical Analysis students equally well as the original German version.

Peter Deuflhard and Andreas Hohmann   Berlin, May 1994

Teaching Hints

The present textbook addresses students of Mathematics, Computer Science and Science covering typical material for introductory courses in Numerical Analysis with clear emphasis towards Scientific Computing.

We start with Gaussian elimination for linear equations as a classical algorithm and discuss additional devices such as pivoting strategies and iterative refinement. Chapter 2 contains the indispensable error analysis based on the fundamental ideas of Wilkinson. The condition of a problem and the stability of algorithms are presented in a unified framework and exemplified by illustrative cases. Only the linearized theory of error analysis is presented — avoiding, however, the typical "ε–battle". Rather, only differentiation is needed as an analytical tool. As a specialty we derive a stability indicator which allows for a rather simple classification of numerical stability. The theory is then worked out for the case of linear equations, thus supplying a posteriori a deeper understanding of Chapter 1. In Chapter 3 we deal with methods of orthogonalization in connection with linear least squares problems. We introduce the extremely useful calculus of pseudoinverses, which is then immediately applied in Chapter 4. There, we consider iterative methods for systems of nonlinear equations (Newton's method), nonlinear least squares problems (Gauss–Newton method) and parameter–dependent problems (continuation methods) in close mutual connection. Special attention is given to the affine invariant form of the convergence theory and the iterative algorithms. A presentation of the power method (direct and inverse) and the QR–algorithm for symmetric eigenvalue problems follow in Chapter 5. The restriction to the real symmetric case is motivated from the beginning by a condition analysis of the general eigenvalue problem. In this context the singular value decomposition fits perfectly, which is so important in applications.

After the first five rather closely connected chapters the remaining four chapters also comprise a closely connected sequence. The sequence begins in Chapter 6 with an extensive treatment of the theory of three–term recurrence relations, which play a key role in the realization of orthogonal projections in function spaces. Moreover, the significant recent spread of symbolic computing has renewed interest in special functions also within Numerical Analysis.

The condition of three–term recurrences is presented via the discrete Green's function. Numerical algorithms for the computation of special functions are exemplified for spherical harmonics and Bessel functions. In Chapter 7 classical interpolation and approximation in the one–dimensional special case are presented first, followed by non–classical methods like Bézier techniques and splines, which nowadays play a central role in CAD (Computer Aided Design) or CAGD (Computer Aided Geometric Design), i.e. special disciplines of computer graphics. Our presentation in Chapter 8, which treats iterative methods for the solution of large symmetric linear equations, is conveniently based on Chapter 6 (three–term recurrences) and Chapter 7 (min–max property of Chebyshev polynomials). The same is true for the Lanczos algorithm for large symmetric eigenvalue problems. The final Chapter 9 turns out to be a bit longer: it carries the bulk of the task to explain principles of the numerical solution of differential equations by means of the simplest problem type, which here is numerical quadrature. After the historical Newton–Cotes formulas and the Gauss quadrature, we progress towards the classical Romberg quadrature as a first example of an adaptive algorithm, which, however, only adapts the approximation order. The formulation of the quadrature problem as an initial value problem opens the possibility of working out a fully adaptive Romberg quadrature (with order and stepsize control) and at the same time a didactic first step into extrapolation methods, which play a prominent role in the solution of ordinary differential equations. The alternative formulation of the quadrature problem as a boundary value problem is exploited for the derivation of an adaptive multigrid algorithm: in this way we once more present an important class of methods for ordinary and partial differential equation in the simplest possible case.

For a typical university term the contents of the book might be too rich. For a possible partitioning of the presented material into two parts we recommend the closely connected sequences Chapter 1 – 5 and Chapter 6 – 9. Of course, different "teaching paths" can be chosen. For this purpose, we give the following connection diagram:

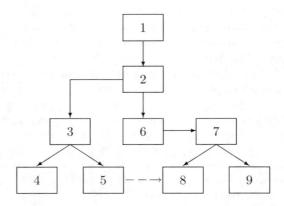

As can be seen from this diagram, the chapters of the last row (Chapters 4, 5, 8, and 9) can be skipped without spoiling the flow of teaching — according to the personal scientific taste. Chapter 4 could be integrated into a course on "Nonlinear optimization", Chapters 5 and 8 into a course on "Numerical linear algebra" or Chapter 9 into "Numerical solution of differential equations".

At the end of each chapter we added exercises. Beyond these explicit exercises further programming exercises may be selected from the numerous algorithms, which are given informally (usually as pseudocodes) throughout the textbook. All algorithms mentioned in the text are internationally accessible via the electronic library *eLib* of the Konrad Zuse Center. In the interactive mode *eLib* can be reached via:

Datex-P:	+45050331033 (WIN) +2043623331033 (IXI)
INTERNET:	elib.ZIB-berlin.de (130.73.108.11)
login:	elib (no password necessary)

In addition, there is the following e-mail access:

X.400:	S=eLib;OU=sc;P=ZIB-Berlin;A=dbp;C=de
INTERNET:	elib@elib.ZIB-Berlin.de
BITNET:	eLib@sc.ZIB-Berlin.dbp.de
UUCP:	unido!sc.ZIB-Berlin.dbp.de!eLib

Especially for users of Internet there is an "anonymous ftp" access (elib.ZIB-Berlin.de - 130.73.108.11) .

Contents

1 Linear Systems

We start with the classical *Gaussian elimination method* for solving systems of linear equations. Carl Friedrich Gauss (1777–1855) describes the method in his 1809 work on celestial mechanics "Theoria Motus Corporum Coelestium" [33] by saying "the values can be obtained with the usual elimination method". The method was used there in connection with the least squares method (cf. Section 3). In fact the method had been used previously by Lagrange in 1759 and had been known in China as early as the first century B.C. The problem is to solve a system of n linear equations

$$
\begin{aligned}
a_{11}x_1 &+ a_{12}x_2 &+ \cdots &+ a_{1n}x_n &=& \; b_1 \\
a_{21}x_1 &+ a_{22}x_2 &+ \cdots &+ a_{2n}x_n &=& \; b_2 \\
&\;\;\vdots & & \quad\vdots & & \;\;\vdots \\
a_{n1}x_1 &+ a_{n2}x_2 &+ \cdots &+ a_{nn}x_n &=& \; b_n
\end{aligned}
$$

or, in short form

$$ Ax = b, $$

where $A \in \mathrm{Mat}_n(\mathbf{R})$ is a real (n, n)-matrix and $b, x \in \mathbf{R}^n$ are real n-vectors. Before starting to compute the solution x, we should ask ourselves whether or not the system is solvable or not? From linear algebra, we know the following result which characterizes solvability in terms of the determinant of the matrix A.

Theorem 1.1 *Let $A \in \mathrm{Mat}_n(\mathbf{R})$ be a real square matrix with $\det A \neq 0$ and $b \in \mathbf{R}^n$. Then there exists a unique $x \in \mathbf{R}^n$ such that $Ax = b$.*

Whenever $\det A \neq 0$, the solution $x = A^{-1}b$ can be computed by Cramer's rule. Here we already see a general property of a "good" algorithm, namely the connection of existence and uniqueness of the solution with a numerical method for computing it. The *cost* of computing $\det A$ amounts to $n \cdot n!$ arithmetic operations when the Leibniz representation

$$ \det A = \sum_{\sigma \in S_n} \mathrm{sgn}\,\sigma \cdot a_{1,\sigma(1)} \cdots a_{n,\sigma(n)} $$

of the determinant as a sum of all permutations $\sigma \in S_n$ of the set $\{1, \ldots, n\}$ is used. Even with the recursive scheme involving development in subdeterminants according to Laplace's rule

$$\det A = \sum_{i=1}^{n} (-1)^{i+1} a_{1i} \det A_{1i}$$

there are 2^n arithmetic operations to be carried out, where $A_{1i} \in \text{Mat}_{n-1}(\mathbf{R})$ is the matrix obtained from A by crossing out the first row and the i-th column. As we will see, all methods to be described in what follows are more efficient than Cramer's rule for $n \geq 3$ so that the latter is interesting only for $n = 2$.

Remark 1.2 Of course, we expect that a good numerical method solves a given problem at minimal cost (in terms of arithmetic operations). Intuitively there is a minimal cost for each problem which is called the *complexity* of the problem. The closer the cost of an algorithm is to the complexity of the problem, the more efficient that algorithm is. The cost of a concrete algorithm is therefore always an *upper bound* for the complexity of the problem it solves. Obtaining *lower bounds* is in general much more difficult — for details see the monograph of TRAUB and WOZNIAKOWSKI [75].

The notation $x = A^{-1}b$ could suggest the idea of computing the solution of $Ax = b$ by first computing the inverse matrix A^{-1} and then applying it to b. However the computation of A^{-1} inherently contains all difficulties related to solving $Ax = b$ for *arbitrary* right hand sides b. We will see in the second chapter that the computation of A^{-1} can be "badly behaved", even when for special b the solution of $Ax = b$ is "well behaved". $x = A^{-1}b$ is therefore meant only as a formal notation which has nothing to do with the actual computation of the solution x. One should therefore avoid talking about "inverting matrices", when in fact one is concerned with "solving linear systems".

Remark 1.3 There has been a long standing bet by an eminent colleague, who wagered a significant amount, that in practice the problem of "inverting a matrix" is always avoidable. As far as we know he has won the bet in all cases.

1.1   Solution of Triangular Systems

In the search for an efficient solution method for arbitrary linear systems we will first consider cases that are particularly easy to solve. Simplest is obviously the case of a diagonal matrix A, where the corresponding system consists of n independent scalar equations. The method that transforms a general system into a diagonal one is called the *Gauss-Jordan method*. However we will omit it here, since it is less efficient than the method described in Section 1.2. Next, in terms of difficulty, is the case of a *triangular system*

$$
\begin{aligned}
r_{11}x_1 \ + \ r_{12}x_2 \ + \ \ldots \ + \ r_{1n}x_n \ &= \ z_1 \\
r_{22}x_2 \ + \ \ldots \ + \ r_{2n}x_n \ &= \ z_2 \\
\ddots \qquad \vdots \qquad &\vdots \\
r_{nn}x_n \ &= \ z_n \ ,
\end{aligned}
$$

and in matrix notation

$$ Rx = z \ , \tag{1.1} $$

where R is an upper triangular matrix , i.e. $r_{ij} = 0$ for all $i > j$. Obviously the components of x can be obtained recursively starting with the n'th row:

$$
\begin{aligned}
x_n \quad &:= \quad z_n/r_{nn} && , \ \ \text{if } r_{nn} \neq 0 \ , \\
x_{n-1} \ &:= \ (z_{n-1} - r_{n-1,n}x_n)/r_{n-1,n-1} && , \ \ \text{if } r_{n-1,n-1} \neq 0 \ , \\
\vdots \ \\
x_1 \quad &:= \quad (z_1 - r_{12}x_2 - \ldots - r_{1n}x_n)/r_{11} && , \ \ \text{if } r_{11} \neq 0 \ .
\end{aligned}
$$

Now, the determinant of the upper triangular matrix R is $\det R = r_{11} \cdots r_{nn}$, and therefore

$$ \det R \neq 0 \iff r_{ii} \neq 0 \text{ for all } i = 1, \ldots, n \ . $$

The above defined algorithm is therefore applicable (as in the case of Cramer's rule) if and only if $\det R \neq 0$, i.e. under the hypothesis of the existence and uniqueness theorem. The computational cost amounts to:

a) for the i-th row: $n - i$ additions and multiplications, and one division

b) for rows n through 1 together:

$$ \sum_{i=1}^{n}(i-1) = \frac{n(n-1)}{2} \doteq \frac{n^2}{2} $$

multiplications and as many additions.

Here the notation "$\doteq$" stands for "equal up to lower order terms", i.e. we consider only the term containing the highest power of n, which dominates the cost for large values of n.

The solution of a triangular system of the form

$$Lx = z, \tag{1.2}$$

with a lower triangular matrix L, is completely analogous, if one starts now with the first row and works through to last one. This way of solving triangular systems is called *backward substitution* in case of (1.1) and *forward substitution* in case of (1.2). The name substitution or replacement is used because each component of the right hand side vector is successively replaced by the solution, as indicated in the following scheme describing the content of the vector stored in the memory of the machine (memory scheme) at each step:

$$(z_1, z_2, \ldots, z_{n-1}, z_n)$$
$$(z_1, z_2, \ldots, z_{n-1}, x_n)$$
$$\vdots$$
$$(z_1, x_2, \ldots, x_{n-1}, x_n)$$
$$(x_1, x_2, \ldots, x_{n-1}, x_n) \, .$$

1.2 Gaussian Elimination

We now return to the general linear system $Ax = b$,

$$
\begin{array}{ccccccc}
a_{11}x_1 & + & a_{12}x_2 & + \ldots + & a_{1n}x_n & = & b_1 \\
a_{21}x_1 & + & a_{22}x_2 & + \ldots + & a_{2n}x_n & = & b_2 \\
\vdots & & \vdots & & \vdots & & \vdots \\
a_{n1}x_1 & + & a_{n2}x_2 & + \ldots + & a_{nn}x_n & = & b_n
\end{array}
\tag{1.3}
$$

and try to transform it into a triangular one. The first row does not have to be changed. We want to manipulate the remaining rows such that the coefficients in front of x_1 vanish, i.e. the variable x_1 from rows 2 through n is *eliminated*. Thus we produce a system of the form

$$
\begin{aligned}
a_{11}x_1 \quad + \quad a_{12}x_2 \quad + \ldots + \quad a_{1n}x_n \quad &= \quad b_1 \\
a_{22}'x_2 \quad + \ldots + \quad a_{2n}'x_n \quad &= \quad b_2' \\
\vdots \qquad\qquad\qquad\quad &\qquad \vdots \\
a_{n2}'x_2 \quad + \ldots + \quad a_{nn}'x_n \quad &= \quad b_n' .
\end{aligned}
\tag{1.4}
$$

Having achieved this we can apply the same procedure to the last $n-1$ rows in order to recursively obtain a triangular system. Therefore it is sufficient to examine the first elimination step from (1.3) to (1.4). We assume that $a_{11} \neq 0$. In order to eliminate the term $a_{i1}x_1$ in row i $(i = 2, \ldots, n)$, we subtract from row i a multiple of row 1 (unaltered), i.e.

$$
\text{new row } i := \text{row } i - l_{i1} \cdot \text{row } 1
$$

or explicitly

$$
\underbrace{(a_{i1} - l_{i1}a_{11})}_{= 0} x_1 + \underbrace{(a_{i2} - l_{i1}a_{12})}_{= a_{i2}'} x_2 + \cdots + \underbrace{(a_{in} - l_{i1}a_{1n})}_{= a_{in}'} x_n = \underbrace{b_i - l_{i1}b_1}_{= b_i'} .
$$

From $a_{i1} - l_{i1}a_{11} = 0$ it follows immediately that $l_{i1} = a_{i1}/a_{11}$. Therefore the first elimination step can be performed under the assumption $a_{11} \neq 0$. The element a_{11} is called a *pivot element* and the first row a *pivot row*. After this first elimination step there remains an $(n-1, n-1)$-submatrix in rows 2 through n. By applying repeatedly the elimination procedure we obtain a sequence

$$
A = A^{(1)} \to A^{(2)} \to \ldots \to A^{(n)} =: R
$$

of matrices of the special form

$$
A^{(k)} =
\begin{bmatrix}
a_{11}^{(1)} & a_{12}^{(1)} & \cdots & \cdots & \cdots & a_{1n}^{(1)} \\
 & a_{22}^{(2)} & \cdots & \cdots & \cdots & a_{2n}^{(2)} \\
 & & \ddots & & & \vdots \\
 & & & a_{kk}^{(k)} & \cdots & a_{kn}^{(k)} \\
 & & & \vdots & & \vdots \\
 & & & a_{nk}^{(k)} & \cdots & a_{nn}^{(k)}
\end{bmatrix}
\tag{1.5}
$$

with an $(n-k+1, n-k+1)$-submatrix, to which we can apply the elimination step

$$l_{ik} \quad := \quad a_{ik}^{(k)}/a_{kk}^{(k)} \qquad \text{for } i = k+1, \ldots, n$$

$$a_{ij}^{(k+1)} \quad := \quad a_{ij}^{(k)} - l_{ik}a_{kj}^{(k)} \qquad \text{for } i, j = k+1, \ldots, n$$

$$b_i^{(k+1)} \quad := \quad b_i^{(k)} - l_{ik}b_k^{(k)} \qquad \text{for } i = k+1, \ldots, n$$

whenever the *pivot* $a_{kk}^{(k)}$ does not vanish. Since every elimination step is a linear operation applied to the rows of A, the transformation from $A^{(k)}$ and $b^{(k)}$ into $A^{(k+1)}$ and $b^{(k+1)}$ can be represented as a *premultiplication* by a matrix $L_k \in \mathrm{Mat}_n(\mathbf{R})$, i.e.

$$A^{(k+1)} = L_k A^{(k)} \quad \text{and} \quad b^{(k+1)} = L_k b^{(k)} .$$

(In case of operations on columns one obtains an analogous *postmultiplication*). The matrix

$$L_k = \begin{bmatrix} 1 & & & & & \\ & \ddots & & & & \\ & & 1 & & & \\ & & -l_{k+1,k} & 1 & & \\ & & \vdots & & \ddots & \\ & & -l_{n,k} & & & 1 \end{bmatrix}$$

is called a *Frobenius matrix*; It has the nice property that its inverse L_k^{-1} is obtained from L_k by changing the signs of the l_{ik}'s. Furthermore the product of the L_k^{-1}'s satisfies

$$L := L_1^{-1} \cdot \ldots \cdot L_{n-1}^{-1} = \begin{bmatrix} 1 & & & & \\ l_{21} & 1 & & & \\ l_{31} & l_{32} & 1 & & \\ \vdots & & \ddots & \ddots & \\ l_{n1} & \cdots & \cdots & l_{n,n-1} & 1 \end{bmatrix} .$$

In this way we have reduced the system $Ax = b$ to the equivalent triangular system $Rx = z$ with

$$R = L^{-1}A \quad \text{and} \quad z = L^{-1}b .$$

A lower (resp. upper) triangular matrix, whose main diagonal elements are all equal to one is a called a *unit* lower (resp. upper) triangular matrix. The above representation $A = LR$ of the matrix A as a product of a unit lower triangular matrix L and an upper triangular matrix R is called the *Gaussian triangular factorization*, or briefly *LR factorization* of A. In the English literature the matrix R is often denoted by U (from *u*pper triangular) and the corresponding Gaussian triangular factorization is called the *LU* factorization. If such a factorization exists, then L and R are uniquely determined (cf. Exercise 1.2).

Algorithm 1.4 *Gaussian Elimination.*

a) $A = LR$ Triangular Factorization, R upper and L lower triangular matrix

b) $Lz = b$ Forward Substitution

c) $Rx = z$ Backward Substitution.

The *memory scheme* for the Gaussian elimination is based upon the representation (1.5) of the matrices $A^{(k)}$. In the remaining memory locations one can store the l_{ik}'s, because the other elements, with values 0 or 1, do not have to be stored. The entire memory cost for Gaussian elimination amounts to $n(n+1)$ memory locations, i.e. as many as needed to define the problem. The cost in terms of number of multiplications is

$\sim \sum_{k=1}^{n-1} k^2 \doteq n^3/3$ for a)

$\sim \sum_{k=1}^{n-1} k \doteq n^2/2$ both for b) and c).

Therefore the main cost comes from the LR-factorization. However, if different right hand sides $b_1, \ldots, b_j$ are considered, then this factorization has to be carried out only once.

1.3   Pivoting Strategies and Iterative Refinement

As seen from the simple example

$$A = \begin{pmatrix} 0 & 1 \\ 1 & 0 \end{pmatrix}, \quad \det A = -1, \quad a_{11} = 0$$

there are cases where the triangular factorization fails even when $\det A \neq 0$. However an interchange of rows leads to the simplest LR-factorization we can imagine, namely

$$A = \begin{pmatrix} 0 & 1 \\ 1 & 0 \end{pmatrix} \quad \longrightarrow \quad \bar{A} = \begin{pmatrix} 1 & 0 \\ 0 & 1 \end{pmatrix} = I = LR \quad \text{with} \quad L = R = I \ .$$

In the numerical implementation of Gaussian Elimination difficulties can arise not only when pivot elements vanish, but also when they are "too small".

Example 1.5 (cf. [30]) We compute the solution of the system

$$(a) \quad 1.00 \cdot 10^{-4}\, x_1 \ + \ 1.00\, x_2 \ = 1.00$$
$$(b) \qquad\qquad 1.00\, x_1 \ + \ 1.00\, x_2 \ = 2.00$$

on a machine, which, for the sake of simplicity, works only with three exact decimal figures. By completing the numbers with zeros, we obtain the "exact" solution with four correct figures

$$x_1 = 1.000 \quad x_2 = 0.9999 \ ,$$

and with three correct figures

$$x_1 = 1.00 \quad x_2 = 1.00 \ .$$

Let us now carry out the Gaussian elimination on our computer, i.e. in three exact decimal figures

$$l_{21} = \frac{a_{21}}{a_{11}} = \frac{1.00}{1.00 \cdot 10^{-4}} = 1.00 \cdot 10^4 \ ,$$

$$(1.00 - 1.00 \cdot 10^4 \cdot 1.00 \cdot 10^{-4})x_1 + (1.00 - 1.00 \cdot 10^4 \cdot 1.00)x_2$$
$$= 2.00 - 1.00 \cdot 10^4 \cdot 1.00 \ .$$

Thus we obtain the upper triangular system

$$1.00 \cdot 10^{-4}\, x_1 \ + \qquad\qquad 1.00\, x_2 \ = \qquad\qquad 1.00$$
$$-1.00 \cdot 10^4\, x_2 \ = \ -1.00 \cdot 10^4$$

and the "solution"

$$x_2 = 1.00 \ (\text{true}) \quad x_1 = 0.00 \ (\text{false!}) \ .$$

However, if before starting the elimination, we interchange the rows

$$(\tilde{a}) \qquad\qquad 1.00\, x_1 \ + \ 1.00\, x_2 \ = \qquad 2.00$$
$$(\tilde{b}) \quad 1.00 \cdot 10^{-4}\, x_1 \ + \ 1.00\, x_2 \ = \qquad 1.00 \ ,$$

then $\tilde{l}_{21} = 1.00 \cdot 10^{-4}$, which yields the upper triangular system

$$
\begin{array}{rcl}
1.00\ x_1 \ + \ 1.00\ x_2 &=& 2.00 \\
1.00\ x_2 &=& 1.00
\end{array}
$$

as well as the "true solution"

$$
x_2 = 1.00 \quad x_1 = 1.00 \ .
$$

By interchanging the rows in the above example we obtain

$$
|\tilde{l}_{21}| < 1 \ \text{ and } \ |\tilde{a}_{11}| \geq |\tilde{a}_{21}| \ .
$$

Thus, the new pivot $\tilde{a}_{11}$ is the largest element, in absolute value, of the first column.

We can deduce the *partial pivoting* or *column pivoting* strategy from the above considerations . This strategy is to choose at each Gaussian elimination step as pivot row the one having the largest element in absolute value within the pivot column. More precisely, we can formulate the following algorithm:

Algorithm 1.6 *Gaussian elimination with column pivoting*

a) In elimination step $A^{(k)} \to A^{(k+1)}$ choose a $p \in \{k, \ldots, n\}$, such that

$$
|a_{pk}^{(k)}| \geq |a_{jk}^{(k)}| \quad \text{for } j = k, \ldots, n
$$

Row p becomes pivot row.

b) Interchange rows p and k

$$
A^{(k)} \to \tilde{A}^{(k)} \ \text{ with } \ \tilde{a}_{ij}^{(k)} = \left\{
\begin{array}{ll}
a_{kj}^{(k)} & , \quad \text{if } i = p \\[2mm]
a_{pj}^{(k)} & , \quad \text{if } i = k \\[2mm]
a_{ij}^{(k)} & , \quad \text{otherwise}
\end{array}
\right. .
$$

Now we have

$$
|\tilde{l}_{ik}| = \left| \frac{\tilde{a}_{ik}^{(k)}}{\tilde{a}_{kk}^{(k)}} \right| = \left| \frac{\tilde{a}_{ik}^{(k)}}{a_{pk}^{(k)}} \right| \leq 1 \ .
$$

c) Perform the next elimination step for $\tilde{A}^{(k)}$, i.e.

$$
\tilde{A}^{(k)} \to A^{(k+1)} \ .
$$

Remark 1.7 Instead of column pivoting with row interchange one can also perform *row pivoting* with column interchange. Both strategies require at most $O(n^2)$ additional operations. If we combine both methods and look at each step for the largest element in absolute value of the entire remaining matrix, then we need $O(n^3)$ additional operations. This *total pivoting* strategy is therefore almost never employed.

In the following formal description of the triangular factorization with partial pivoting we use *permutation matrices* $P \in \mathrm{Mat}_n(\mathbf{R})$. For each permutation $\pi \in S_n$ we define the corresponding matrix

$$P_\pi = \left[e_{\pi(1)} \cdots e_{\pi(n)} \right] ,$$

where $e_j = (\delta_{1j}, \ldots, \delta_{nj})^T$ is the j-th unit vector. A permutation π of the rows of the matrix A can be expressed as a premultiplication by P_π

$$\text{Permutation of rows } \pi \colon A \longrightarrow P_\pi A.$$

and analogously a permutation π of the columns as a postmultiplication

$$\text{Permutation of columns } \pi \colon A \to A P_\pi .$$

It is known from linear algebra that the mapping

$$\pi \longmapsto P_\pi$$

is a group homeomorphism $S_n \to \mathbf{O}(n)$ of the symmetric group S_n into the orthogonal group $\mathbf{O}(n)$. In particular we have

$$P^{-1} = P^T .$$

The determinant of the permutation matrix is just the sign of the corresponding permutation

$$\det P_\pi = \mathrm{sgn}\, \pi \in \{\pm 1\} ,$$

i.e. it is equal to $+1$, if π consists of an even number of transpositions, and -1 otherwise. The following proposition shows that, *theoretically*, the triangular factorization with partial pivoting fails only when the matrix A is singular.

Theorem 1.8 *For every invertible matrix A there exists a permutation matrix P such that a triangular factorization of the form*

$$PA = LR$$

is possible. Here P can be chosen so that all elements of L are less than or equal to one in absolute value, i.e.

$$|L| \leq 1 .$$

Proof. We employ the LR-factorization algorithm with column pivoting. Since $\det A \neq 0$, there is a transposition $\tau_1 \in S_n$ such that the first diagonal element $a_{11}^{(1)}$ of the matrix

$$A^{(1)} = P_{\tau_1} A$$

is different from zero and is also the largest element in absolute value in the first column, i.e.

$$0 \neq |a_{11}^{(1)}| \geq |a_{i1}^{(1)}| \quad \text{for} \quad i = 1, \ldots, n .$$

After eliminating the remaining elements of the first column we obtain the matrix

$$A^{(2)} = L_1 A^{(1)} = L_1 P_{\tau_1} A = \left[\begin{array}{c|ccc} a_{11}^{(1)} & * & \cdots & * \\ \hline 0 & & & \\ \vdots & & B^{(2)} & \\ 0 & & & \end{array} \right] ,$$

where all elements of L_1 are less than or equal to one in absolute value, i.e. $|L_1| \leq 1$, and $\det L_1 = 1$. The remaining matrix $B^{(2)}$ is again invertible since $|a_{11}^{(1)}| \neq 0$ and

$$0 \neq \text{sgn}\,(\tau_1) \det A = \det A^{(2)} = a_{11}^{(1)} \det B^{(2)} .$$

Now by we can proceed by induction and obtain

$$R = A^{(n)} = L_{n-1} P_{\tau_{n-1}} \cdots L_1 P_{\tau_1} A , \qquad (1.6)$$

where $|L_k| \leq 1$, and τ_k is either the identity or the transposition of two numbers $\geq k$. If $\pi \in S_n$ only permutes numbers $\geq k+1$, then the Frobenius matrix

$$L_k = \left[\begin{array}{ccccccc} 1 & & & & & & \\ & \ddots & & & & & \\ & & 1 & & & & \\ & & -l_{k+1,k} & 1 & & & \\ & & \vdots & & \ddots & & \\ & & -l_{n,k} & & & 1 \end{array} \right]$$

satisfies

$$\hat{L}_k = P_\pi L_k P_\pi^{-1} = \begin{bmatrix} 1 & & & & & & \\ & \ddots & & & & & \\ & & 1 & & & & \\ & & -l_{\pi(k+1),k} & 1 & & & \\ & & \vdots & & \ddots & & \\ & & -l_{\pi(n),k} & & & 1 \end{bmatrix} . \qquad (1.7)$$

Therefore we can separate Frobenius matrices L_k and permutations P_{τ_k} by inserting in (1.6) the identities $P_{\tau_k}^{-1} P_{\tau_k}$ i.e.

$$R = L_{n-1} P_{\tau_{n-1}} L_{n-2} P_{\tau_{n-1}}^{-1} P_{\tau_{n-1}} P_{\tau_{n-2}} L_{n-3} \cdots L_1 P_{\tau_1} A .$$

Hence we obtain

$$R = \hat{L}_{n-1} \cdots \hat{L}_1 P_{\pi_0} A \quad \text{with} \quad \hat{L}_k = P_{\pi_k} L_k P_{\pi_k}^{-1} ,$$

where $\pi_{n-1} := \mathrm{id}$ and $\pi_k = \tau_{n-1} \cdots \tau_{k+1}$ for $k = 0, \ldots, n-2$. Since the permutation π_k interchanges in fact only numbers $\geq k+1$, the matrices $\hat{L}_k$ are of the form (1.7). Consequently

$$P_{\pi_0} A = LR$$

with $L := \hat{L}_1^{-1} \cdots \hat{L}_{n-1}^{-1}$ or explicitly

$$L = \begin{bmatrix} 1 & & & & \\ l_{\pi_1(2),1} & 1 & & & \\ l_{\pi_1(3),1} & l_{\pi_2(3),2} & 1 & & \\ \vdots & & \ddots & \ddots & \\ l_{\pi_1(n),1} & & \cdots & l_{\pi_{n-1}(n),n-1} & 1 \end{bmatrix} ,$$

and therefore $|L| \leq 1$. $\qquad\qquad\qquad\qquad\qquad\qquad\qquad\qquad\qquad\qquad\quad$ □

Note that we have used the Gaussian elimination algorithm with column pivoting to constructively prove an existence theorem.

Remark 1.9 Let us also note that the *determinant* of A can be easily computed by using the $PA = LR$ factorization of Proposition 1.8 via the formula

$$\det A = \det(P) \cdot \det(LR) = \mathrm{sgn}\,(\pi_0) \cdot r_{11} \cdots r_{nn}$$

A warning should be made against the naive computation of determinants! As is well known, multiplication of a linear system by an arbitrary scalar α results in

$$\det(\alpha A) = \alpha^n \det A \ .$$

This trivial transformation may be used to convert a "small" determinant into an arbitrarily "large" one and the other way around. The only invariants under this class of trivial transformations are the Boolean quantities $\det A = 0$ or $\det A \neq 0$; for an odd n we have additionally $\mathrm{sgn}\,(\det A)$. The above noted theoretical difficulty will lead later on to a completely different characterization of the solvability of linear systems.

Furthermore, it is apparent that the pivoting strategy can be arbitrarily changed by multiplying different rows by different scalars. This observation leads to the question of *scaling*. By row scaling we mean premultiplication of A by a diagonal matrix

$$A \rightarrow D_r A \ , \quad D_r \ \text{ diagonal matrix}$$

and analogously, by column scaling we mean postmultiplication by a diagonal matrix

$$A \rightarrow A D_c \ , \quad D_c \ \text{ diagonal matrix} \ .$$

(As we have already seen in the context of Gaussian elimination, linear operations on the rows of a matrix can be expressed by premultiplication with suitable matrices and correspondingly operations on columns are represented by postmultiplication.) Mathematically speaking scaling changes the length of the basis vectors of the range (row scaling) and of the domain (column scaling) of the linear mapping defined by the matrix A, respectively. If this mapping models a physical phenomenon then we can interpret scaling as a change of unit, or gauge transformation (e.g. from Å to km). In order to make the solution of the linear system $Ax = b$ independent of the choice of unit we have to appropriately scale the system by pre– or postmultiplying the matrix A by suitable diagonal matrices:

$$A \rightarrow \tilde{A} := D_r A D_c \ ,$$

where

$$D_r = \mathrm{diag}(\sigma_1, \ldots, \sigma_n) \ \text{ and } \ D_c = \mathrm{diag}\,(\tau_1, \ldots, \tau_n) \ .$$

At first glance the following three strategies seem to be reasonable:

a) *Row equilibration* of A with respect to a vector norm $\| \cdot \|$. Let A^i be the i-th row of A and assume that there are no zero rows. By setting $D_s := I$ and

$$\sigma_i := \|A^i\|^{-1} \text{ for } i = 1, \ldots, n,$$

we make all rows of $\tilde{A}$ have norm one.

b) *Column equilibration.* Suppose that there are no columns A_j of A equal to zero. By setting $D_z := I$ and

$$\tau_j := \|A_j\|^{-1} \text{ for } j = 1, \ldots, n,$$

we make all columns of $\tilde{A}$ have norm one.

c) Following a) and b) it is natural to require that all rows of A have the same norm and at the same time that all columns of A have the same norm. In order to determine σ_i and τ_j up to a mutual common factor one has to solve a *nonlinear* system with $2n - 2$ unknowns. This obviously requires a great deal more effort than solving the original problem. As will be seen in the fourth chapter the solution of this nonlinear system requires the solution of a sequence of linear systems, now in $2n - 2$ unknowns, for which the problem of scaling has to be addressed again.

Because of this dilemma, most programs (e.g. LINPACK [26]) leave the scaling issue to the user.

The pivoting strategies discussed above cannot prevent the possibility of computing a rather inaccurate solution $\tilde{x}$. How can one improve the accuracy of $\tilde{x}$ without too much effort? Of course we can simply discard the solution $\tilde{x}$ altogether and try to compute a "better" solution by using a higher machine precision. However in this way all information obtained in computing $\tilde{x}$ is lost. This is avoided in the so called *iterative refinement* method by explicitly evaluating the *residual*

$$r(y) := b - Ay = A(x - y)$$

The absolute error $\Delta x_0 := x - x_0$ of $x_0 := \tilde{x}$ satisfies the equation

$$A\Delta x_0 = r(x_0) . \tag{1.8}$$

In solving this *corrector equation* (1.8), we obtain an approximate correction $\tilde{\Delta} x_0 \neq \Delta x_0$ which is again afflicted by rounding errors. In spite of this fact we expect that the approximate solution

$$x_1 := x_0 + \tilde{\Delta} x_0$$

is "better" than x_0. The idea of iterative refinement consists in repeating this process until the approximate solution x_i is "accurate enough". We should remark that the linear system (1.8) differs from the original linear system only by the right hand side, so that the computation of the corrections Δx_i requires little effort. In Section 2.4.3 we will make precise the meaning of the terms "better approximate solution" and "accurate enough". In fact iterative refinement works excellently in conjunction with Gaussian elimination. In Section 2.4.3 we will state the substantial result of SKEEL [70] that for triangular factorization with column pivoting, a single refinement step is sufficient for obtaining a suitably accurate solution of the given problem.

1.4   Cholesky's Method for Symmetric Positive Definite Matrices

We want now to apply Gaussian elimination to the special class of systems of equations with symmetric positive definite matrices. It will become clear that in this case, the triangular factorization can be substantially simplified. We recall that a symmetric matrix $A = A^T \in \mathrm{Mat}_n(\mathbf{R})$ is *positive definite* if and only if

$$\langle x, Ax \rangle > 0 \ \ \text{for all } x \neq 0 \,. \tag{1.9}$$

We call such matrices for short *spd-matrices*.

Theorem 1.10 *For any spd-matrix $A \in \mathrm{Mat}_n(\mathbf{R})$ we have:*

   i) *A is invertible.*

   ii) *$a_{ii} > 0$ for $i = 1, \ldots, n$.*

   iii) *$\max_{i,j=1,\ldots,n} |a_{ij}| = \max_{i=1,\ldots,n} a_{ii}$.*

   iv) *Each rest matrix obtained during Gaussian elimination without pivoting is also symmetric positive definite.*

Obviously iii) and iv) say that row or column pivoting is not necessary for LR factorization, in fact even absurd because it might destroy the structure of A. In particular iii) means that total pivoting can be reduced to diagonal pivoting.

Proof.   The invertibility of A follows immediately from (1.9). If we put in (1.9) a basis vector e_i instead of x, it follows immediately that $a_{ii} = \langle e_i, Ae_i \rangle > 0$ and therefore the second claim is proven. The third statement is proved similarly, cf. Exercise 1.7. In order to prove statement iv) we write

$A = A^{(1)}$ as

$$
A^{(1)} =
\left[
\begin{array}{c|c}
a_{11} & z^T \\
\hline
z & B^{(1)}
\end{array}
\right]
\tag{1.10}
$$

where $z = (a_{12}, \ldots, a_{1n})^T$ and after one elimination step we obtain

$$
A^{(2)} = L_1 A^{(1)} =
\left[
\begin{array}{c|c}
a_{11} & z^T \\
0 & \\
\vdots & B^{(2)} \\
0 &
\end{array}
\right]
\quad \text{with} \quad
L_1 =
\left[
\begin{array}{cccc}
1 & & & \\
-l_{21} & 1 & & \\
\vdots & & \ddots & \\
-l_{n1} & & & 1
\end{array}
\right] .
$$

Now if we premultiply $A^{(2)}$ with L_1^T, then z^T in the first row is also eliminated and and the submatrix $B^{(2)}$ remains unchanged, i.e.

$$
L_1 A^{(1)} L_1^T =
\left[
\begin{array}{c|ccc}
a_{11} & 0 & \cdots & 0 \\
\hline
0 & & & \\
\vdots & & B^{(2)} & \\
0 & & &
\end{array}
\right] .
$$

The operation $A \rightarrow L_1 A L_1^T$ describes a change of basis for the bilinear form defined by the symmetric matrix A. According to the inertia theorem of Sylvester, $L_1 A^{(1)} L_1^T$ and with it $B^{(2)}$ remain positive definite. □

Together with the LR factorization we can deduce now the *rational Cholesky factorization* for symmetric positive definite matrices.

Theorem 1.11 *For every symmetric positive definite matrix A there exists a uniquely determined factorization of the form*

$$
A = LDL^T ,
$$

where L is a unit lower triangular matrix and D a positive diagonal matrix.

Proof. We continue the construction from the proof of Theorem 1.10 for $k = 2, \ldots, n-1$ and obtain immediately L as the product of $L_1^{-1}, \ldots, L_{n-1}^{-1}$ and D as the diagonal matrix of the pivots. □

Corollary 1.12 *Since $D = \text{diag}(d_i)$ is positive, the square root $D^{\frac{1}{2}} = \text{diag}(\sqrt{d_i})$ exists and with it the* Cholesky factorization

$$A = \bar{L}\bar{L}^T , \qquad (1.11)$$

where $\bar{L}$ is the lower triangular matrix $\bar{L} := LD^{\frac{1}{2}}$.

The matrix $\bar{L} = (l_{ij})$ can be computed by using *Cholesky's method*, :

Algorithm 1.13 *Cholesky's method.*

> **for** $k := 1$ **to** n **do**
> $\qquad l_{kk} := (a_{kk} - \sum_{j=1}^{k-1} l_{kj}^2)^{1/2};$
> $\qquad$ **for** $i := k + 1$ **to** n **do**
> $\qquad\qquad l_{ik} = (a_{ik} - \sum_{j=1}^{k-1} l_{ij}l_{kj})/l_{kk};$
> $\qquad$ **end for**
> **end for**

The derivation of this algorithm is nothing more than the element-wise evaluation of equation (1.11)

$$\begin{bmatrix} l_{11} & & \\ \vdots & \ddots & \\ l_{n1} & \cdots & l_{nn} \end{bmatrix} \begin{bmatrix} l_{11} & \cdots & l_{n1} \\ & \ddots & \vdots \\ & & l_{nn} \end{bmatrix} = \begin{bmatrix} a_{11} & \cdots & a_{1n} \\ \vdots & & \vdots \\ a_{n1} & \cdots & a_{nn} \end{bmatrix}$$

$$i = k : a_{kk} = l_{k1}^2 + \cdots + l_{k,k-1}^2 + l_{kk}^2$$
$$i > k : a_{ik} = l_{i1}l_{k1} + \cdots + l_{i,k-1}l_{k,k-1} + l_{ik}l_{kk} .$$

The sophistication of the method is contained in the sequence of computations for the elements of $\bar{L}$. As for the computational cost we have

$$\sim \frac{1}{6}n^3 \text{ multiplications and } n \text{ square roots .}$$

In contrast, the rational Cholesky factorization requires no square roots, but only rational operations (whence the name). By smart programming the cost can be kept here also to $\sim \frac{1}{6}n^3$. An advantage of the rational Cholesky factorization is that almost singular matrices D can be recognized. Also the method can be extended to symmetric indefinite matrices ($x^T Ax \neq 0$ for all x).

Remark 1.14 The supplemental spd property has obviously led to a sensible reduction of the computational cost. At the same time, this property forms the basis of completely different types of solution methods that will be described in Section 8.

1.5   Exercises

Exercise 1.1 Give an example of a full nonsingular (3,3)-matrix for which Gaussian elimination without pivoting fails.

Exercise 1.2   a) Show that the unit (nonsingular) lower (upper) triangular matrices form a subgroup of $GL(n)$.

   b) Apply a) to show that the representation

$$A = LR$$

     of a nonsingular matrix $A \in GL(n)$ as the product of a unit lower triangular matrix L and a nonsingular upper triangular matrix R is unique, provided it exists.

   c) If $A = LR$ as in b), then L and R can be computed by Gaussian triangular factorization. Why is this another proof of b) ? Hint: use induction.

Exercise 1.3 A matrix $A \in \mathrm{Mat}_n(\mathbf{R})$ is called strictly diagonally dominant if

$$|a_{ii}| > \sum_{\substack{i=1 \\ j \neq i}}^{n} |a_{ij}| \;\; \text{for } i = 1, \ldots, n.$$

Show that Gaussian triangular factorization can be performed for any matrix $A \in \mathrm{Mat}_n(\mathbf{R})$ with a strictly diagonally dominant transpose A^T. In particular any such A is invertible. Hint: use induction.

Exercise 1.4 The *numerical range* $W(A)$ of a matrix $A \in \mathrm{Mat}_n(\mathbf{R})$ is defined as the set

$$W(A) := \{\langle Ax, x \rangle \mid \langle x, x \rangle = 1, \; x \in \mathbf{R}^n\}$$

Here $\langle \cdot, \cdot \rangle$ is the Euclidean scalar product on $\mathbf{R}^n$.

   a) Show that the matrix $A \in \mathrm{Mat}_n(\mathbf{R})$ has an LR factorization (L unit lower triangular, R upper triangular) if and only if the origin is not contained in the numerical range of A, i.e.

$$0 \notin W(A) .$$

   Hint: use induction.

b) Use a) to show that the matrix

$$
\begin{bmatrix}
1 & 2 & 3 \\
2 & 4 & 7 \\
3 & 5 & 3
\end{bmatrix}
$$

has no LR factorization.

Exercise 1.5 Program the Gaussian triangular factorization. The program should read data A and b from a data file and should be tested on the following examples:

a) with the matrix from Example 1.1,

b) with $n = 1$, $A = 25$ and $b = 4$,

c) with $a_{ij} = i^{j-1}$ and $b_i = i$ for $n = 7$, 15 and 50.

Compare in each case the computed and the exact solutions.

Exercise 1.6 Gaussian factorization with column pivoting applied to the matrix A delivers the factorization $PA = LR$, where P is the permutation matrix produced during elimination. Show that:

a) Gaussian elimination with column pivoting is invariant with respect to

i) Permutation of rows of A (with the trivial exception that there are several elements of equal absolute value per column)

ii) Multiplication of the matrix by a number $\sigma \neq 0$, $A \longrightarrow \sigma A$.

b) If D is a diagonal matrix, then Gaussian elimination with column pivoting applied to $\bar{A} := AD$ delivers the factorization $P\bar{A} = L\bar{R}$ with $\bar{R} = RD$.

Consider the corresponding behavior for a row pivoting strategy with column interchange as well as for total pivoting with row and column interchange.

Exercise 1.7 Let the matrix $A \in \mathrm{Mat}_n(\mathbf{R})$ be symmetric positive definite.

a) Show that

$$
|a_{ij}| \;\leq\; \sqrt{a_{ii} a_{jj}} \;\leq\; \frac{1}{2}(a_{ii} + a_{jj}) \quad \text{for all } i, j = 1, \ldots, n.
$$

Hint: show first that the matrix $\begin{pmatrix} a_{ii} & a_{ij} \\ a_{ji} & a_{jj} \end{pmatrix}$ is symmetric positive definite for all i, j.

b) Deduce from a) that

$$\max_{i,j} |a_{ij}| = \max_i a_{ii} \ .$$

Interpret the result in the context of pivoting strategies.

Exercise 1.8 Show that for any $u, v \in \mathbf{R}^n$ we have:

a) $(I + uv^T)^{-1} = I - \dfrac{uv^T}{1 + v^T u}$,   whenever $u^T v \neq -1$

b) $I + uv^T$ is singular whenever $u^T v = -1$.

Exercise 1.9 The linear system $Ax = b$ with matrix

$$A = \left[\begin{array}{c|c} R & v \\ \hline u^T & 0 \end{array} \right] ,$$

is to be solved, where $R \in \mathrm{Mat}_n(\mathbf{R})$ is an invertible upper triangular matrix, $u, v \in \mathbf{R}^n$ and $x, b \in \mathbf{R}^{n+1}$.

a) Specify the triangular factorization of A.

b) Show that A is nonsingular if and only if

$$u^T R^{-1} v \neq 0 \ .$$

c) Formulate an economical algorithm for solving the above linear system and determine its computational cost.

Exercise 1.10 In the context of probability distributions one encounters matrices $A \in \mathrm{Mat}_n(\mathbf{R})$ with the following properties:

i) $\sum_{i=1}^n a_{ij} = 0$ for $j = 1, \ldots, n$;

ii) $a_{ii} < 0$ and $a_{ij} \geq 0$ for $i = 1, \ldots, n$ and $j \neq i$.

Let $A = A^{(1)}, A^{(2)}, \ldots, A^{(n)}$ be produced during Gaussian elimination. Show that:

a) $|a_{11}| \geq |a_{i1}|$ for $i = 2, \ldots, n$;

b) $\sum_{i=2}^n a_{ij}^{(2)} = 0$ for $j = 2, \ldots, n$;

c) $a_{ii}^{(1)} \leq a_{ii}^{(2)} \leq 0$ for $i = 2, \ldots, n$;

d) $a_{ij}^{(2)} \geq a_{ij}^{(1)} \geq 0$ for $i, j = 2, \ldots, n$ and $j \neq i$;

e) If the diagonal elements produced successively during the first $n - 2$ Gaussian elimination steps are all nonzero (i.e. $a_{ii}^{(i)} < 0$ for $i = 1, \ldots, n-1$) then $a_{nn}^{(n)} = 0$.

Exercise 1.11 A problem from astrophysics ("cosmic maser") can be formulated as a system of $(n + 1)$ linear equations in n unknowns of the form

$$\begin{pmatrix} & & \\ & A & \\ & & \\ 1 & \cdots & 1 \end{pmatrix} \begin{pmatrix} x_1 \\ \vdots \\ x_n \end{pmatrix} = \begin{pmatrix} 0 \\ \vdots \\ 0 \\ 1 \end{pmatrix}$$

where A is the matrix from Exercise 1.10. In order to solve this system we apply Gaussian elimination on the matrix A with the following two additional rules, where the matrices produced during elimination are denoted again by $A = A^{(1)}, \ldots, A^{(n-1)}$ and the relative machine precision is denoted by eps.

a) If during the algorithm $|a_{kk}^{(k)}| \leq |a_{kk}|$eps for some $k < n$, then shift simultaneously column k and row k to the end and the other columns and rows towards the front (rotation of rows and columns).

b) If $|a_{kk}^{(k)}| \leq |a_{kk}|$ eps for all remaining $k < n - 1$, then terminate the algorithm.

Show that:

i) If the algorithm does not terminate in b) then after $n - 1$ elimination steps it delivers a factorization of A as $PAP = LR$, where P is a permutation and $R = A^{(n-1)}$ is an upper triangular matrix with $r_{nn} = 0$, $r_{ii} < 0$ for $i = 1, \ldots, n-1$ and $r_{ij} \geq 0$ for $j > i$.

ii) The system has in this case a unique solution x, and all components of x are nonnegative (interpretation: probabilities).

Give a simple scheme for computing x.

Exercise 1.12 Program the algorithm developed in Exercise 1.11 for solving the special system of equations and test the program on two examples of your choice of dimensions $n = 5$ and $n = 7$, as well as on the matrix

$$
\begin{pmatrix}
-2 & 2 & 0 & 0 \\
2 & -4 & 1 & 1 \\
0 & 2 & -1 & 1 \\
0 & 0 & 0 & -2
\end{pmatrix}.
$$

Exercise 1.13 Let a linear system $Cx = b$ be given, where C is an invertible $(2n, 2n)$-matrix of the following special form:

$$
C = \begin{bmatrix} A & B \\ B & A \end{bmatrix}, \quad A, B \text{ invertible}.
$$

a) Let C^{-1} be partitioned as C:

$$
C^{-1} = \begin{bmatrix} E & F \\ G & H \end{bmatrix}.
$$

Prove SCHUR's identity:

$$
E = H = (A - BA^{-1}B)^{-1} \quad \text{and} \quad F = G = (B - AB^{-1}A)^{-1}.
$$

b) Let $x = (x_1, x_2)^T$ and $b = (b_1, b_2)^T$ be likewise partitioned and

$$
(A + B)y_1 = b_1 + b_2, \quad (A - B)y_2 = b_1 - b_2.
$$

Show that

$$
x_1 = \frac{1}{2}(y_1 + y_2), \quad x_2 = \frac{1}{2}(y_1 - y_2).
$$

Numerical advantage?

2 Error Analysis

In the last chapter, we introduced a class of methods for the numerical solution of linear systems. There, from a given input (A, b) we computed the solution $f(A, b) = A^{-1}b$. In a more abstract formulation the problem consists in evaluating a mapping $f : U \subset X \to Y$ at a point $x \in U$. The numerical solution of such a problem (f, x) computes the result $f(x)$ from the input x by means of an algorithm that eventually produces some intermediate values as well.

In this chapter we want to see how errors arise in this process and in particular to see if Gaussian elimination is indeed a dependable method. The errors in the numerical result arise from *errors in the data* or *input errors* as well as from *errors in the algorithm*.

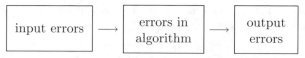

In principle we are powerless against the former, as they belong to the given problem and at best they can be avoided by changing the setting of the problem. The situation appears to be different with the errors caused by the algorithm. Here we have the chance to avoid, or to diminish, errors by changing the method. The distinction between the two kind of errors will lead us in what follows to the notions of *condition of a problem* and *stability of an algorithm*. First we want to discuss the possible sources of errors.

2.1 Sources of Errors

Even when input data are considered to be given exactly, errors in the data may still occur because of the machine representation of non-integer numbers. With today's usual *floating point representation*, a number z of "real type" is represented as $z = ad^e$, where the *basis* d is a power of two (as a rule $2, 8$ or 16) and the *exponent* e is an integer of a given maximum number of binary positions,

$$e \in \{e_{\min}, \ldots, e_{\max}\} \subset \mathbf{Z} \,.$$

The so called *mantissa* a is either 0 or a number satisfying $d^{-1} \le |a| < 1$ and has the form

$$a = v \sum_{i=1}^{l} a_i d^{-i},$$

where $v \in \{\pm 1\}$ is the *sign*, $a_i \in \{0, \ldots, d-1\}$ are the *digits* (it is assumed that $a = 0$ or $a_1 \ne 0$), and l is the *length of the mantissa*. The numbers that are representable in this way form a subset

$$\mathbf{F} := \{x \in \mathbf{R} \mid \text{ there is } a, e \text{ as above, so that } \; x = ad^e\}$$

of real numbers. The range of the exponent e defines the largest and smallest number that can be represented on the machine (by which we mean the processor together with the compiler). The *length of the mantissa* is responsible for the *relative precision* of the representation of real numbers on the given machine. Every number $x \ne 0$ with

$$d^{e_{\min}-1} \le |x| \le d^{e_{\max}}(1 - d^{-l})$$

is represented as a *floating point number* by rounding to the closest *machine number* whose relative error is estimated by

$$\frac{|x - \mathrm{fl}(x)|}{|x|} \le \mathrm{eps} := d^{1-l}/2$$

Here we use for division the convention $0/0 = 0$ and $x/0 = \infty$ for $x > 0$. We say that we have an *underflow* when $|x|$ is smaller than the smallest machine number $d^{e_{\min}-1}$ and, an *overflow* when $|x| > d^{e_{\max}}(1 - d^{-l})$. We call eps the *relative machine precision* or the *machine epsilon*. In the literature this quantity is also denoted by $\mathbf{u}$ for "unit roundoff" or "unit round". For *single precision* in FORTRAN, or *float* in C, we have usually eps $\approx 10^{-7}$.

Let us imagine that we wanted to enter in the machine a mathematically exact real number x, for example

$$x = \pi = 3.141592653589\ldots \,,$$

It is known theoretically that π as an irrational number cannot be represented with a finite mantissa and therefore it is a quantity affected by errors on any computer, e.g. for eps $= 10^{-7}$

$$\pi \mapsto \mathrm{fl}(\pi) = 3.141593 \,, \quad |\mathrm{fl}(\pi) - \pi| \le \mathrm{eps}\, \pi \,.$$

In the above it is essential to note that the number x after being introduced in the machine is indistinguishable from other numbers $\tilde{x}$ that are rounded to the same floating point number $\mathrm{fl}(x) = \mathrm{fl}(\tilde{x})$. In particular a real number that is obtained by appending zeros to a machine number is by no means distinguished. Therefore it is absurd to aim at an "exact" input. This insight will decisively influence the following error analysis.

A further important source of errors for the input data are the *measurement* errors, when input quantities are obtained from experimental data. Such data x are usually given together with the *absolute* error δx, the so called *tolerance*: the distance from x to the "true" value $\bar{x}$ can be estimated component-wise by

$$|x - \bar{x}| \leq \delta x .$$

In many important practical situations the *relative* precision $|\delta x/x|$ lies in between 10^{-2} and 10^{-3} — a quantity that in general outweighs by far the rounding of the input data. One often uses in this context the term *technical precision*.

Let us go now to the second group of error sources, the *errors in the algorithm*. The realization of an elementary operation

$$\circ \in \{+, -, \cdot, /\}$$

by the corresponding floating point operation $\hat{\circ} \in \{\hat{+}, \hat{-}, \hat{\cdot}, \hat{/}\}$ does not avoid the *rounding errors*. The relative error here is less than or equal to the machine precision, i.e. for $x, y \in \mathbf{F}$, we have

$$x \hat{\circ} y = (x \circ y)(1 + \varepsilon) \quad \text{for an} \quad \varepsilon = \varepsilon(x, y) \quad \text{with} \quad |\varepsilon| \leq \text{eps} .$$

One should notice that in general the operation $\hat{\circ}$ is not associative (see Exercise 2.1), so that within $\mathbf{F}$ the order sequence of the operations to be executed is very important.

Besides rounding errors, an algorithm may also produce *approximation errors* . They always appear when a function cannot be calculated exactly, but must be approximated. This happens for example when computing sine by a truncated power series or, to mention a more complex problem, in the solution of a differential equation. In this chapter we will essentially limit ourselves to the treatment of rounding errors. Approximation errors will be studied in the context of particular algorithms in later chapters.

2.2 Condition of Problems

Now we ask ourselves the question: How do perturbations of input variables influence the result *independently* of the choice of algorithm? We have seen in the above description of input errors that the input x is logically indistinguishable from all inputs $\tilde{x}$ that are within the range of a given precision. Instead of the "exact" input x we should instead consider an *input set E*, that contains all perturbed inputs $\tilde{x}$ (see figure 2.1). A machine number x

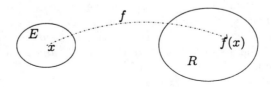

Figure 2.1: Input and output sets

represents therefore the input set

$$E = \{\tilde{x} \in \mathbf{R} \mid |\tilde{x} - x| \le \text{eps } |x|\} \ .$$

If we know that the input x is given with absolute tolerance δx, then the defining input quantities form the set

$$E = \{\tilde{x} \in \mathbf{R} \mid |\tilde{x} - x| \le \delta x\} \ .$$

The function f defining our problem, maps this set into an *output set* $R = f(E)$. In view of the same argument that led us from the supposedly exact input x to the input set E, it makes sense, when analyzing the input error, to study instead of the *pointwise* mapping $f : x \mapsto f(x)$ the *set valued* mapping $f : E \mapsto R = f(E)$. The effects of perturbations of input quantities on the output can be read only through the ratio between the input and output sets. We call *condition* of the problem described by (f, x) a certain characterization of the ratio between E and R.

Example 2.1 In order to develop a feeling for this concept that has not yet been defined precisely, let us examine beforehand the geometrical problem of determining graphically the intersection point r of two lines g and h in the plane (see Figure 2.2). When solving this problem graphically it is not possible to represent the lines g and h exactly. The question is then how strongly does the constructed intersection point depend on the drawing error (or: input error). The input set E in our example consists of all lines $\tilde{g}$ and $\tilde{h}$ laying within drawing precision from g and h respectively, and the output set R of the corresponding intersection points $\tilde{r}$. We see at once that the ratio between the input and the output sets depends strongly on the angle $\measuredangle (g, h)$ at which g and h intersect. If g and h are nearly perpendicular then

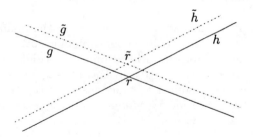

Figure 2.2: Nearly perpendicular intersection of two lines g, h (well-conditioned)

Figure 2.3: Small angle intersection of two lines g, h (ill-conditioned)

the intersection point $\tilde{r}$ varies about the same as the lines $\tilde{g}$ and $\tilde{h}$. However if the angle $\measuredangle\,(g, h)$ is small, i.e. the lines g and h are *nearly parallel* then one has great difficulty to locate precisely the intersection point with the naked eye. (see Figure 2.3). Actually, the intersection point $\tilde{r}$ moves several times more than the small perturbation of the lines. We can therefore say that the determination of the intersection point is well conditioned in the first case, and ill-conditioned in the second case.

We arrive thus at a mathematical definition of the concept of conditioning. For the sake of simplicity we assume that the problem (f, x) is given by a mapping

$$f : U \subset \mathbf{R}^n \to \mathbf{R}^m$$

from a open subset $U \subset \mathbf{R}^n$ into $\mathbf{R}^m$, a point $x \in U$ and a (relative or absolute) precision δ of the input data. The precision δ can be given either through a *norm* $\|\cdot\|$ on $\mathbf{R}^n$

$$\|\tilde{x} - x\| \leq \delta \ \ \text{(absolute) or} \ \ \|\tilde{x} - x\| \leq \delta \,\|x\| \ \ \text{(relative)} \qquad (2.1)$$

or *component-wise*

$$|\tilde{x}_i - x_i| \leq \delta \ \ \text{(absolute) or} \ \ |\tilde{x}_i - x_i| \leq \delta \,|x_i| \ \ \text{(relative)} \qquad (2.2)$$

for $i = 1, \ldots, n$. Correspondingly we measure the output error $f(\tilde{x}) - f(x)$ either norm-wise or component-wise.

2.2.1   Norm-wise condition analysis

In order to keep the calculations involved with the quantitative condition
analysis manageable, one usually chooses a sufficiently small input error δ
and examines the asymptotic ratio defining the condition in the so called
linearized error theory as $\delta \to 0$. In this setting the following shortened
notation is extremely useful: Two functions $g, h : \mathbf{R}^n \to \mathbf{R}^m$ are equal *to a
first approximation* or *in their leading order for $x \to x_0$*, in short

$$g(x) \doteq h(x) \quad \text{for } x \to x_0,$$

if

$$g(x) = h(x) + o(\|h(x)\|) \quad \text{for } \; x \to x_0 \; ,$$

where the Landau symbol '$o(\|h(x)\|)$ for $x \to x_0$' denotes a generic function
$\varphi(x)$ having the property

$$\lim_{x \to x_0} \frac{\|\varphi(x)\|}{\|h(x)\|} = 0.$$

Thus for a differentiable function f we have

$$f(\tilde{x}) - f(x) \doteq f'(x)(\tilde{x} - x) \quad \text{for } \; \tilde{x} \to x \; .$$

Analogously we define '$g(x) \stackrel{\cdot}{\leq} h(x)$ for $x \to x_0$' (component-wise) by '$g(x) \leq
h(x) + o(\|h(x)\|)$ for $x \to x_0$'.

   We can characterize the norm-wise ratio between the input and output
sets, without direct recourse to the derivative of f, via the following *asymptotic Lipschitz condition*.

Definition 2.2 The *absolute norm-wise condition number* of the problem
(f, x) is the smallest number $\kappa_{\mathrm{abs}} \geq 0$, such that

$$\|f(\tilde{x}) - f(x)\| \stackrel{\cdot}{\leq} \kappa_{\mathrm{abs}} \|\tilde{x} - x\| \quad \text{for } \tilde{x} \to x.$$

The problem (f, x) is *ill-posed*, if such a number does not exist (formally:
$\kappa_{\mathrm{abs}} = \infty$). Analogously the *relative norm-wise condition number* of (f, x)
is the smallest number $\kappa_{\mathrm{rel}} \geq 0$, such that

$$\frac{\|f(\tilde{x}) - f(x)\|}{\|f(x)\|} \stackrel{\cdot}{\leq} \kappa_{\mathrm{rel}} \frac{\|\tilde{x} - x\|}{\|x\|} \quad \text{for } \tilde{x} \to x.$$

   Thus κ_{abs} and κ_{rel} describe the increase of the absolute and the relative errors respectively. A problem (f, x) is *well-conditioned* if its condition
number is small and *ill-conditioned* if it is large. Naturally, the meaning of

"small" and "large" has to be considered separately for each problem. For the relative condition number, unity serves for orientation: it corresponds to the case of pure rounding of the result (see the discussion in Section 2.1). Further below we will show in a sequence of illustrative examples what "small" and "large" means.

If f is differentiable in x, then according to the mean value theorem, we can express the condition numbers in terms of the derivative:

$$\kappa_{\text{abs}} = \|f'(x)\| \quad \text{and} \quad \kappa_{\text{rel}} = \frac{\|x\|}{\|f(x)\|}\|f'(x)\| , \tag{2.3}$$

where $\|f'(x)\|$ is the norm of the Jacobian $f'(x) \in \text{Mat}_{m,n}(\mathbf{R})$ in the subordinate (or operator) norm

$$\|A\| := \sup_{x \neq 0} \frac{\|Ax\|}{\|x\|} = \sup_{\|x\|=1} \|Ax\| \quad \text{for} \quad A \in \text{Mat}_{m,n}(\mathbf{R})$$

For illustration let us compute the condition numbers of some simple problems.

Example 2.3 *Condition of addition (resp. subtraction).* Addition is a linear mapping

$$f : \mathbf{R}^2 \to \mathbf{R} , \quad \begin{pmatrix} a \\ b \end{pmatrix} \mapsto f(a,b) := a + b$$

with derivative $f'(a,b) = (1,1) \in \text{Mat}_{1,2}(\mathbf{R})$. If we choose the 1-norm on $\mathbf{R}^2$

$$\|(a,b)^T\| = |a| + |b|$$

and the absolute value on $\mathbf{R}$, then it follows that the subordinate matrix norm (see Exercise 2.8) is

$$\|f'(a,b)\| = \|(1,1)\| = 1 .$$

Therefore the condition numbers of addition are

$$\kappa_{\text{abs}} = 1 \quad \text{and} \quad \kappa_{\text{rel}} = \frac{|a| + |b|}{|a + b|} .$$

Hence for the *addition* of two numbers of the same sign we have $\kappa_{\text{rel}} = 1$. On the other hand it turns out that the *subtraction* of two nearly equal numbers is ill-conditioned according to the relative condition number, because in this case we have

$$|a + b| \ll |a| + |b| \iff \kappa_{\text{rel}} \gg 1 .$$

The phenomenon is called *cancellation of significant digits* or *loss of significance error*.

Example 2.4 *Unavoidable cancellation.* For illustration we give the following example with eps $= 10^{-7}$:

$$
\begin{aligned}
a &= 0.\,1234\,67* &&\leftarrow \text{ perturbation at position 7}\\
b &= 0.\,1234\,56* &&\leftarrow \text{ perturbation at position 7}\\
a - b &= 0.\underbrace{0000}11*\underbrace{000} &&\leftarrow \text{ perturbation at position 3.}
\end{aligned}
$$

$\quad\quad\quad\quad\quad\quad$ leading zeroes $\;$ shifted zeroes

An error in the seventh significant decimal digit of the input data a, b leads to an error in the third significant decimal digit of the result $a-b$ i.e. $\kappa_{\mathrm{rel}} \approx 10^4$.

Be aware of the fact that the cancellation of a digit of the result given by a computer cannot be noticed afterwards. The shifted zeroes are zeroes in the *binary* representation and they are lost through transformation to the decimal system. Therefore one has the following rule:

Avoid whenever possible subtraction of nearly equal numbers!

A further rule will be derived for unavoidable subtractions in Section 2.3.

Example 2.5 *Quadratic Equations.* A really classical example of an avoidable loss of significance error arises in the solution of the quadratic equation (see also Chapter 4)

$$x^2 - 2px + q = 0 \, ,$$

whose solution is usually given by

$$x_{1,2} = p \pm \sqrt{p^2 - q} \, .$$

In this form loss of significance errors occur when a solution is close to zero. However, this cancellation of significant digits is avoidable, because by Vieta's theorem, q is the product of the roots and therefore we can use

$$x_1 = p + \mathrm{sgn}\,(p)\sqrt{p^2 - q}, \quad x_2 = q/x_1 \, ,$$

to compute both solutions in a stable manner.

Example 2.6 Often loss of significance errors can be avoided by using power series expansions. For example let us consider the function

$$\frac{1 - \cos(x)}{x} = \frac{1}{x}\left(1 - [1 - \frac{x^2}{2} + \frac{x^4}{24} \pm \cdots]\right) = \frac{x}{2}\left(1 - \frac{x^2}{12} \pm \cdots\right) \, .$$

For $x = 10^{-4}$ we have $x^2/12 < 10^{-9}$ and therefore, according to Leibniz's theorem on alternating power series, $x/2$ is an approximation of $(1 - \cos x)/x$ correct up to eight decimal digits.

Example 2.7 *Condition of a linear system* $Ax = b$. If in the solution of the linear system $Ax = b$ we consider as input only the vector $b \in \mathbf{R}^n$, then the problem can be described by the mapping

$$f : \mathbf{R}^n \to \mathbf{R}^n \ , \ b \mapsto f(b) := A^{-1}b$$

It is linear in b. Its derivative is clearly $f'(b) = A^{-1}$, so that the condition numbers of the problem (A^{-1}, b) are given by

$$\kappa_{\mathrm{abs}} = \|A^{-1}\| \ \text{ and } \ \kappa_{\mathrm{rel}} = \frac{\|b\|}{\|A^{-1}b\|} \|A^{-1}\| = \frac{\|Ax\|}{\|x\|} \|A^{-1}\|$$

We can take into account perturbations in A as well. To do it we consider the matrix A as input quantity

$$f : GL(n) \subset \mathrm{Mat}_n(\mathbf{R}) \to \mathbf{R}^n \ , \ \ A \mapsto f(A) = A^{-1}b$$

and keep b fixed. This mapping is nonlinear in A. However it is differentiable. This follows for example from Cramer's rule and the fact that the determinant of a matrix is a polynomial in the entries of the matrix (see also Remark 2.9). For the computation of the derivative we need the following Lemma.

Lemma 2.8 *The mapping* $g : \mathrm{GL}(n) \subset \mathrm{Mat}_n(\mathbf{R}) \to \mathrm{GL}(n)$ *with* $g(A) = A^{-1}$ *is differentiable, and*

$$g'(A)C = -A^{-1}CA^{-1} \ \ \textit{for all} \ \ C \in \mathrm{Mat}_n(\mathbf{R}) \ . \tag{2.4}$$

Proof. We differentiate with respect to t the equation $(A+tC)(A+tC)^{-1} = I$ and obtain

$$0 = C(A + tC)^{-1} + (A + tC)\frac{d}{dt}(A + tC)^{-1} \ .$$

In particular for $t = 0$ it follows that

$$g'(A)C = \frac{d}{dt}(A + tC)^{-1}\Big|_{t=0} = -A^{-1}CA^{-1} \ .$$

$\square$

Remark 2.9 The differentiability of the inverse follows easily also from the *Neumann series*. If $C \in \mathrm{Mat}_n(\mathbf{R})$ is a matrix with $\|C\| < 1$, then $I - C$ is invertible and

$$(I - C)^{-1} = \sum_{k=0}^{\infty} C^k = I + C + C^2 + \cdots$$

This fact is proved as for the summation formula $\sum_{k=0}^{\infty} q^k = 1/(1 - q)$ of a geometrical progression with $|q| < 1$. Hence, for a matrix $C \in \mathrm{Mat}_n(\mathbf{R})$ with $\|A^{-1}C\| < 1$ it follows that

$$\begin{aligned}
(A + C)^{-1} &= (A(I + A^{-1}C))^{-1} = (I + A^{-1}C)^{-1}A^{-1} \\
&= (I - A^{-1}C + o(\|C\|))\, A^{-1} = A^{-1} - A^{-1}CA^{-1} + o(\|C\|)
\end{aligned}$$

for $\|C\| \to 0$ and therefore (2.4) holds. This argument remains valid for bounded linear operators in Banach spaces as well.

From Lemma 2.8 it follows that the derivative with respect to A of the solution $f(A) = A^{-1}b$ of the linear system satisfies

$$f'(A)\, C = -A^{-1}CA^{-1}b = -A^{-1}Cx \quad \text{for} \quad C \in \mathrm{Mat}_n(\mathbf{R}) \ .$$

In this way we get the condition numbers

$$\kappa_{\mathrm{abs}} = \|f'(A)\| = \sup_{\|C\|=1} \|A^{-1}Cx\| \le \|A^{-1}\|\, \|x\|,$$

$$\kappa_{\mathrm{rel}} = \frac{\|A\|}{\|x\|}\|f'(A)\| \le \|A\|\, \|A^{-1}\| \ .$$

The relative condition number corresponding to the input b can also be estimated by

$$\kappa_{\mathrm{rel}} \le \|A\|\, \|A^{-1}\|$$

because of the submultiplicativity $\|Ax\| \le \|A\|\, \|x\|$ of the matrix norm. The quantity

$$\kappa(A) := \|A\|\, \|A^{-1}\|$$

will therefore be called the *condition number* of the matrix A. It describes in particular the relative condition number of a linear system $Ax = b$ for *all* possible right hand sides $b \in \mathbf{R}^n$. Another representation for $\kappa(A)$ (see Exercise 2.12) is

$$\kappa(A) := \frac{\max_{\|x\|=1} \|Ax\|}{\min_{\|x\|=1} \|Ax\|} \in [0, \infty] \ . \tag{2.5}$$

It has the advantage that it is well defined for noninvertible and rectangular matrices as well. With this representation we immediately verify the following three properties of $\kappa(A)$:

i) $\kappa(A) \geq 1$

ii) $\kappa(\alpha A) = \kappa(A)$ for all $\alpha \in \mathbf{R}$, $\alpha \neq 0$

iii) $A \neq 0$ is singular if and only if $\kappa(A) = \infty$.

We see that the condition number $\kappa(A)$, as opposed to the determinant $\det A$, is *invariant* under the scalar transformation $A \to \alpha A$. Together with properties i) and iii) this favors the use of condition numbers rather than determinants for characterizing the solvability of a linear system. We will go deeper into this subject in Section 2.4.1.

Example 2.10 *Condition of nonlinear systems.* Assume that we want to solve a nonlinear system $f(x) = y$, where $f : \mathbf{R}^n \to \mathbf{R}^n$ is a continuously differentiable function and $y \in \mathbf{R}^n$ is the input quantity (mostly $y = 0$). We see immediately that the problem is well defined only if the derivative $f'(x)$ is invertible. In this case, according to the Inverse Function Theorem, f is also invertible in a neighborhood of y, i.e. $x = f^{-1}(y)$. The derivative satisfies

$$(f^{-1})'(y) = f'(x)^{-1} .$$

The condition numbers of the problem (f^{-1}, y) are therefore

$$\kappa_{\mathrm{abs}} = \|(f^{-1})'(y)\| = \|f'(x)^{-1}\| \quad \text{and} \quad \kappa_{\mathrm{rel}} = \frac{\|f(x)\|}{\|x\|} \|f'(x)^{-1}\| .$$

The conclusion clearly agrees with the geometrical determination of the intersection point of two lines. If κ_{abs} or κ_{rel} are large we have a situation similar to the *small angle intersection of two lines* (see. Figure 2.4).

2.2.2 Component-wise condition analysis

After being convinced of the the value of the norm-wise condition analysis, we want to introduce similar concepts for component-wise error analysis. The latter is often proves beneficial because all individual components are afflicted by some relative errors, and therefore some phenomena cannot be explained by norm-wise considerations. Furthermore, a norm-wise consideration does not take into account any eventual special structure of the matrix A, but rather it analyses the behavior relative to arbitrary perturbations δA, i.e. including those that do not preserve this special structure.

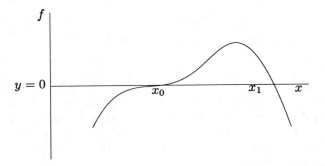

Figure 2.4: Ill-conditioned zero at x_0, well-conditioned zero at x_1.

Example 2.11 The solution of a linear system $Ax = b$ with a diagonal matrix

$$A = \begin{pmatrix} 1 & 0 \\ 0 & \varepsilon \end{pmatrix} \ , \quad A^{-1} = \begin{pmatrix} 1 & 0 \\ 0 & \varepsilon^{-1} \end{pmatrix} \ ,$$

is obviously a well-conditioned problem, because the equations are completely independent of each other (one says also *decoupled*). Here we implicitly assume that the admissible perturbations preserve the diagonal form. The norm-wise condition number $\kappa_\infty(A)$,

$$\kappa_\infty(A) = \|A^{-1}\|_\infty \, \|A\|_\infty = \frac{1}{\varepsilon} \ ,$$

becomes however arbitrary large for small $\varepsilon \leq 1$. It describes the condition for *arbitrary* perturbations of the matrix.

The example suggests that the notion of condition defined in Section 2.2.1 turns out to be deficient in some situations. Intuitively, we expect that the condition number of a diagonal matrix, i.e. of a completely decoupled linear system, is equal to one, as in the case of a scalar linear equation. The following component-wise analysis will lead us to such a condition number. If we want to carry over the concept of Section 2.2.1 to the component-wise setting, we have to replace norms with absolute values of the components. We will work this out in detail for the relative error concept.

Definition 2.12 The (prototype) *component-wise condition number* of the problem (f, x) is the smallest number $\kappa_{\text{rel}} \geq 0$, such that

$$\frac{\|f(\tilde{x}) - f(x)\|_\infty}{\|f(x)\|_\infty} \overset{\cdot}{\leq} \kappa_{\text{rel}} \max_i \frac{|\tilde{x}_i - x_i|}{|x_i|} \quad \text{for } \tilde{x} \to x.$$

Remark 2.13 Alternatively we can define the *relative component-wise condition number* also by

$$\max_i \frac{|f_i(\tilde{x}) - f_i(x)|}{|f_i(x)|} \stackrel{.}{\leq} \kappa_{\text{rel}} \max_i \frac{|\tilde{x}_i - x_i|}{|x_i|} \quad \text{for } \tilde{x} \to x.$$

The condition number defined in this way is even *submultiplicative*, i.e.

$$\kappa_{\text{rel}}(g \circ h, x) \leq \kappa_{\text{rel}}(g, h(x)) \cdot \kappa_{\text{rel}}(h, x) .$$

By analogy to (2.3) we can compute this condition number for differentiable mappings via the derivative at x. Application of the mean value theorem

$$f(\tilde{x}) - f(x) = \int_{t=0}^1 f'(x + t(\tilde{x} - x))(\tilde{x} - x) \, dt$$

gives component-wise

$$|f(\tilde{x}) - f(x)| \leq \int_{t=0}^1 |f'(x + t(\tilde{x} - x))| \, |(\tilde{x} - x)| \, dt .$$

and hence

$$\kappa_{\text{rel}} = \frac{\| \, |f'(x)| \, |x| \, \|_\infty}{\|f(x)\|_\infty} .$$

We want to compute also the component-wise condition number for some problems.

Example 2.14 *Condition of multiplication.* The multiplication of two real numbers x, y is described by the mapping

$$f : \mathbf{R}^2 \to \mathbf{R} , \quad (x, y)^T \mapsto f(x, y) = xy$$

It is differentiable with $f'(x, y) = (y, x)$, and the relative component-wise condition number becomes

$$\kappa_{\text{rel}} = \frac{\| \, |f'(x, y)| \, |(x, y)^T| \, \|_\infty}{\|f(x, y)\|_\infty} = 2 \frac{|xy|}{|xy|} = 2 .$$

Therefore multiplication can be considered as well-conditioned.

Example 2.15 *Condition of scalar products.* When computing a scalar product $\langle x, y \rangle = \sum_{i=1}^n x_i y_i$ we evaluate the mapping

$$f : \mathbf{R}^{2n} \to \mathbf{R} , \quad (x, y) \mapsto \langle x, y \rangle$$

at (x, y). Since f is differentiable with $f'(x, y) = (y^T, x^T)$, it follows that the component-wise relative condition number is

$$\kappa_{\text{rel}} = \frac{\| \, |(y^T, x^T)| \, |(x, y)| \, \|_\infty}{\|\langle x, y \rangle\|_\infty} = 2 \frac{\langle |x|, |y| \rangle}{|\langle x, y \rangle|} .$$

Example 2.16 *Component-wise condition of a linear system (Skeel's condition number).* If we consider, as in Example 2.7, the problem (A^{-1}, b) of a linear system with b as input, then we obtain the following value of the component-wise relative condition number:

$$\kappa_{\mathrm{rel}} = \frac{\| \, |A^{-1}| \, |b| \, \|_\infty}{\|A^{-1}b\|_\infty} = \frac{\| \, |A^{-1}| \, |b| \, \|_\infty}{\|x\|_\infty} \, .$$

This number was introduced by SKEEL [69]. With it the error $\tilde{x} - x$, $\tilde{x} = A^{-1}\tilde{b}$ can be estimated by

$$\frac{\|\tilde{x} - x\|_\infty}{\|x\|_\infty} \le \kappa_{\mathrm{rel}}\varepsilon \quad \text{for} \quad |\tilde{b} - b| \le \varepsilon |b|.$$

The ideas of Example 2.7 can be transferred for perturbations in A. We already know that the mapping $f : GL(n) \to \mathbf{R}^n$, $A \mapsto f(A) = A^{-1}b$, is differentiable with

$$f'(A)\, C = -A^{-1}CA^{-1}b = -A^{-1}Cx \quad \text{for} \quad C \in \mathrm{Mat}_n(\mathbf{R}) \, .$$

It follows that (see Exercise 2.14) the component-wise relative condition number is given by

$$\kappa_{\mathrm{rel}} = \frac{\| \, |f'(A)| \, |A| \, \|_\infty}{\|f(A)\|_\infty} = \frac{\| \, |A^{-1}| \, |A| \, |x| \, \|_\infty}{\|x\|_\infty} \, .$$

If we bring the results together and consider perturbations in both A and b, then the relative condition numbers add up and we obtain as condition number for the combined problem

$$\kappa_{\mathrm{rel}} = \frac{\| \, |A^{-1}| \, |A| \, |x| + |A^{-1}| \, |b| \, \|_\infty}{\|x\|_\infty} \le 2\frac{\| \, |A^{-1}| \, |A| \, |x| \, \|_\infty}{\|x\|_\infty} \, .$$

Taking for x the vector $e = (1, \ldots, 1)$, yields the following characterization of the component-wise condition of $Ax = b$ for arbitrary right hand sides b

$$\frac{1}{2}\kappa_{\mathrm{rel}} \le \frac{\| \, |A^{-1}| \, |A| \, |e| \, \|_\infty}{\|e\|_\infty} = \| \, |A^{-1}| \, |A| \, \|_\infty$$

in terms of the so called *Skeel's condition number*

$$\kappa_C(A) := \| \, |A^{-1}| \, |A| \, \|_\infty$$

of A. This condition number $\kappa_C(A)$ satisfies, the same as $\kappa(A)$, properties i) through iii) of Example 2.7. Moreover $\kappa_C(A)$ has the property $\kappa_C(D) = 1$

for any diagonal matrix D, that we have been intuitively looking for from the beginning. It is even invariant under row scaling i.e.

$$\kappa_C(DA) = \kappa_C(A) \, ,$$

because

$$|(DA)^{-1}| \, |DA| = |A^{-1}| \, |D^{-1}| \, |D| \, |A| = |A^{-1}| \, |A| \, .$$

Example 2.17 *Component-wise condition of nonlinear systems.* Let us compute the component-wise condition number of the problem (f^{-1}, y) of a nonlinear system $f(x) = y$, where $f : \mathbf{R}^n \to \mathbf{R}^n$ is as in Example 2.10 a continuously differentiable function. It follows in a completely analogous manner that

$$\kappa_{\text{rel}} = \frac{\| \, |f'(x)^{-1}| \, |f(x)| \|_\infty}{\|x\|_\infty} \, .$$

The expression $|f'(x)^{-1}| \, |f(x)|$ strongly resembles the correction

$$\Delta x = -f'(x)^{-1} f(x)$$

of Newton's method that we are going to meet in Section 4.2.

2.3 Stability of Algorithms

In this section we turn to the second group of errors, the errors in the algorithm. The mapping f describing the given problem (f, x) is realized in the numerical solution by a mapping $\tilde{f}$. This mapping contains all rounding and approximation errors and gives instead of $f(x)$ a perturbed output $\tilde{f}(x)$. Now the problem of *stability of the algorithm* reads:

'Is it acceptable to have the output $\tilde{f}(x)$ instead of $f(x)$?'

To answer it we must first think about how to characterize the perturbed mapping $\tilde{f}$. We have seen above that the errors in performing a floating point operation $\circ \in \{+, -, \cdot, /\}$ can be estimated by

$$a \hat{\circ} b = (a \circ b)(1 + \varepsilon) \, , \quad \varepsilon = \varepsilon(a, b) \quad \text{with} \quad |\varepsilon| \le \text{eps}. \tag{2.6}$$

Here it does not make too much sense (even if it may be possible in principle) to determine $\varepsilon = \varepsilon(a, b)$ for all values a and b on a given computer. In this respect our algorithm has to deal not with a single mapping $\tilde{f}$, but with a whole class $\{\tilde{f}\}$, containing all mappings characterized by estimates of the form (2.6). This class also contains the given problem $f \in \{\tilde{f}\}$.

The estimate of the error of a floating point operation (2.6) was derived in Section 2.1 only for machine numbers. Because we want to study the relation of the whole class of mappings we can allow arbitrary real numbers as arguments. In this way we put the mathematical tools of calculus at our disposal. Our model of the algorithm consists therefore of mappings operating on real numbers and satisfying estimates of the form (2.6).

In order to avoid unwieldy notation, let us denote the family $\{\tilde{f}\}$ by $\tilde{f}$ as well, i.e. $\tilde{f}$ stands for the whole family or for a representative according to the context. Statements on such an algorithm $\tilde{f}$ (for example error estimates) are always appropriately interpreted for all the mappings in the family. In particular we define the image $\tilde{f}(E)$ of a set E as the union

$$\tilde{f}(E) := \bigcup_{\phi \in \tilde{f}} \phi(E) \, .$$

We are left with the question of how to assess the error $\tilde{f}(x) - f(x)$. In our condition analysis we have seen that input data are always (at least for floating point numbers) affected by input errors that lead through the condition of the problem to unavoidable errors in the output. We cannot expect from our algorithm to accomplish more than allowed by the problem itself. Therefore we are happy when its error $\tilde{f}(x) - f(x)$ lies within reasonable bounds of the error $f(\tilde{x}) - f(x)$ caused by the input error. There are essentially two approaches for realising these considerations: the *forward analysis* and the *backward analysis*. They will be treated in what follows.

2.3.1   Stability concepts

With *forward analysis* we analyze the set

$$\tilde{R} := \tilde{f}(E)$$

of all outputs, as perturbed by input errors *and* errors in the algorithm. Because $f \in \tilde{f}$, the set $\tilde{R}$ contains in particular $R = f(E)$. The enlargement from R to $\tilde{R}$ identifies the stability in the sense of forward analysis. If the measure of $\tilde{R}$ has the same order of magnitude as that of R, then we say that the algorithm is *stable* in the sense of forward analysis.

The idea of *backward analysis* introduced by Wilkinson consists of passing the errors of the algorithm back to the starting point and interpreting them as input errors. To do this we formulate the perturbed output $\tilde{y} = \tilde{f}(\tilde{x})$ as the exact result $\tilde{y} = f(\hat{x})$ corresponding to the perturbed input quantity $\hat{x}$. Obviously this can be done only if $\tilde{f}(E)$ lies in the image of f. If this is not the case, a backward analysis is not possible and the algorithm is regarded

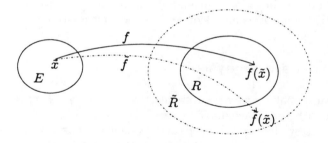

Figure 2.5: Input and output sets in forward analysis

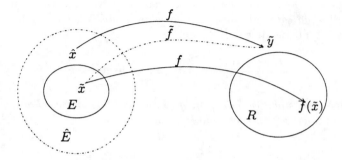

Figure 2.6: Input and output sets in backward analysis

as *unstable* in the sense of backward analysis. For non injective mappings f this is expressed in the form $\hat{x} \in f^{-1}(\tilde{y})$. In the concept of backward analysis $\hat{x}$ is chosen as the *least* deviation of the input, i.e.

$$\|\hat{x} - \tilde{x}\| = \min .$$

(As f was supposed to be continuous, there is at least one such $\hat{x}$.) If we construct the set

$$\hat{E} := \left\{\hat{x} \mid f(\hat{x}) = \tilde{f}(\tilde{x}) \text{ and } \|\hat{x} - \tilde{x}\| = \min \text{ for some } \tilde{x} \in E\right\}$$

of all those perturbed input quantities $\hat{x}$ of minimal distance, then any perturbed result $\tilde{y} \in \tilde{f}(E)$ can be interpreted as the exact result $f(\hat{x})$ corresponding to an input $\hat{x} \in \hat{E}$. The ratio between the sets $\hat{E}$ and E is here the measure of the stability of the algorithm in the sense of backward analysis.

In the following quantitative description of stability we will limit ourselves to roundoff errors both in input and algorithm and therefore we choose the relative error concept as an adequate basis for our analysis.

2.3.2   Forward analysis

With forward analysis we have to relate the error $\tilde{f}(x) - f(x)$ caused by the algorithm with the unavoidable error which, according to Section 2.2, can be estimated by the product κ eps of the relative condition number κ and the input error eps. We define therefore a *stability indicator* σ as the factor by which the algorithm enlarges the unavoidable error, κ eps.

Definition 2.18 Let $\tilde{f}$ be the floating point realization of an algorithm for solving the problem (f, x) with (norm-wise) relative condition number κ_{rel}. The *stability indicator of norm-wise forward analysis* is the smallest number $\sigma \geq 0$, such that for all $\tilde{x} \in E$

$$\frac{\|\tilde{f}(\tilde{x}) - f(\tilde{x})\|}{\|f(\tilde{x})\|} \;\overset{.}{\leq}\; \sigma \, \kappa_{\mathrm{rel}} \, \mathrm{eps} \quad \text{for eps} \to 0 \; .$$

Analogously the *stability indicator of component-wise forward analysis* is the smallest number $\sigma \geq 0$, such that

$$\max_i \frac{|\tilde{f}_i(\tilde{x}) - f_i(\tilde{x})|}{|f_i(\tilde{x})|} \;\overset{.}{\leq}\; \sigma \, \bar{\kappa}_{\mathrm{rel}} \, \mathrm{eps} \quad \text{for eps} \to 0 \; ,$$

where $\bar{\kappa}_{\mathrm{rel}}$ is the component-wise relative condition number of (f, x) (see remark 2.13).

The algorithm $\tilde{f}$ is called *stable* in the sense of forward analysis, if σ is smaller than the number of successively performed elementary operations

Lemma 2.19 *For the elementary operations* $\{+, -, *, /\}$ *and their floating point realizations* $\{\hat{+}, \hat{-}, \hat{*}, \hat{/}\}$ *we have*

$$\sigma \kappa \;\leq\; 1 \; .$$

Proof.   For any elementary operation $\hat{\circ} \in \{\hat{+}, \hat{-}, \hat{*}, \hat{/}\}$ we have $a \hat{\circ} b = (a \circ b)(1 + \varepsilon)$ for some ε with $|\varepsilon| \leq \mathrm{eps}$ and hence

$$\frac{|a \hat{\circ} b - a \circ b|}{|a \circ b|} = \frac{|(a \circ b)(1 + \varepsilon) - a \circ b|}{|a \circ b|} = |\varepsilon| \leq \mathrm{eps} \; .$$

$\square$

Example 2.20 *Subtraction.* We see in particular that in the case of cancellation $\kappa \gg 1$ so that the stability indicator is very small, $\sigma \ll 1$. The subtraction is outstandingly stable in this case and in the case of total cancellation is indeed error free $a\hat{-}b = a - b$.

When computing the stability indicators it is advantageous to first analyze the algorithm step by step and from there to obtain the stability of the whole algorithm in a recursive way. We will clarify this process for algorithms that can be split in two steps. To do this we divide the problem (f,x) in two partial problems (g,x) and $(h, g(x))$, where

$$f = h \circ g, \quad \mathbf{R}^n \xrightarrow{g} \mathbf{R}^l \xrightarrow{h} \mathbf{R}^n ,$$

and we assume that the stability indicators σ_g, σ_h for the partial algorithms $\tilde{g}$ and $\tilde{h}$ that implement g and h are known. How can we assess from here the stability of the composed algorithm

$$\tilde{f} = \tilde{h} \circ \tilde{g} := \{\psi \circ \phi \mid \phi \in \tilde{g}, \ \psi \in \tilde{h}\} \quad ?$$

Lemma 2.21 *The stability indicator σ_f of the composed algorithm $\tilde{f}$ satisfies*

$$\sigma_f \kappa_f \ \leq \ \sigma_h \kappa_h + \sigma_g \kappa_g \kappa_h \tag{2.7}$$

for both norm- and component-wise approaches.

Proof. We work out the proof for the norm-wise approach. Let $\tilde{g}$ and $\tilde{h}$ be arbitrary representatives of the algorithms for g and h as well as $\tilde{f} = \tilde{h} \circ \tilde{g}$ for $f = h \circ g$. Then

$$
\begin{aligned}
\|\tilde{f}(x) - f(x)\| &= \|\tilde{h}(\tilde{g}(x)) - h(g(x))\| \\
&\dot{\leq} \|\tilde{h}(\tilde{g}(x)) - h(\tilde{g}(x))\| + \|h(\tilde{g}(x)) - h(g(x))\| \\
&\dot{\leq} \sigma_h \kappa_h \, \mathrm{eps} \, \|h(\tilde{g}(x))\| + \kappa_h \frac{\|\tilde{g}(x) - g(x)\|}{\|g(x)\|} \|h(g(x))\| \\
&\dot{\leq} \sigma_h \kappa_h \, \mathrm{eps} \, \|h(\tilde{g}(x))\| + \kappa_h \sigma_g \kappa_g \, \mathrm{eps} \, \|h(g(x))\| \\
&\dot{\leq} (\sigma_h \kappa_h + \sigma_g \kappa_g \kappa_h) \, \mathrm{eps} \, \|f(x)\|
\end{aligned}
$$

$\square$

This lemma has an immediate important consequence. Because for each elementary operation we have $\sigma \kappa \leq 1$, the stability of the composed mapping $f = h \circ g$ is dangerously threatened whenever a subtraction arises in the second mapping h so that $\kappa_h \gg 1$. An illustrative example is the computation of the variance, see Exercise 2.16. We therefore derive the principle (compare with page 30):

Put unavoidable subtractions as close to the beginning of the algorithm as possible.

We now analyse via the recursive application of formula (2.7) the floating point realization of the sum of n real numbers and the scalar product.

Example 2.22 *Summation.* The simplest algorithm for the sum

$$s_n : \mathbf{R}^n \longrightarrow \mathbf{R}, \quad (x_1, \ldots, x_n) \longmapsto \sum_{i=1}^{n} x_i$$

is by recursive computation in accordance with $s_n = s_{n-1} \circ \alpha_n$ for $n > 2$ and $s_2 = \alpha_2$, where

$$\alpha_n : \mathbf{R}^n \longrightarrow \mathbf{R}^{n-1}, \quad (x_1, \ldots, x_n) \longmapsto (x_1 + x_2, x_3 \ldots, x_n)$$

denotes the addition of the first two components. We want to examine this "algorithm" component-wise. The condition number and stability indicator for α_n coincide with those for the addition of two numbers, i.e. $\kappa_{\alpha_n} = \kappa_+$ and $\sigma_{\alpha_n} = \sigma_+$. With the notation $\kappa_j := \kappa_{s_j}$ and $\sigma_j := \sigma_{s_j}$ we have by virtue of Lemma 2.21 that

$$\sigma_n \kappa_n \leq (\sigma_{n-1} + \sigma_+ \kappa_+)\kappa_{n-1} \leq (1 + \sigma_{n-1})\kappa_{n-1} \ .$$

According to Example 2.3, the condition number κ_n satisfies

$$\kappa_n = \frac{\sum_{i=1}^{n} |x_i|}{|\sum_{i=1}^{n} x_i|} \geq 1 \quad \text{and} \quad \kappa_{n-1} = \frac{\sum_{i=3}^{n} |x_i| + |x_1 + x_2|}{|\sum_{i=1}^{n} x_i|} \leq \kappa_n$$

and therefore $\sigma_n \leq 1 + \sigma_{n-1}$. Since $\sigma_2 = \sigma_+ \leq 1/\kappa_+ \leq 1$, we obtain for the stability indicator that

$$\sigma_n \leq n - 1 \ .$$

Hence the naive summation algorithm is numerically stable for the required $n - 1$ elementary operations.

Example 2.23 *Implementation of scalar products.* We subdivide the computation of the scalar product $f(x, y) = \langle x, y \rangle$ for $x, y \in \mathbf{R}^n$ into the component-wise multiplication

$$p : \mathbf{R}^n \times \mathbf{R}^n \longrightarrow \mathbf{R}^n, \quad ((x_i), (y_i)) \longmapsto (x_i y_i) \ ,$$

followed by the summation, $f = s_n \circ p$, analyzed in the last example. According to Lemmas 2.21 and 2.19 together with the estimation of the stability indicator of the last example we have

$$\sigma_f \kappa_f \leq (\sigma_n + \sigma_p \kappa_p)\kappa_n \leq (1 + \sigma_n)\kappa_n \leq n\kappa_n$$

and therefore

$$\sigma_f \leq n \frac{\kappa_n}{\kappa_f} = n \frac{\sum_{i=1}^{n} |x_i y_i| / |\sum_{i=1}^{n} x_i y_i|}{2 \sum_{i=1}^{n} |x_i y_i| / |\sum_{i=1}^{n} x_i y_i|} = n/2 . \tag{2.8}$$

At $2n - 1$ elementary operations, this algorithm for the scalar product is also numerically stable.

Actually, this estimate proves to be as a rule too pessimistic. Frequently one observes a factor $\sqrt{n}$ rather than n in (2.8).

Remark 2.24 Some computers have a 'scalar product function' with variable mantissa length, the so called *dot-product* $\odot$. Here the mantissas are enlarged in such a way that the additions can be performed in fixed point arithmetic. Thus one achieves the same stability as for addition, i.e. $\sigma \approx 1$.

In the following we will see how to carry out the forward analysis for scalar functions. In this special case we have a simplified version of Lemma 2.21.

Lemma 2.25 *If the functions g and h of Lemma 2.21 are scalar and differentiable then the stability indicator σ_f of the combined algorithm $\tilde{f} = \tilde{h} \circ \tilde{g}$ satisfies*

$$\sigma_f \leq \frac{\sigma_h}{\kappa_g} + \sigma_g .$$

Proof. In this special case the condition number of the combined problem is the product of the condition numbers of the parts

$$\kappa_f = \frac{|x| |f'(x)|}{|f(x)|} = \frac{|g(x)| |h'(g(x))| |g'(x)| |x|}{|h(g(x))| |g(x)|} = \kappa_h \, \kappa_g .$$

Hence from Lemma 2.21 it follows that

$$\sigma_f \, \kappa_h \, \kappa_g \leq \sigma_h \, \kappa_h + \sigma_g \, \kappa_h \, \kappa_g .$$

$\square$

If the condition number κ_g of the first partial problem is very small, $\kappa_g \ll 1$, then the algorithm becomes unstable. A small condition number can also be interpreted as *loss of information*: A change in the input has almost no influence on the output. Such a loss of information at the beginning of the algorithm leads therefore to instability. Moreover we see that an instability in the beginning of the algorithm (large σ_g) fully affects the composed algorithm. For example let us analyze the recursive method for computing $\cos mx$ and an intermediary result.

Example 2.26 For the mapping $f(x) = \cos x$ we have

$$\kappa_{\mathrm{abs}} = \sin x \quad \text{and} \quad \kappa_{\mathrm{rel}} = x \tan x \; .$$

If $x \to 0$ then $\kappa_{\mathrm{rel}} \doteq x^2 \to 0$. The evaluation of the function alone is extremely well conditioned for small x. However, if the information of x were to be *subsequently* used, then it becomes inaccessible through this intermediary step. We will come back to this phenomenon in the following example:

Example 2.27 Now we can analyze the recursive computation of $\cos mx$. It is important for example in the *Fourier-synthesis*, i.e. in the evaluation of trigonometric polynomials of the form

$$f(x) = \sum_{k=1}^{N} a_k \cos kx + b_k \sin kx \; .$$

On the basis of the addition theorem

$$\cos(k+1)x = 2 \cos x \cdot \cos kx - \cos(k-1)x$$

we can compute $c_m := f(x) := \cos mx$ from $c_0 = 1$ and $c_1 = \cos x$ by means of the three-term recurrence relation

$$c_{k+1} = 2 \cos x \cdot c_k - c_{k-1} \quad \text{for} \quad k = 1, 2, \ldots \tag{2.9}$$

Is this a stable algorithm? A crucial role is played by the evaluation of $g(x) := \cos x$, which occurs in each step of the three-term recurrence. We have just seen in example 2.26 that for small x information is lost when computing $\cos x$ and this can lead to instability. The corresponding stability indicator contains the factor

$$\frac{1}{\kappa(x)} = \left| \frac{g(x)}{g'(x)x} \right| = \left| \frac{1}{x \, \tan x} \right|$$

in each term of the sum. Since

$$x \to 0 \quad \Rightarrow \quad \frac{1}{\kappa(x)} \longrightarrow \frac{1}{x^2} \to \infty$$

$$x \to \pi \quad \Rightarrow \quad \frac{1}{\kappa(x)} \longrightarrow \frac{1}{\pi(x - \pi)} \to \infty \; ,$$

the recurrence is unstable for both limit cases $x \to 0, \pi$ with the former case $x \to 0$ being the more critical one. If we have a relative machine precision $\mathrm{eps} = 5 \cdot 10^{-12}$ and we compute the value $\cos mx$ for $m = 1240$ and $x = 10^{-4}$

according to (2.9), we obtain for example a relative error of 10^{-8}. With the condition number

$$\kappa = |mx \tan mx| \approx 1.5 \cdot 10^{-2}$$

it follows that $\sigma > 1.3 \cdot 10^5$, i.e. the calculation is clearly unstable. There is however a stable recurrence relation for computing $\cos mx$ for small values of x, developed by C. REINSCH [64]. It is based upon introducing differences $\Delta c_k := c_{k+1} - c_k$ and transforming the three-term recurrence relation (2.9) into a system of two-term recurrence relation

$$\begin{aligned} \Delta c_k &= -4 \sin^2(x/2) \cdot c_k + \Delta c_{k-1} \\ c_{k+1} &= c_k + \Delta c_k \end{aligned}$$

for $k = 1, 2, \ldots$ with starting values $c_0 = 1$ and $\Delta c_0 = -2 \sin^2(x/2)$. The evaluation of $h(x) := \sin^2(x/2)$ is stable for small $x \in [-\pi/2, \pi/2]$, since

$$\frac{1}{\kappa(h,x)} = \left| \frac{h(x)}{h'(x)x} \right| = \left| \frac{\tan(x/2)}{x} \right| \to \frac{1}{2} \quad \text{for} \quad x \to 0 .$$

For the above numerical example this recurrence yields an essentially better solution with a relative error of $1.5 \cdot 10^{-11}$. The recurrence for $x \to \pi$ can be stabilized in a similar way. (It turns out that these stabilizations lead ultimately to usable results only because the three-term recurrence relation (2.9) is well conditioned— see Section 6.2.1)

2.3.3 Backward analysis

For the quantitative description of the backward analysis we have to relate the errors in the algorithm, passed back to the input side with the original input errors.

Definition 2.28 The *norm-wise backward error* of the algorithm $\tilde{f}$ for solving the problem (f, x), is the smallest number $\eta \geq 0$, having the property that for any $\tilde{x} \in E$ there is $\hat{x}$ such that

$$\frac{\|\hat{x} - \tilde{x}\|}{\|\tilde{x}\|} \overset{.}{\leq} \eta \quad \text{for eps} \to 0 .$$

The *component-wise backward error* is defined analogously by replacing the inequality with

$$\max_i \frac{|\hat{x}_i - \tilde{x}_i|}{|\tilde{x}_i|} \overset{.}{\leq} \eta \quad \text{for eps} \to 0 .$$

The algorithm is called *stable* with respect to the relative input error δ, if

$$\eta < \delta \, .$$

For the input error $\delta = $ eps caused by roundoff we define the *stability indicator of the backward analysis* as the quotient

$$\sigma_R := \eta/\text{eps} \, .$$

As we see the condition of the problem does not appear in the definition. Also, in contrast with forward analysis the backward analysis does not require a beforehand condition analysis of the problem. Furthermore, the results are easily interpreted by comparing the input error and the backward error. Due to these properties the backward analysis is preferable, especially in case of complex algorithms. All the stability results for Gaussian elimination collected in the next section are related to backward analysis.

The two stability indicators σ and σ_R are not identical in general. The concept of backward analysis is rather stronger, as shown by the following lemma.

Lemma 2.29 *The stability indicators σ and σ_R of the forward and backward analysis satisfy*

$$\sigma \leq \sigma_R \, .$$

In particular backward stability implies forward stability.

Proof. From the definition of the backward error it follows that for any $\tilde{x} \in E$ there is a $\hat{x}$, such that $f(\hat{x}) = \tilde{f}(\tilde{x})$ and

$$\frac{\|\hat{x} - \tilde{x}\|}{\|\tilde{x}\|} \ \dot{\leq} \ \eta = \sigma_R \, \text{eps} \ \text{ for eps} \to 0 \, .$$

It follows that the relative error of the results satisfies

$$\frac{\|\tilde{f}(\tilde{x}) - f(\tilde{x})\|}{\|f(\tilde{x})\|} = \frac{\|f(\hat{x}) - f(\tilde{x})\|}{\|f(\tilde{x})\|} \ \dot{\leq} \ \kappa \, \frac{\|\hat{x} - \tilde{x}\|}{\|\tilde{x}\|} \ \dot{\leq} \ \kappa \, \sigma_R \, \text{eps}$$

for eps $\to 0$. $\square$

As an example of the backward analysis let us look again to the scalar product $\langle x, y \rangle$ for $x, y \in \mathbf{R}^n$ in its floating point implementation

$$\langle x, y \rangle := x_n y_n + \langle x^{n-1}, y^{n-1} \rangle \, , \tag{2.10}$$

where

$$x^{n-1} := (x_1, \ldots, x_{n-1})^T \ \text{ and } \ y^{n-1} := (y_1, \ldots, y_{n-1})^T \, .$$

This is the form used on sequential machines. We concentrate on the stability corresponding to the input x, because we need it in Section 2.4.2 for the analysis of the backward substitution.

Lemma 2.30 *The floating point implementation of the scalar product in accordance with (2.10) computes for $x, y \in \mathbf{R}^n$ a solution $\langle x, y \rangle_{fl}$, such that*

$$\langle x, y \rangle_{fl} = \langle \hat{x}, y \rangle$$

for an $\hat{x} \in \mathbf{R}^n$ with

$$|x - \hat{x}| \;\dot{\le}\; n \operatorname{eps} |x| \,,$$

i.e. the relative component-wise backward error amounts to

$$\eta \;\le\; n \operatorname{eps}$$

and the scalar product is (with $2n - 1$ elementary operations) stable in the sense of backward analysis

Proof. The recursive formulation (2.10) naturally suggests an inductive proof. For $n = 1$ the assertion is clear. Therefore let $n > 1$ and the assertion already proved for $n - 1$. For the floating point implementation of the recurrence relation (2.10) we have:

$$\langle x, y \rangle_{fl} = (x_n y_n (1 + \delta) + \langle x^{n-1}, y^{n-1} \rangle_{fl})(1 + \varepsilon) \,,$$

where δ and ε with $|\delta|, |\varepsilon| \le \operatorname{eps}$ characterize the relative errors of multiplication and addition respectively. Furthermore, according to the induction hypothesis we have

$$\langle x^{n-1}, y^{n-1} \rangle_{fl} = \langle z, y^{n-1} \rangle$$

for a $z \in \mathbf{R}^{n-1}$ with

$$|x^{n-1} - z| \;\le\; (n - 1) \operatorname{eps} |x^{n-1}| \,.$$

If we set $\hat{x}_n := x_n (1 + \delta)(1 + \varepsilon)$ and $\hat{x}_k := z_k (1 + \varepsilon)$ for $k = 1, \dots, n - 1$, it follows that

$$\langle x, y \rangle = x_n y_n (1 + \delta)(1 + \varepsilon) + \langle z(1 + \varepsilon), y \rangle = \hat{x}_n y_n + \langle \hat{x}^{n-1}, y \rangle = \langle \hat{x}, y \rangle \,,$$

where $|x_n - \hat{x}_n| \;\dot{\le}\; 2 \operatorname{eps} |x_n| \;\le\; n \operatorname{eps} |x_n|$ and

$$
\begin{aligned}
|x_k - \hat{x}_k| \;&\le\; |x_k - z_k| + |z_k - \hat{x}_k| \\
&\dot{\le}\; (n - 1) \operatorname{eps} |x_k| + \operatorname{eps} |\hat{x}_k| \;\dot{\le}\; n \operatorname{eps} |x_k|
\end{aligned}
$$

for $k = 1, \dots, n - 1$, hence $|x - \hat{x}| \;\dot{\le}\; n \operatorname{eps} |x|$. □

Remark 2.31 If we pass back the errors in x and y in an equally distributed way, then we deduce as in the case of forward analysis that the stability indicator satisfies $\sigma_R \le n/2$.

2.4   Application to Linear Systems

In what follows the above concepts of condition and stability will be discussed
again and deepened in the context of linear systems.

2.4.1   A closer look at solvability

As we did before in Chapter 1, we ask ourselves again the question: When
is a linear system $Ax = b$ solvable? In contrast to Chapter 1, where this
question was answered on the fictive basis of real numbers, we want to use
now the above derived error theory. Accordingly, the characteristic quantity
$\det A$ will be replaced by the condition number. We restate the results of
Section 2.2.1

$$\frac{\|\tilde{x} - x\|}{\|x\|} \overset{.}{\leq} \kappa_{\mathrm{rel}}\delta \,, \tag{2.11}$$

where δ is the relative input error of A (or b) . By norm-wise examination
we obtained for the relative condition number

$$\kappa_{\mathrm{rel}} = \|A^{-1}\|\frac{\|Ax\|}{\|x\|} \leq \|A^{-1}\| \, \|A\| = \kappa(A)$$

and by component-wise examination

$$\kappa_{\mathrm{rel}} = \frac{\| \, |A^{-1}| \, |A| \, |x| \, \|_\infty}{\|x\|_\infty} \leq \| \, |A^{-1}| \, |A| \, \|_\infty = \kappa_C(A) \,.$$

As seen in Example 2.7 for a nonzero matrix $A \in \mathrm{Mat}_n(\mathbf{R})$ we have

$$\det A = 0 \iff \kappa(A) = \infty \,. \tag{2.12}$$

If $\kappa(A) < \infty$, then the linear system is in principle uniquely solvable for
any right hand side. On the other hand, according to (2.11) we expect a
numerically usable result if $\kappa_{\mathrm{rel}}\delta$ is sufficiently small. In addition to that we
only have seen that a point condition of type (2.12) makes no sense because
instead of an individual matrix A we have to consider the set of all matrices
that are indistinguishable from A , for example

$$E := \{\tilde{A} \mid \|\tilde{A} - A\| \leq \delta\|A\|\}$$

It is therefore suitable to call a matrix A with relative precision δ "almost
singular" , or "numerically singular" whenever the corresponding input set
contains an (exactly) singular matrix. This leads to the following definition:

Definition 2.32 A matrix A is called *almost singular* or *numerically singular* with respect to the condition number $\kappa(A)$, whenever

$$\delta\kappa(A) \geq 1,$$

where δ is the relative precision of the matrix A.

For the rounding error caused when inputting the matrix A into a computer we assume, for example, that $\delta = \mathrm{eps}$. With experimental data δ is taken as the largest tolerance. Nevertheless we would like to stress the fact that linear systems with almost singular matrices may also be numerically "well behaved" – a fact that can be interpreted through the x-dependency of the relative condition number κ_{rel}.

Example 2.33 As an example we examine the linear system $Ax = b_1$,

$$A := \begin{pmatrix} 1 & 1 \\ 0 & \varepsilon \end{pmatrix}, \quad b_1 := \begin{pmatrix} 2 \\ \varepsilon \end{pmatrix},$$

where we regard $0 < \varepsilon \ll 1$ as input. The matrix A and the right hand side b_1 have the same common input variable ε; they are connected with each other. Therefore the condition number of the matrix

$$\kappa(A) = \|A^{-1}\|_\infty \|A\|_\infty = \frac{1}{\varepsilon} \gg 1$$

is not meaningful for the solution of the problem (f_1, ε),

$$f_1(\varepsilon) = A^{-1} b_1 = \begin{pmatrix} 1 \\ 1 \end{pmatrix}.$$

The solution is indeed independent of ε, i.e. $f_1'(\varepsilon) = 0$, and consequently the problem is well conditioned (relatively and absolutely) for any ε. However, if we examine a right hand side independent of ε, $b_2 := (0, 1)^T$, then we obtain the solution

$$f_2(\varepsilon) = A^{-1} b_2 = x_2 = \begin{pmatrix} -1/\varepsilon \\ 1/\varepsilon \end{pmatrix}$$

and the component-wise condition numbers

$$\kappa_{\mathrm{abs}} = \|f_2'(\varepsilon)\| = \frac{1}{\varepsilon^2} \quad \text{and} \quad \kappa_{\mathrm{rel}} = \frac{\| \, |f_2'(\varepsilon)| \, |\varepsilon| \, \|}{\|f_2(\varepsilon)\|} = 1 \, .$$

In this case only the direction of the solution is well-conditioned, which is reflected in the relative condition number. We will encounter a situation of this type again in Sections 3.1.2 and 5.2

2.4.2   Backward analysis of Gaussian elimination

We have already seen in Lemma 2.30 that the floating point computation of the scalar product $f(x,y) = \langle x,y \rangle$ is stable in the sense of the backward analysis. The algorithm for solving a linear system $Ax = b$ discussed in the first chapter requires only the evaluation of scalar products of certain rows and columns of matrices, so that on the basis of this consideration it is possible to perform a backward analysis of the Gaussian elimination. The technicality of the proof for backward analysis certainly increases with the complexity of the algorithm, and therefore we will carry it out explicitly here only for the forward substitution, while for the Gaussian triangular factorization we will just give the statement of the results that can be essentially found in WILKINSON [79] and [80]. A nice overview is also offered by the paper of HIGHAM [48].

Theorem 2.34 *The floating point implementation of the forward substitution for the solution of a triangular system $Lx = b$ computes a solution $\hat{x}$, such that there is a lower triangular matrix $\hat{L}$ with*

$$\hat{L}\hat{x} = b \quad \text{and} \quad |L - \hat{L}| \dot{\leq} n \operatorname{eps} |L| \,,$$

(i.e. for the component-wise relative backward error we have $\eta \leq n \cdot \operatorname{eps}$) so that the forward substitution is stable in the sense of backward analysis.

Proof. Similar to the scalar product (see Lemma 2.30), the forward substitution algorithm may also be recursively formulated as

$$l_{kk}x_k = b_k - \langle l^{k-1}, x^{k-1} \rangle \tag{2.13}$$

for $k = 1, \ldots, n$, where

$$x^{k-1} = (x_1, \ldots, x_{k-1})^T \quad \text{and} \quad l^{k-1} = (l_{k1}, \ldots, l_{k,k-1})^T \,.$$

Floating point implementation turns (2.13) into the recurrence relation

$$l_{kk}(1 + \delta_k)(1 + \varepsilon_k)\hat{x}_k = b_k - \langle l^{k-1}, \hat{x}^{k-1} \rangle_{fl} \,,$$

where δ_k and ε_k with $|\delta_k|, |\varepsilon_k| \leq \operatorname{eps}$ describe the relative error of the multiplication and addition respectively. For the floating point implementation of the scalar product we know already from Lemma 2.30 that

$$\langle l^{k-1}, \hat{x}^{k-1} \rangle_{fl} = \langle \hat{l}^{k-1}, \hat{x}^{k-1} \rangle$$

for some vector $\hat{l}^{k-1} = (\hat{l}_{k1}, \ldots, \hat{l}_{k,k-1})^T$ with

$$|l^{k-1} - \hat{l}^{k-1}| \leq (k-1)\operatorname{eps} |l^{k-1}| \,.$$

By setting also $\hat{l}_{kk} := l_{kk}(1 + \delta_k)(1 + \varepsilon_k)$, we get, as asserted,

$$\hat{L}x = b \quad \text{and} \quad |L - \hat{L}| \le n\,\text{eps}\,|L| \,.$$

$\square$

As a first result of backward analysis we assess the quality of the LR-factorization.

Lemma 2.35 *Let A possess an LR-factorization. Then the Gaussian elimination computes $\hat{L}$ and $\hat{R}$, such that*

$$\hat{L}\hat{R} = \hat{A}$$

for a matrix $\hat{A}$ with

$$|A - \hat{A}| \overset{.}{\le} n|\hat{L}|\,|\hat{R}|\,\text{eps} \,.$$

Proof. A simple inductive proof for the weaker statement with $4n$ instead of n is found in the book of GOLUB and VAN LOAN [39], Theorem 3.3.1. $\square$

The following estimation of the component-wise backward error of the Gaussian elimination was proved by SAUTTER. [68] The weaker statement with $3n$ instead of $2n$ follows easily from Theorem 2.34 and Lemma 2.35.

Theorem 2.36 *Let A possess a LR-factorization. Then the Gaussian elimination for the linear system $Ax = b$ computes a solution $\hat{x}$ with*

$$\hat{A}\hat{x} = b$$

for a matrix $\hat{A}$ with

$$|A - \hat{A}| \overset{.}{\le} 2n|\hat{L}|\,|\hat{R}|\,\text{eps} \,.$$

From here we can derive the following statement on the norm-wise backward error of the Gaussian elimination, that goes back to Wilkinson.

Theorem 2.37 *The Gaussian elimination with column pivoting for the linear system $Ax = b$ computes an $\hat{x}$, such that*

$$\hat{A}\hat{x} = b$$

for a matrix $\hat{A}$ with

$$\frac{\|A - \hat{A}\|_\infty}{\|A\|_\infty} \overset{.}{\le} 2n^3 \rho_n(A)\text{eps} \,,$$

where

$$\rho_n(A) := \frac{\alpha_{\max}}{\max_{i,j} |a_{ij}|}$$

and $\alpha_{\max}$ is the largest absolute value of an element of the remainder matrices $A^{(1)} = A$ through $A^{(n)} = R$ appearing during elimination.

Proof. We denote by $\hat{P}$, $\hat{L}$, $\hat{R}$ and $\hat{x}$ the quantities computed during the Gaussian factorization with column pivoting $PA = LR$. Then $\hat{P}A$ possesses a LR-factorization and according to Theorem 2.36 there is a matrix $\hat{A}$, such that $\hat{A}\hat{x} = b$ and

$$|A - \hat{m}A| \stackrel{.}{\leq} 2n\,\hat{P}^T\,|\hat{L}|\,|\hat{R}|\,\text{eps}\ .$$

Since $\|\hat{P}\|_\infty = 1$, it follows that

$$\|A - \hat{A}\|_\infty \stackrel{.}{\leq} 2n\,\|\hat{L}\|_\infty\,\|\hat{R}\|_\infty\,\text{eps}\ . \qquad (2.14)$$

The column pivoting strategy takes care that all the components of $\hat{L}$ are less than or equal to one in absolute value , i.e.

$$\|\hat{L}\|_\infty \leq n\ .$$

The norm of $\hat{R}$ can be estimated by

$$\|\hat{R}\|_\infty \leq n\,\max_{i,j}|\hat{r}_{ij}| \leq n\,\alpha_{\max}\ .$$

The statement follows therefore from (2.14), because $\max_{i,j}|a_{ij}| \leq \|A\|_\infty$. $\square$

So what is the stability of Gaussian elimination? This question is not clearly answered by Theorem 2.37. Whether the matrix is suitable for Gaussian elimination depends obviously on the number $\rho_n(A)$. In general this quantity can be estimated by

$$\rho_n(A) \leq 2^{n-1},$$

where the estimate is sharp because the bound is attained by the (pathological) matrix given by Wilkinson (see Exercise 2.20).

$$A_W = \begin{bmatrix} 1 & & & 1 \\ -1 & \ddots & & \vdots \\ \vdots & \ddots & \ddots & \vdots \\ -1 & \cdots & -1 & 1 \end{bmatrix}$$

Therefore Gaussian elimination with column pivoting is not stable for the whole class of invertible matrices. However for special classes of matrices the situation looks considerably better. In Table 2.1 we have listed some classes of matrices and the corresponding estimates for ρ_n. The Gaussian

Table 2.1: Stability of Gaussian elimination for special matrices

Type of matrix	column pivoting	$\rho_n \leq$
invertible	yes	2^{n-1}
upper Hessenberg	yes	n
A or A^T strictly diagonally dominant	superfluous	2
tridiagonal	yes	2
symmetric positive definite	no	1
random	yes	$n^{2/3}$ (average)

elimination is stable for symmetric positive definite matrices (Cholesky factorization) as well as for diagonally dominant matrices, i.e.

$$|a_{ii}| > \sum_{\substack{j=1 \\ i \neq j}}^{n} |a_{ij}| \text{ for } i = 1, \ldots, n,$$

Furthermore one could state with a clear conscience that Gaussian elimination is stable "as a rule", i.e. is stable for matrices usually encountered in practice. This statement is also supported by the probabilistic considerations carried out by TREFETHEN and SCHREIBER [76] — (see the last row of Table 2.1).

2.4.3   Assessment of approximate solutions

Now let us assume that an approximate solution $\tilde{x}$ of a given linear system $Ax = b$ is already available and we want to answer the question: How good is an approximate solution $\tilde{x}$? The fact is that x is the exact solution of the linear system $Ax = b$ exactly when the corresponding *residual* vanishes,

$$r(x) := b - Ax = 0 \, .$$

From a "good" approximation one would expect the residual $r(\tilde{x})$ to be as small as possible. The question is then 'how small is small' ? In looking for a sensible limit we run into the problem that the norm $\|r(\tilde{x})\|$ can be arbitrarily changed by a row scaling

$$Ax = b \rightarrow (D_r A)x = D_r b$$

although the problem itself is not changed by it. This can be also read from the invariance of Skeel's condition number $\kappa_C(A)$ with respect to row scaling. Therefore the residual can be invoked at best when it possesses a problem specific meaning. Better suited is the concept of backward analysis introduced in Definition 2.28. The norm-wise and component-wise backward errors of an approximate solution $\tilde{x}$ of a linear system can be given directly.

Theorem 2.38 *The norm-wise relative backward error of an approximate solution $\tilde{x}$ of a linear system $f(A, b) = f^{-1}b$ relative to $\|(A, b)\| := \|A\| + \|b\|$ is*

$$\eta_N(\tilde{x}) = \frac{\|A\tilde{x} - b\|}{\|A\| \, \|\tilde{x}\| + \|b\|} \; .$$

Proof. See RIGAL and GACHES [65].   □

The following theorem goes back to PRAGER and OETTLI [62] .

Theorem 2.39 *The component-wise relative backward error of an approximate solution $\tilde{x}$ of $Ax = b$ is*

$$\eta_C(\tilde{x}) = \max_i \frac{|A\tilde{x} - b|_i}{(|A| \, |\tilde{x}| + |b|)_i} \; . \tag{2.15}$$

Proof. Let

$$\eta_C = \min\{w \mid \text{there is } \tilde{A}, \tilde{b} \text{ with } \tilde{A}\tilde{x} = \tilde{b}, \, |\tilde{A} - A| \leq w|A|, \, |\tilde{b} - b| \leq w|b|\}.$$

We set

$$\Theta := \max_i \frac{|r(\tilde{x})|_i}{(\, |A| \, |\tilde{x}| + |b|)_i}$$

and we have to prove that $\eta_C(\tilde{x}) = \Theta$. Half of the statement, $\eta_C(\tilde{x}) \geq \Theta$, follows from the fact that for any solution $\tilde{A}, \tilde{b}, w$ we have

$$|A\tilde{x} - b| = |(A - \tilde{A})\tilde{x} + (\tilde{b} - b)| \leq |A - \tilde{A}| \, |\tilde{x}| + |\tilde{b} - b| \leq w(|A| \, |\tilde{x}| + |b|) \; ,$$

and therefore $w \geq \Theta$. For the other half of the statement $\eta_C(\tilde{x}) \leq \Theta$, we write the residual $\tilde{r} = b - A\tilde{x}$ as

$$\tilde{r} = D(|A| \, |\tilde{x}| + |b|) \quad \text{with} \quad D = \text{diag}(d_1, \dots, d_n) \; ,$$

where

$$d_i = \frac{\tilde{r}_i}{(|A| \, |\tilde{x}| + |b|)_i} \leq \Theta \; .$$

Now by setting

$$\tilde{A} := A + D|A| \, \text{diag}(\text{sgn}\,(\tilde{x})) \quad \text{and} \quad \tilde{b} := b - D|b| \; ,$$

we get $|\tilde{A} - A| \leq \Theta|A|$, $|\tilde{b} - b| \leq \Theta|b|$ and

$$\tilde{A}\tilde{x} - \tilde{b} = A\tilde{x} + D|A||\tilde{x}| - b + D|b| = -\tilde{r} + D(|A| \, |\tilde{x}| + |b|) = 0.$$

$\square$

Remark 2.40 Naturally, formula (2.15) for the backward error is usable in practice only when the error occurring in the evaluation does not alter the meaningfulness of the result. SKEEL showed that the value $\tilde{\eta}$ computed in floating point arithmetic differs from the actual value η by at most $n \cdot$ eps. Thus formula (2.15) can really be invoked to assess the quality of an approximate solution $\tilde{x}$.

In Chapter 1, the iterative refinement was introduced as a possibility of "improving" approximate solution. The following result proved by SKEEL [70] shows the effectiveness of the iterative refinement. For Gaussian elimination one refinement step already implies component-wise stability.

Theorem 2.41 *If*

$$K_n \kappa_C(A^{-1}) \, \sigma(A, x)\text{eps} < 1,$$

where $\kappa_C(A^{-1}) = \| \, |A| \, |A^{-1}| \, \|_\infty$ *is Skeel's condition number,*

$$\sigma(A, x) := \frac{\max_i(|A||x|)_i}{\min_i(|A||x|)_i}$$

and K_n *is a constant close to* n, *then Gaussian elimination with column pivoting followed by one refinement step has a component-wise stability indicator*

$$\sigma \leq n + 1 \,,$$

i.e this form of Gaussian elimination is stable.

The quantity $\sigma(A, x)$ is a measure of the quality of the scaling (see [70]).

Example 2.42 Hamming's example

$$A = \begin{bmatrix} 3 & 2 & 1 \\ 2 & 2\varepsilon & 2\varepsilon \\ 1 & 2\varepsilon & -\varepsilon \end{bmatrix} , \quad b = \begin{pmatrix} 3 + 3\varepsilon \\ 6\varepsilon \\ 2\varepsilon \end{pmatrix} , \quad x = \begin{pmatrix} \varepsilon \\ 1 \\ 1 \end{pmatrix}$$

We take as ε the relative machine precision, here $\varepsilon := \text{eps} = 3 \cdot 10^{-10}$. We obtain for Skeel's condition number

$$\kappa_{\text{rel}} = \frac{\| \, |A^{-1}| \, |A| \, |x| + |A^{-1}| \, |b| \, \|_\infty}{\|x\|_\infty} \doteq 6 \,,$$

i.e. this system of equations is well conditioned. By using Gaussian elimination with column pivoting we obtain the solution

$$\tilde{x} = x_0 = (6.20881717 \cdot 10^{-10}, 1, 1)^T$$

with the component-wise backward error (according to Theorem 2.39) $\eta(x_0) = 0.15$. From the quantity $\sigma(A, x) = 2.0 \cdot 10^9$ we can read that the problem is extremely badly scaled. In spite of that, after one refinement step we obtain the solution $x_1 = x$ with the computed backward error $\eta(x_1) = 1 \cdot 10^{-9}$.

2.5   Exercises

Exercise 2.1 Show that the elementary operation $\hat{+}$ is not associative in floating point arithmetic.

Exercise 2.2 Determine through a short program the relative machine precision of the computer you use.

Exercise 2.3 The zeros of the cubic polynomial

$$z^3 + 3qz - 2r = 0, \quad r, q > 0$$

are to be found. According to Cardano-Tartaglia a real root is given by

$$z = \left(r + \sqrt{q^3 + r^2}\right)^{\frac{1}{3}} + \left(r - \sqrt{q^3 + r^2}\right)^{\frac{1}{3}}$$

This formula is numerically inappropriate because two cubic roots must be computed and total cancellation occurs for $r \to 0$. Give a cancellation free formula for z that requires the computation of only one cubic and one square root.

Exercise 2.4 Determination of zeros of polynomials in coefficient representation is in general an ill-conditioned problem. For illustration consider the following polynomial:

$$P(x) = x^4 - 4x^3 + 6x^2 - 4x + a, \quad a = 1.$$

How do the roots x_i of P, change when the coefficient a is distorted to $\bar{a} := a - \varepsilon, 0 < \varepsilon \le \text{eps}$? Give the condition number of the problem. Specify the condition of the problem. With how many exact digits can the solution be computed if the relative machine precision is $\text{eps} = 10^{-16}$?

Exercise 2.5 Determine

$$\alpha = \left(\frac{1 + \frac{1}{2n}}{1 - \frac{1}{2n}}\right)^n - e\left(1 + \frac{1}{12n^2}\right) > 0$$

for $n = 10^6$ up to three exact digits.

Exercise 2.6 Compute the condition number of the evaluation of a polynomial that is given by its coefficients $a_0, \ldots, a_n$

$$p(x) = a_n x^n + a_{n-1} x^{n-1} + \cdots + a_1 x + a_0$$

at the point x, first with respect to perturbation of the coefficients $a_i \rightarrow \tilde{a}_i = a_i(1 + \varepsilon_i)$, and then with respect to perturbations $x \rightarrow \tilde{x} = x(1 + \varepsilon)$ of x. Consider in particular the polynomial

$$p(x) = 8118x^4 - 11482x^3 + x^2 + 5741x - 2030$$

at the point $x = 0.707107$. The exact result is

$$p(x) = -1.9152732527082 \cdot 10^{-11}.$$

Suppose that a computer delivers the value

$$\tilde{p}(x) = -1.9781509763561 \cdot 10^{-11}.$$

Assess the solution by using the condition of the problem.

Exercise 2.7 Let $\| \cdot \|_1$ and $\| \cdot \|_2$ be norms on $\mathbf{R}^n$ and $\mathbf{R}^m$ respectively. Show that

$$\|A\| := \sup_{x \neq 0} \frac{\|Ax\|_2}{\|x\|_1}$$

defines a norm on the space $\text{Mat}_{m,n}(\mathbf{R})$ of all real (m, n) matrices.

Exercise 2.8 The p-norms on $\mathbf{R}^n$ are defined for $1 \leq p < \infty$ by

$$\|x\|_p := \left(\sum_{i=1}^n |x_i|^p\right)^{\frac{1}{p}}$$

and for $p = \infty$ by

$$\|x\|_\infty := \max_{i=1,\ldots,n} |x_i|.$$

Show that the corresponding matrix norms

$$\|A\|_p := \sup_{x \neq 0} \frac{\|Ax\|_p}{\|x\|_p}$$

satisfy:

a) $\|A\|_1 = \max_{j=1,\ldots,n} \sum_{i=1}^{m} |a_{ij}|$

b) $\|A\|_\infty = \max_{i=1,\ldots,m} \sum_{j=1}^{n} |a_{ij}|$

c) $\|A\|_2 \le \sqrt{\|A\|_1 \|A\|_\infty}$

d) $\|AB\|_p \le \|A\|_p \|B\|_p$ for $1 \le p \le \infty$.

Exercise 2.9 (Spectral norm and Frobenius norm) Let $A \in \mathrm{Mat}_n(\mathbf{R})$.

a) Show that $\|A\|_2 = \|A^T\|_2$.

b) Show that $\|A\|_F = \sqrt{tr(A^T A)} = \|\sigma(A)\|_2$, where $\sigma(A) = [\sigma_1, \ldots, \sigma_n]^T$ is the vector of singular values of A .

c) Deduce from b), that $\|A\|_2 \le \|A\|_F \le \sqrt{n}\|A\|_2$.

d) Prove the submultiplicativity of the Frobenius norm $\| \cdot \|_F$.

e) Show that the Frobenius norm $\| \cdot \|_F$ is not an operator norm.

Exercise 2.10 Let A be a matrix with columns A_j, $j = 1, \ldots, n$. We define a matrix norm by
$$\|A\|_\square := \max_j \|A_j\|_2$$

Show that:

a) $\| \cdot \|_\square$ defines indeed a norm on $\mathrm{Mat}_{m,n}(\mathbf{R})$.

b) $\|Ax\|_2 \le \|A\|_\square \|x\|_1$ and $\|A^T x\|_\infty \le \|A\|_\square \|x\|_2$.

c) $\|A\|_F \le \sqrt{n}\|A\|_\square$ and $\|AB\|_\square \le \|A\|_F \|B\|_\square$.

Exercise 2.11 Let A be a nonsingular matrix, that is transformed into $\tilde{A} := A + \delta A$ by a small perturbation δA, and let $\| \cdot \|$ be a submultiplicative norm with $\|I\| = 1$. Show that:

a) If $\|B\| < 1$, then there exists $(I - B)^{-1} = \sum_{k=0}^{\infty} B^k$, and the following inequality holds
$$\|(I - B)^{-1}\| \le \frac{1}{1 - \|B\|} .$$

b) If $\|A^{-1}\delta A\| \le \varepsilon < 1$, then
$$\kappa(\tilde{A}) \le \frac{1+\varepsilon}{1-\varepsilon}\kappa(A) .$$

Exercise 2.12 Show that the following property holds for any invertible matrix $A \in \mathrm{GL}(n)$

$$\kappa(A) = \|A\| \, \|A^{-1}\| = \frac{\max_{\|x\|=1} \|Ax\|}{\min_{\|x\|=1} \|Ax\|} \ .$$

Exercise 2.13 Let $A \in \mathrm{GL}(n)$ be symmetric positive definite. Show that in this case

$$\kappa(A) = \frac{\lambda_{\max}(A)}{\lambda_{\min}(A)} \ .$$

Exercise 2.14 Show that the mapping $f : \mathrm{GL}(n) \to \mathbf{R}^n$, defined by $f(A) = A^{-1}b$, satisfies

$$|f'(A)| \, |C| = |A^{-1}| \, |C| \, |A^{-1}b| \ .$$

Exercise 2.15 Determine the component-wise relative condition number of the sum $\sum_{i=1}^{n} x_i$ of n real number $x_1, \ldots, x_n$ and compare it with the norm-wise relative condition number relative to the norm $\| \cdot \|_1$.

Exercise 2.16 There are two formulas for computing the variance of a vector $x \in \mathbf{R}^n$:

$$a) \qquad S^2 = \frac{1}{n-1} \sum_{i=1}^{n} (x_i - \bar{x})^2$$

$$b) \qquad S^2 = \frac{1}{n-1} \left(\sum_{i=1}^{n} x_i^2 - n\bar{x}^2 \right) ,$$

where $\bar{x} = \frac{1}{n} \sum_{i=1}^{n} x_i$ is the mean value of x_i. Which of the two formulas for computing S^2 is numerically more stable, and therefore preferable? Support your assertion on the stability indicator, and illustrate your choice with a numerical example.

Exercise 2.17 We consider the approximation of the exponential $\exp(x)$ by the truncated Taylor series

$$\exp(x) \approx \sum_{k=0}^{N} \frac{x^k}{k!} \ . \qquad (2.16)$$

Compute approximate values for $\exp(x)$ for $x = -5.5$ of the following three types with $N = 3, 6, 9, \ldots, 30$:

a) With the help of formula (2.16).

b) With $\exp(-5.5) = 1/\exp(5.5)$ and formula (2.16) for $\exp(5.5)$.

c) With $\exp(-5.5) = (\exp(0.5))^{-11}$ and formula (2.16) for $\exp(0.5)$.

The "exact" value is

$$\exp(-5.5) = 0.0040867714\ldots.$$

How are the results interpreted?

Exercise 2.18 Let there be given a matrix

$$A_\varepsilon = \begin{pmatrix} 1 & 1 \\ 1 & 1+\varepsilon \end{pmatrix}$$

and two right hand sides

$$b_1^T = (1,1), \quad b_2^T = (-1,1) \ .$$

Compute the two solutions x_1 and x_2 of the equations $Ax_i = b_i$, the condition number $\kappa_\infty(A_\varepsilon)$ and Skeel's condition numbers $\kappa_{\mathrm{rel}}(A_\varepsilon, x_1)$ and $\kappa_{\mathrm{rel}}(A_\varepsilon, x_2)$. How are these results interpreted?

Exercise 2.19 Let be given a linear system $Ax = b$ with

$$A = \begin{pmatrix} 0.780 & 0.563 \\ 0.913 & 0.659 \end{pmatrix}, \quad b = \begin{pmatrix} 0.217 \\ 0.254 \end{pmatrix}$$

with relative precision eps $= 10^{-4}$. Assess the precision of the two approximate solutions

$$\tilde{x}_1 = \begin{pmatrix} 0.999 \\ -1.001 \end{pmatrix} \quad \text{and} \quad \tilde{x}_2 = \begin{pmatrix} 0.341 \\ -0.087 \end{pmatrix}.$$

with the help of the Theorem of Prager and Oettli (Theorem 2.39).

Exercise 2.20 Show that in the pathological example of Wilkinson

$$A = \begin{bmatrix} 1 & & & 1 \\ -1 & \ddots & & \vdots \\ \vdots & \ddots & \ddots & \vdots \\ -1 & \cdots & -1 & 1 \end{bmatrix}$$

the maximal absolute value of a pivot in column pivoting is

$$|\alpha_{\max}| = 2^{n-1}$$

Exercise 2.21 According to R. Skeel (1979) the component-wise backward error η in Gaussian elimination with column pivoting for solving $Ax = b$, satisfies

$$\eta \le \chi_n \, \sigma(A, x) \text{ eps },$$

where χ_n is a constant depending only on n (as a rule $\chi_n \approx n$), and

$$\sigma(A, x) = \frac{\max_i(|A| \, |x|)_i}{\min_i(|A| \, |x|)_i} .$$

Specify a row scaling $A \to DA$ of the matrix A with a diagonal matrix $D = \mathrm{diag}(d_1, \dots, d_n)$, $d_i > 0$, such that

$$\sigma(DA, x) = 1.$$

Why is this an impractical stabilization method?

Exercise 2.22 Let

$$A = \begin{bmatrix} 1 & 1 & -1 & -1 \\ 0 & \varepsilon & 0 & 0 \\ 0 & 0 & \varepsilon & 0 \\ 1 & 0 & 0 & 1 \end{bmatrix}, \quad b = \begin{bmatrix} 0 \\ 1 \\ 1 \\ 2 \end{bmatrix}.$$

The solution of the linear system $Ax = b$ is $x = [1, \varepsilon^{-1}, \varepsilon^{-1}, 1]^T$.

a) Show that this system is well conditioned but badly scaled, by computing the condition number $\kappa_C(A) = \|\, |A|^{-1}|A| \,\|_\infty$ and the scaling quantity $\sigma(A, x)$ (see Exercise 2.21). What do you expect from Gaussian elimination, when ε is substituted by the relative machine precision eps ?

b) Solve the system by a Gaussian elimination program with column pivoting for $\varepsilon = $ eps. How big is the computed backward error $\hat{\eta}$?

c) Convince yourself that one single refinement step delivers a stable result.

3 Linear Least Squares Problems

This chapter deals with the solution of overdetermined linear systems by means of the *linear least squares method*, also known as *maximum likelihood method*. Because of the invariance of this type of problem, orthogonal transformations are very well suited for its solution. We give in Section 3.2 the description of the so called *orthogonalization method* that can be used for a stable solution of the linear least squares problem. Associated with it is a (somewhat more expensive) alternative to Gaussian elimination for solving linear systems.

3.1 Least Squares Method of Gauss

Gauss first described the method in 1809 in 'Theoria Motus Corporum Coelestium', as part of the solution algorithm (see Section 3.1.4), as he also did with the elimination method discussed in the first chapter. At the same time in this work he also studies fundamental questions from probability theory–obviously celestial mechanics appears to be an extremely stimulative application problem.

3.1.1 Formulation of the problem

We start with the following problem formulation: Let

$$(t_i, b_i), \quad t_i, b_i \in \mathbf{R}, \quad i = 1, \ldots, m ,$$

be m given points, where b_i may describe for example the position of an object at time t_i. We assume that these measurements are in conformity to a natural law, so that the dependence of b on t can be expressed by a *model function* φ

$$b(t) = \varphi(t; x_1, \ldots, x_n)$$

where the model function contains n unknown parameters x_i.

Example 3.1 We consider as an example Ohm's law $b = xt = \varphi(t; x)$, where t is the intensity of the current, b the voltage and x the resistance. The task is to draw a line through the origin that is 'as close as possible' to the measurements.

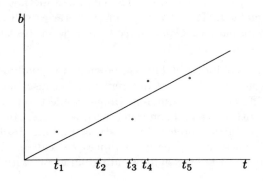

Figure 3.1: Linear least squares computation for Ohm's law.

If there were no measurement errors, then the model would describe exactly the situation and the parameters $x_1, \ldots, x_n$ would need to be determined such that

$$b_i = b(t_i) = \varphi(t_i; x_1, \ldots, x_n) \quad \text{for} \quad i = 1, \ldots, m \ .$$

The measurements are actually corrupted by errors and the model function describes only approximately the reality. So Ohm's law holds approximately only in the middle of a temperature range, a fact that will become clear at least when a wire burns out. Therefore we can only require that

$$b_i \approx \varphi(t_i; x_1, \ldots, x_n) \quad \text{for} \quad i = 1, \ldots, m \ .$$

There are several possibilities of weighing the individual deviations

$$\Delta_i := b_i - \varphi(t_i; x_1, \ldots, x_n) \quad \text{for} \quad i = 1, \ldots, m \ .$$

Gauss chose first to minimize *even powers* of the deviations. Based on considerations from the theory of probability he finally chose the *squares* Δ_i^2. This leads to the problem of determining the parameters $x_1, \ldots, x_n$ such that $\sum_{i=1}^m \Delta_i^2$ becomes minimal. We denote it simply by

$$\Delta^2 := \sum_{i=1}^m \Delta_i^2 = \min \ . \tag{3.1}$$

Remark 3.2 The relation between the linear least squares problems and probability theory is reflected in the equivalence of the minimization problem (3.1) with the maximization problem

$$\exp(-\Delta^2) = \max \ .$$

The exponential term characterizes here a probabilistic distribution, the so called *Gaussian normal distribution*. The complete method is called *maximum likelihood method*. The attentive contemporaries will have surely noticed that the accompanying *Gaussian bell curve* decorates the new German ten mark bill introduced in 1991.

In (3.1) the errors of individual measurements are equally weighted. However, the measurements (t_i, b_i) are different just because the measuring apparatus works differently over different ranges while the measurements are taken sometime with more and sometime with less care. To any individual measurement b_i pertains therefore in a natural way an absolute measuring precision, or tolerance, δb_i. These tolerances δb_i can be included in the problem formulation (3.1) by weighing individual errors with tolerances, i.e.

$$\sum_{i=1}^{m} \left(\frac{\Delta_i}{\delta b_i} \right)^2 = \min \ .$$

This form of minimization has also a sensible statistical interpretation (somewhat similar to standard deviation).

We consider here at first only the special case when the model function φ is *linear* in x, i.e.

$$\varphi(t; x_1, \ldots, x_n) = a_1(t)x_1 + \cdots + a_n(t)x_n \ ,$$

where $a_1, \ldots, a_n : \mathbf{R} \to \mathbf{R}$ are arbitrary functions. The nonlinear case will be considered in Section 4.3. In this section $\| \cdot \|$ denotes the Euclidean norm $\| \cdot \| = \| \cdot \|_2$. In the linear case the least squares problem can be written in short form as

$$\|b - Ax\| = \min$$

where $b = (b_1, \ldots, b_m)^T$, $x = (x_1, \ldots, x_n)^T$ and

$$A = (a_{ij}) \in \text{Mat}_{m,n}(\mathbf{R}) \quad \text{with} \quad a_{ij} := a_j(t_i) \ .$$

As a further restriction we will consider here only the overdetermined case $m \geq n$, i.e. there are more data than parameters to be determined, which sounds sensible from a statistical point of view. We obtain therefore the framework of the *linear least squares problem*: For given $b \in \mathbf{R}^m$ and $A \in Mat_{m,n}(\mathbf{R})$ with $m \geq n$ find an $x \in \mathbf{R}^n$, such that

$$\|b - Ax\| = \min \ . \tag{3.2}$$

Remark 3.3 By replacing the 2-norm with the 1-norm one obtains the standard *linear optimization* problem that will not be treated here by us because it appears rather infrequently in the natural sciences and engineering. It plays however a very important role in economics and management science. With the maximum norm instead of the 2-norm we come across the so called *Chebyshev approximation* problem. This problem occurs more often in the natural sciences, although not as frequently as the Gaussian least squares problem. Because the Euclidean norm is induced by a scalar product the latter offers more ground for geometric insight.

3.1.2   Normal equations

Geometrically speaking the solution of the linear least squares problem consists of finding a point $z = Ax$ in the range space $R(A)$ of A, which has the smallest distance to the given point b. For $m = 2$ and $n = 1$ $R(A) \subset \mathbf{R}^2$ is either the origin or a straight line through the origin (see Figure 3.2). It is graphically clear that the the distance $\|b - Ax\|$ is minimal exactly where the difference $b - Ax$ is perpendicular to the subspace $R(A)$. In other words: Ax is the *orthogonal projection* of b onto the subspace $R(A)$. As we want to fall back on this result later, we will formulate it in a somewhat more abstract form.

Theorem 3.4 *Let V be a finite dimensional Euclidean vector space with scalar product $\langle \cdot, \cdot \rangle$, $U \subset V$ a subspace and*

$$U^{\perp} = \left\{ v \in V \mid \langle v, u \rangle = 0 \text{ for all } u \in U \right\}$$

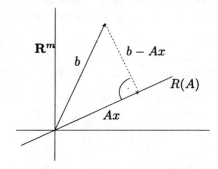

Figure 3.2: Projection on the range space $R(A)$

its orthogonal complement in V. Then for all $v \in V$ we have the following
property regarding the norm $\|v\| = \sqrt{\langle v, v \rangle}$ induced by the scalar product:

$$\|v - u\| = \min_{u' \in U} \|v - u'\| \iff v - u \in U^\perp .$$

Proof. Let $u \in U$ be the (uniquely determined) point such that $v - u \in U^\perp$.
Then for all $u' \in U$ we have

$$
\begin{aligned}
\|v - u'\|^2 &= \|v - u\|^2 + 2\langle v - u, u - u' \rangle + \|u - u'\|^2 \\
&= \|v - u\|^2 + \|u - u'\|^2 \geq \|v - u\|^2 ,
\end{aligned}
$$

where equality holds if and only if $u = u'$.   □

Remark 3.5 With this the solution $u \in U$ of $\|v - u\| = \min$ is uniquely
determined and is called the *orthogonal projection* of v onto U. The mapping

$$P : V \longrightarrow U, \quad v \longmapsto Pv \quad \text{with} \quad \|v - Pv\| = \min_{u \in U} \|v - u\|$$

is linear and is called the *orthogonal projection* from V onto U.

Remark 3.6 The theorem generally holds also when U is replaced by an
affine subspace $W = w_0 + U \subset V$, where $w_0 \in V$ and U is a subspace of V
parallel to W. Then for all $v \in V$ and $w \in W$ it follows

$$\|v - w\| = \min_{w' \in W} \|v - w'\| \iff v - w \in U^\perp .$$

This defines, as in Remark 3.5, a function

$$P : V \longrightarrow W, \quad v \longmapsto Pv \quad \text{with} \quad \|v - Pv\| = \min_{w \in W} \|v - w\|$$

which is an affine mapping called the *orthogonal projection* of V onto the
affine subspace W. This consideration will prove to be quite useful in Chapter 8.

With Theorem 3.4 we can easily prove a statement on the existence and
uniqueness of the solution of the linear least squares problem.

Theorem 3.7 *The vector $x \in \mathbf{R}^n$ is a solution of the linear least squares*
problem $\|b - Ax\| = \min$, if and only if it satisfies the so called normal
equations

$$A^T A x = A^T b . \tag{3.3}$$

In particular, the linear least squares problem is uniquely solvable if and only
if the rank of A is maximal, i.e. rank $A = n$.

Proof. By applying Theorem 3.4 to $V = \mathbf{R}^m$ and $U = R(A)$ we get

$$
\begin{aligned}
\|b - Ax\| = \min \quad &\Longleftrightarrow \quad \langle b - Ax, Ax' \rangle = 0 \ \text{ for all } x' \in \mathbf{R}^n \\
&\Longleftrightarrow \quad \langle A^T(b - Ax), x' \rangle = 0 \ \text{ for all } x' \in \mathbf{R}^n \\
&\Longleftrightarrow \quad A^T(b - Ax) = 0 \\
&\Longleftrightarrow \quad A^T A x = A^T b
\end{aligned}
$$

and therefore the first statement. The second part follows from the fact that $A^T A$ is invertible if and only if rank $A = n$. $\qquad\square$

Remark 3.8 Geometrically, the normal equations mean precisely that $b - Ax$ is normal to $R(A) \subset \mathbf{R}^m$, hence the name.

3.1.3 Condition

We begin our condition analysis with the orthogonal projection $P : \mathbf{R}^m \to V$, $b \mapsto Pb$, onto a subspace V of $\mathbf{R}^m$ (see Figure 3.3). Clearly the relative

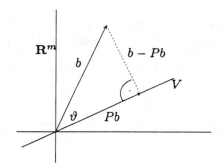

Figure 3.3: Projection onto the subspace V

condition number of the projection problem (P, b) corresponding to the input b depends strongly on the angle ϑ of intersection between b and the subspace V. If the angle is small, i.e. $b \approx Pb$, then perturbations of b leave the result Pb nearly unchanged. On the other hand if b is almost perpendicular to V then small perturbations of b produce relatively large variations of Pb. These observations are reflected in the following lemma.

Lemma 3.9 *Let $P : \mathbf{R}^m \to V$ be the orthogonal projection onto a subspace V of $\mathbf{R}^n$. For an input b let ϑ denote the angle between b and V, i.e.*

$$\sin \vartheta = \frac{\|b - Pb\|_2}{\|b\|_2} .$$

Then the relative condition number of the problem (P, b) corresponding to the Euclidean norm satisfies

$$\kappa = \frac{1}{\cos \vartheta} \|P\|_2 .$$

Proof. According to the Pythagorean theorem $\|Pb\|^2 = \|b\|^2 - \|b - Pb\|^2$ and therefore

$$\frac{\|Pb\|^2}{\|b\|^2} = 1 - \sin^2 \vartheta = \cos^2 \vartheta .$$

Because P is linear it follows that the relative condition number of (P, b) satisfies, as stated

$$\kappa = \frac{\|b\|}{\|Pb\|} \|P'(b)\| = \frac{\|b\|}{\|Pb\|} \|P\| = \frac{1}{\cos \vartheta} \|P\| .$$

$\square$

For the next theorem we also need the following relationship between the condition numbers of A and $A^T A$ corresponding to the Euclidean norm.

Lemma 3.10 *For a matrix $A \in Mat_{m,n}(\mathbf{R})$ of maximal rank $p = n$ we have*

$$\kappa_2(A^T A) = \kappa_2(A)^2 .$$

Proof. According to Definition (2.5) the condition number of a rectangular matrix satisfies

$$
\begin{aligned}
\kappa_2(A)^2 &= \frac{\max_{\|x\|_2=1} \|Ax\|^2}{\min_{\|x\|_2=1} \|Ax\|^2} \\
&= \frac{\max_{\|x\|_2=1} \langle A^T Ax, x \rangle}{\min_{\|x\|_2=1} \langle A^T Ax, x \rangle} = \frac{\lambda_{\max}(A^T A)}{\lambda_{\min}(A^T A)} = \kappa_2(A^T A) .
\end{aligned}
$$

$\square$

With these preparations the following result on the condition of a linear least squares problem no longer comes as a surprise.

Theorem 3.11 *Let $A \in \mathrm{Mat}_{m,n}(\mathbf{R})$, $m \geq n$, be a matrix of full column rank, $b \in \mathbf{R}^n$, and x the (unique) solution of the linear least squares problem*

$$\|b - Ax\|_2 = \min .$$

We assume that $x \neq 0$ and we denote by ϑ the angle between b and the range space $R(A)$ of A, i.e.

$$\sin \vartheta = \frac{\|b - Ax\|_2}{\|b\|_2} = \frac{\|r\|_2}{\|b\|_2}$$

with residual $r = b - Ax$. Then the relative condition number of x in the Euclidean norm satisfies:

a) corresponding to perturbations of b

$$\kappa \leq \frac{\kappa_2(A)}{\cos \vartheta} \tag{3.4}$$

b) corresponding to perturbations of A

$$\kappa \leq \kappa_2(A) + \kappa_2(A)^2 \tan \vartheta . \tag{3.5}$$

Proof.   a) The solution x is given through the normal equations by the linear mapping

$$x = \phi(b) = (A^T A)^{-1} A^T b$$

so that

$$\kappa = \frac{\|\phi'(b)\|_2 \|b\|_2}{\|x\|_2} = \frac{\|A\|_2 \|(A^T A)^{-1} A^T\|_2 \|b\|_2}{\|A\|_2 \|x\|_2} .$$

It is easily seen that for a full column rank matrix A the condition number $\kappa_2(A)$ is precisely

$$\kappa_2(A) = \frac{\max_{\|x\|_2=1} \|Ax\|_2}{\min_{\|x\|_2=1} \|Ax\|_2} = \|A\|_2 \|(A^T A)^{-1} A^T\|_2 .$$

Now, as in Lemma 3.9, the assertion follows from

$$\kappa = \frac{\|b\|_2}{\|A\|_2 \|x\|_2} \kappa_2(A) \leq \frac{\|b\|_2}{\|Ax\|_2} \kappa_2(A) = \frac{\kappa_2(A)}{\cos \vartheta} .$$

b) Here we consider $x = \phi(A) = (A^T A)^{-1} A^T b$ as a function of A. Because the matrices of rank n form an open subset of $\mathrm{Mat}_{m,n}(\mathbf{R})$, ϕ is differentiable in a neighborhood of A. We construct the directional derivative $\phi'(A)C$

for a matrix $C \in \mathrm{Mat}_{m,n}(\mathbf{R})$, by differentiating the equation characterizing $\phi(A + tC)$

$$(A + tC)^T (A + tC)\phi(A + tC) = (A + tC)^T b,$$

with respect to t at the point $t = 0$. It follows that

$$C^T Ax + A^T Cx + A^T A\phi'(A)C = C^T b \,,$$

i.e.

$$\phi'(A)C = (A^T A)^{-1}(C^T (b - Ax) - A^T Cx) \,.$$

From it we can estimate the derivative of ϕ by

$$\|\phi'(A)\| \ \leq \ \|(A^T A)^{-1}\|_2 \, \|r\|_2 + \|(A^T A)^{-1} A^T\|_2 \, \|x\|_2 \,,$$

so that

$$
\begin{aligned}
\kappa \ &= \ \frac{\|\phi'(A)\|_2 \, \|A\|_2}{\|x\|_2} \\[2mm]
&\leq \ \underbrace{\|A\|_2 \, \|(A^T A)^{-1} A^T\|_2}_{=\kappa_2(A)} + \underbrace{\|A\|_2^2 \|(A^T A)^{-1}\|_2}_{=\kappa_2(A^T A)=\kappa_2(A)^2} \frac{\|r\|_2}{\|A\|_2 \, \|x\|_2}\,.
\end{aligned}
$$

Now the assertion follows as in a) from

$$\frac{\|r\|_2}{\|A\|_2 \, \|x\|_2} = \frac{\|r\|_2}{\|b\|_2} \cdot \frac{\|b\|_2}{\|A\|_2 \, \|x\|_2} \leq \sin\vartheta \cdot \frac{1}{\cos\vartheta} = \tan\vartheta \,.$$

$\square$

If the residual $r = b - Ax$ of the linear least squares problem is small compared with the input b, then we have $\cos\vartheta \approx 1$ and $\tan\vartheta \approx 0$. In this case (which should be the normal case for linear least squares problems) the problem behaves as a linear system from the point of view of condition. For large residuals, i.e. $\cos\vartheta \ll 1$ and $\tan\vartheta > 1$, the estimate (3.5) contains the quantity $\kappa_2(A)$, which is relevant for linear systems, as well as its square $\kappa_2(A)^2$. Hence for large residuals the linear least squares problem behaves essentially differently from linear systems.

3.1.4   Solution of normal equations

We switch now to the solution of the normal equations. We start from the assumption that the linear least squares problem is uniquely solvable, i.e. rank $A = n$, so that $A^T A$ is an spd-matrix. According to Section 1.4 this gives the possibility of using the Cholesky factorization. As for the cost of solving the linear least squares problem with the help of the normal equations we have (in number of multiplications):

a) Computation of $A^T A$: $\sim \frac{1}{2}n^2 m$

b) Cholesky factorization of $A^T A$: $\sim \frac{1}{6}n^3$.

For $m \gg n$ part a) predominates, so that we have altogether a cost of

$$\sim \frac{1}{2}n^2 m \ \text{ for } m \gg n \ \text{ and } \ \sim \frac{2}{3}n^3 \ \text{ for } m \approx n$$

The numerical treatment described above requires in the first step the computation of $A^T A$, that is of numerous scalar products of rows of A. The numerical intuition developed in the second chapter (see Example 2.33) makes this appear dubious. The errors that may arise in performing each additional step will be propagated further until the final solution. Therefore in most cases it is better to look for an efficient 'direct' method operating *only on A itself*. A further point of criticism is the fact that the errors in $A^T b$ are amplified in the solution of the linear system $A^T A x = A^T b$ by a factor close to the condition

$$\kappa_2(A^T A) = \kappa_2(A)^2$$

(see Lemma 3.10). For large residuals this agrees with the condition number of the linear least squares problem (3.5). However for small residuals the condition number of the latter problem is described instead by $\kappa(A)$, so that passing to the normal equations means a considerable worsening of the condition number. In addition the matrices usually arising in linear least squares problems are already badly conditioned so that they require extreme attention and further worsening of the condition by passing to $A^T A$ is not manageable. Hence the solution of linear least squares problems via normal equations with Cholesky factorization can be recommended only for problems with large residuals.

Remark 3.12 Based on the properties of the linear least squares problem discussed in the last section one should perform a residual correction for problems with matrices of maximal rank and large residuals. In the iterative refinement the residual is introduced as an explicit variable (see BJØRCK [4] and Problem 3.2). With this variable we write the normal equations as an equivalent symmetric system

$$\begin{bmatrix} I & A \\ A^T & 0 \end{bmatrix} \begin{pmatrix} r \\ x \end{pmatrix} = \begin{pmatrix} b \\ 0 \end{pmatrix} \tag{3.6}$$

of double dimension with the input quantities A and b appearing directly. This formulation is also especially appropriate for large sparse matrices A.

3.2   Orthogonalization Methods

Each elimination process for linear systems, as for example the Gaussian elimination from Section 1.2, can be formally represented as

$$A \xrightarrow{f_1} B_1 A \xrightarrow{f_2} B_2 B_1 A \xrightarrow{f_3} \cdots \xrightarrow{f_k} B_k \cdots B_1 A = R$$

where the matrices B_j describe the operations on the matrix A. We have seen in the recursive stability analysis from Lemma 2.21 that the stability indicators of the partial steps of an algorithm are increased by the condition of all successive partial steps. For an elimination process as described above they are the condition numbers of the matrices B_j. These condition numbers , e.g. that of the elimination matrix $B_j = L_j$ of the Gaussian triangular factorization, are in general not bounded from above, so that we may encounter instabilities. On the other hand if we choose instead of L_j orthogonal transformations Q_j for the elimination, then we have in Euclidean norm

$$\kappa(Q_j) = \|Q_j\|_2 \|Q_j^{-1}\| = \|Q_j\|_2 \|Q_j^T\|_2 = 1 \ .$$

Therefore these *orthogonalization methods* are always stable. Unfortunately the stability comes together with a somewhat higher cost than, for example, Gaussian elimination as will be seen in the following. In addition there is a further reason that makes the orthogonalization method for linear least squares problems a good alternative to the solution of the normal equations. Actually, it is the invariance of the Euclidean norm with respect to orthogonal transformations that imposes the application of orthogonalization methods for the solution of linear systems. Let us assume that we have brought a matrix $A \in \mathrm{Mat}_{m,n}$ with $m \geq n$ to the upper triangular form R by means of an orthogonal matrix $Q \in \mathbf{O}(m)$

$$Q^T A = \begin{bmatrix} * & \cdots & * \\ & \ddots & \vdots \\ & & * \end{bmatrix} = \begin{bmatrix} R \\ 0 \end{bmatrix} \ .$$

Then, as an alternative to solving the normal equations, we can determine the solution of the linear least squares problem $\|b - Ax\| = \min$ as follows:

Theorem 3.13 *Let* $A \in \mathrm{Mat}_{m,n}(\mathbf{R})$, $m \geq n$ *be a matrix of maximal rank,* $b \in \mathbf{R}^m$ *a vector, and* $Q \in \mathbf{O}(m)$ *an orthogonal matrix with*

$$Q^T A = \begin{pmatrix} R \\ 0 \end{pmatrix} \quad and \quad Q^T b = \begin{pmatrix} b_1 \\ b_2 \end{pmatrix},$$

where $b_1 \in \mathbf{R}^n$, $b_2 \in \mathbf{R}^{m-n}$ *and* $R \in \mathrm{Mat}_n(\mathbf{R})$ *is an (invertible) upper triangular matrix. Then* $x = R^{-1}b_1$ *is the solution of the linear least squares problem* $\|b - Ax\| = \min$.

Proof. Because $Q \in \mathbf{O}(m)$, we have for all $x \in \mathbf{R}^m$

$$
\begin{aligned}
\|b - Ax\|^2 &= \|Q^T(b - Ax)\|^2 = \left\| \begin{pmatrix} b_1 - Rx \\ b_2 \end{pmatrix} \right\|^2 \\
&= \|b_1 - Rx\|^2 + \|b_2\|^2 \geq \|b_2\|^2 .
\end{aligned}
$$

As $\operatorname{rank} A = \operatorname{rank} R = n$, R is invertible. The term $\|b_1 - Rx\|^2$ vanishes precisely for $x = R^{-1}b_1$. Observe that the residual $r := b - Ax$ does not vanish in general and that $\|r\| = \|b_2\|$. $\square$

For $m = 2$ the orthogonal transformations in question can be derived geometrically as rotations, and reflections, see Figure 3.4. If we want to map

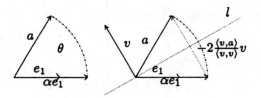

Figure 3.4: Rotation and reflection of a over αe_1

the vector $a \in \mathbf{R}^2$ via an orthogonal transformation into a multiple αe_1 of the first unit vector it follows that $\alpha = \|a\|$. The first possibility is to rotate a with an angle θ over αe_1. We obtain

$$a \longmapsto \alpha e_1 = Qa \quad \text{with} \quad Q := \begin{pmatrix} \cos\theta & \sin\theta \\ -\sin\theta & \cos\theta \end{pmatrix} .$$

As a second possibility we can reflect a with respect to a vector l which is perpendicular on v, i.e.

$$a \longmapsto \alpha e_1 = a - 2\frac{\langle v, a \rangle}{\langle v, v \rangle}v ,$$

where v is collinear to the difference $a - \alpha e_1$. The numerical handling by means of rotations will be introduced in Section 3.2.1 and that by means of reflections in Section 3.2.2.

3.2.1   Givens rotations

The name *Givens rotation* (GIVENS 1953) is used to denote matrices of the form

$$
\Omega_{kl} := \begin{bmatrix} I & & & & \\ & c & & s & \\ & & I & & \\ & -s & & c & \\ & & & & I \end{bmatrix} \begin{matrix} \\ \leftarrow k \\ \\ \leftarrow l \\ \end{matrix} \in \mathrm{Mat}_m(\mathbf{R}) \ ,
$$

where I is the identity matrix of corresponding dimension and $c^2 + s^2 = 1$. Here c and s should remind us naturally of $\cos\theta$ and $\sin\theta$. Geometrically the matrix describes a rotation of angle θ in the (k, l)-plane. If we apply Ω_{kl} to a vector $x \in \mathbf{R}^m$ it follows that

$$
x \longmapsto y = \Omega_{kl}x \quad \text{with} \quad y_i = (\Omega_{kl}x)_i = \begin{cases} cx_k + sx_l & \text{if } i = k \\ sx_k + cx_l & \text{if } i = l \\ x_i & \text{if } i \neq k, l \end{cases} \quad . \tag{3.7}
$$

If we premultiply a matrix

$$
A = [A_1, \ldots, A_n] \in \mathrm{Mat}_{m,n}(\mathbf{R})
$$

with columns $A_1, \ldots, A_n \in \mathbf{R}^m$ by Ω_{kl}, then the Givens rotation operates on the columns, i.e.

$$
\Omega_{kl}A = [\Omega_{kl}A_1, \ldots, \Omega_{kl}A_n] \ .
$$

According to (3.7) only rows k and l of the matrix A are changed. This is especially important when the sparsity structure is to be preserved as much as possible by the transformation.

Now how can we determine the coefficients c and s in order to eliminate a component x_l of the vector x? As Ω_{kl} operates only on the (k, l)-plane it is sufficient to clarify the principle in case $m = 2$. From $x_k^2 + x_l^2 \neq 0$ and $s^2 + c^2 = 1$ it follows that

$$
\begin{pmatrix} c & s \\ -s & c \end{pmatrix}\begin{pmatrix} x_k \\ x_l \end{pmatrix} = \begin{pmatrix} r \\ 0 \end{pmatrix} \iff r = \pm\sqrt{x_k^2 + x_l^2}, \ c = x_k/r \text{ and } s = x_l/r.
$$

Actually c and s are more conveniently computed with the formula (τ stands for $\tan\theta$ and $\cot\theta$ respectively)

$$\tau := x_k/x_l \ , \quad s := 1/\sqrt{1+\tau^2} \ , \quad c := s\tau$$

if $|x_l| > |x_k|$ and

$$\tau := x_l/x_k \ , \quad c := 1/\sqrt{1+\tau^2} \ , \quad s := c\tau$$

if $|x_k| \geq |x_l|$. With this overflow of the exponent is also avoided.

Now we can bring a given matrix $A \in \mathrm{Mat}_{m,n}$ to an upper triangular form R where $r_{ij} = 0$ for $i > j$ with the help of Givens rotations by eliminating column by column the nonzero elements below the diagonal. For example the algorithm can be illustrated on a full $(5,4)$-matrix as follows (the index pair over the arrows give the indices of the Givens rotation Ω_{kl} performed at that step):

$$A = \begin{bmatrix} * & * & * & * \\ * & * & * & * \\ * & * & * & * \\ * & * & * & * \\ * & * & * & * \end{bmatrix} \xrightarrow{(5,4)} \begin{bmatrix} * & * & * & * \\ * & * & * & * \\ * & * & * & * \\ * & * & * & * \\ 0 & * & * & * \end{bmatrix} \xrightarrow{(4,3)} \cdots \xrightarrow{(2,1)} \begin{bmatrix} * & * & * & * \\ 0 & * & * & * \\ 0 & * & * & * \\ 0 & * & * & * \\ 0 & * & * & * \end{bmatrix}$$

$$\xrightarrow{(5,4)} \begin{bmatrix} * & * & * & * \\ 0 & * & * & * \\ 0 & * & * & * \\ 0 & * & * & * \\ 0 & 0 & * & * \end{bmatrix} \xrightarrow{(4,3)} \cdots \xrightarrow{(5,4)} \begin{bmatrix} * & * & * & * \\ 0 & * & * & * \\ 0 & 0 & * & * \\ 0 & 0 & 0 & * \\ 0 & 0 & 0 & 0 \end{bmatrix} .$$

After carefully counting the operations we obtain the cost of the QR factorization of a full matrix $A \in \mathrm{Mat}_{m,n}$:

a) $\sim n^2/2$ square roots and $\sim 4n^3/3$ multiplications, if $m \approx n$

b) $\sim mn$ square roots and $\sim 2mn^2$ multiplications, if $m \gg n$

For $m = n$ we obtain an alternative to the Gaussian triangular factorization of Section 1.2. The better stability is bought at a considerable higher cost of $\sim 4n^3/3$ multiplications, versus $\sim n^3/3$ for the Gaussian elimination. However one should observe that for sparse matrices the comparison turns out to be essentially more favorable. Thus only $n-1$ Givens rotations are needed in order to bring to upper triangular form a *Hessenberg matrix*

$$
A = \begin{bmatrix}
* & \cdots & \cdots & \cdots & * \\
* & \ddots & & & \vdots \\
0 & \ddots & \ddots & & \vdots \\
\vdots & \ddots & \ddots & \ddots & \vdots \\
0 & \cdots & 0 & * & *
\end{bmatrix}.
$$

which is almost upper triangular and has nonzero components only in the first subdiagonal. With Gaussian elimination the pivot search eventually doubles the subdiagonal band.

Remark 3.14 If A is stored with a row scaling DA, then the Givens rotations can be realized (similarly to the rational Cholesky factorization) without evaluating square roots. In 1973 GENTLEMAN [35] and HAMMARLING [46] developed a variant, the so called *fast Givens* or *rational Givens*. This type of factorization is invariant with respect to column scaling, i. e.

$$A = QR \implies AD = Q(RD) \text{ for a diagonal matrix } D.$$

3.2.2 Householder reflections

In 1958 HOUSEHOLDER [49] introduced matrices $Q \in Mat_n(\mathbf{R})$ of the form

$$Q = I - 2\frac{vv^T}{v^Tv} \tag{3.8}$$

with $v \in \mathbf{R}^n$. Today they are called *Householder reflections*. Such matrices describe exactly the reflections on the plane perpendicular to v (compare Figure 3.4). In particular Q depends only on the direction of v. The Householder reflections Q have the following properties:

a) Q is *symmetric*, i.e. $Q^T = Q$.

b) Q is *orthogonal*, i.e. $QQ^T = Q^TQ = I$.

c) Q is *idempotent*, i.e. $Q^2 = I$.

If we apply Q to a vector $y \in \mathbf{R}^n$ we get

$$y \longmapsto Qy = \left(I - 2\frac{vv^T}{v^Tv}\right) y = y - 2\frac{\langle v, y \rangle}{\langle v, v \rangle} v \ .$$

If y is to be mapped on a multiple of the unit vector e_1,

$$\alpha e_1 = y - 2\frac{\langle v, y \rangle}{\langle v, v \rangle} v \in \text{ span}(e_1) \,,$$

then

$$|\alpha| = \|y\|_2 \quad \text{and} \quad v \in \text{span}(y - \alpha e_1) \,.$$

From it we determine Q by

$$v := y - \alpha e_1 \quad \text{with} \quad \alpha = \pm\|y\|_2 \,.$$

In order to avoid cancelling in computing $v = (y_1 - \alpha, y_2, \ldots, y_n)^T$ we choose $\alpha := -\text{sgn}(y_1)\|y\|_2$. Because

$$\langle v, v \rangle = \langle y - \alpha e_1, y - \alpha e_1 \rangle = \|y\|_2^2 - 2\alpha\langle y, e_1 \rangle + \alpha^2 = -2\alpha(y_1 - \alpha)$$

we can compute Qx for arbitrary $x \in \mathbf{R}^n$ by the very simple formula

$$Qx = x - 2\frac{\langle v, x \rangle}{\langle v, v \rangle} v = x + \frac{\langle v, x \rangle}{\alpha(y_1 - \alpha)} v \,.$$

With the help of Householder reflections we can transform a matrix $A = [A_1, \ldots, A_n] \in \text{Mat}_{m,n}(\mathbf{R})$ to upper triangular form as well, by eliminating successively the elements below the main diagonal. In the first step we "shorten" the column A_1 and obtain

$$A \to A' := Q_1 A = \begin{bmatrix} \alpha_1 & & & \\ 0 & & & \\ \vdots & A_2' \cdots A_n' \\ 0 & & & \end{bmatrix},$$

where

$$Q_1 = I - 2\frac{v_1 v_1^T}{v_1^T v_1} \quad \text{with } v_1 := A_1 - \alpha_1 e_1 \text{ and } \alpha_1 := -\text{sgn}(a_{11})\|A_1\|_2.$$

After the k-th step the output matrix is brought to upper triangular form except for a remainder matrix $T^{(k+1)} \in \mathrm{Mat}_{m-k,n-k}(\mathbf{R})$

$$
A^{(k)} = \begin{bmatrix}
* & \cdots & \cdots & & \cdots & & * \\
 & \ddots & & & & & \vdots \\
 & & * & & \cdots & & * \\
 & & 0 & & & & \\
 & & \vdots & & T^{(k+1)} & & \\
 & & 0 & & & &
\end{bmatrix} .
$$

Now let us build an orthogonal matrix

$$
Q_{k+1} = \left[\begin{array}{c|c}
I_k & 0 \\
\hline
0 & \bar{Q}_{k+1}
\end{array} \right] ,
$$

where $\bar{Q}_{k+1} \in \mathbf{O}(m-k)$ is constructed as in the first step with $T^{(k+1)}$ instead of A. Altogether after $p = \min(m-1, n)$ steps we obtain the upper triangular matrix

$$
R = Q_p \cdots Q_1 A
$$

and from here, because $Q_i^2 = I$, the factorization

$$
A = QR \quad \text{with} \quad Q = Q_1 \cdots Q_p.
$$

If we calculate the solution of the linear least squares problem $\|b - Ax\| = \min$ as in Theorem 3.13, by computing the QR factorization of the matrix $A \in \mathrm{Mat}_{m,n}(\mathbf{R})$, $m \geq n$, with the help of Householder reflections $Q_j \in \mathbf{O}(m)$ then we arrive at the following method:

1) $A = QR$,   QR factorization with Householder reflections;

2) $(b_1, b_2)^T = Q^T b$ with $b_1 \in \mathbf{R}^n$ and $b_2 \in \mathbf{R}^{m-n}$,   Transformation of b;

3) $Rx = b_1$,   Solution upper triangular system.

In a computer implementation we have to store the Householder vectors $v_1, \ldots, v_p$ as well as the upper triangular matrix R . The diagonal elements

$$
r_{ii} = \alpha_i \quad \text{for } i = 1, \ldots, p
$$

are stored in a separate vector, so that the Householder vectors $v_1, \ldots, v_p$ find a place in the lower half of A (see Figure 3.5). Another possibility is to

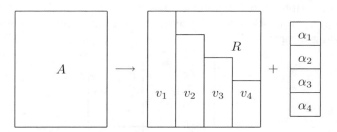

Figure 3.5: Memory partition for QR-factorization with Householder reflections for $m = 5$ and $n = 4$.

normalize the Householder vectors in such a way that the first component $\langle v_i, e_i \rangle$ is always 1 and therefore does not need to be stored.

For the *cost* of this method we obtain:

a) $\sim 2n^2 m$ multiplications, if $m \gg n$,

b) $\sim \frac{2}{3} n^3$ multiplications, if $m \approx n$.

For $m \approx n$ we have about the same cost as for the Cholesky method for the normal equations. For $m \gg n$ the cost is worse by a factor of two but the method has the stability advantage discussed above.

As in the case of Gaussian elimination there is also a pivoting strategy for the QR factorization, the *column exchange strategy* of BUSINGER and GOLUB [10]. In contrast to Gaussian elimination this strategy is of minor importance for the numerical stability of the algorithm. If one pushes the column with maximal 2-norm to the front, so that after the change we have

$$\|T_1^{(k)}\|_2 = \max_j \|T_j^{(k)}\|_2,$$

then the diagonal elements r_{kk} of R satisfy

$$|r_{kk}| = \|T^{(k)}\|_\square \quad \text{and} \quad |r_{k+1,k+1}| \leq |r_{kk}|,$$

for the matrix norm $\|A\|_\square := \max_j \|A_j\|_2$. If $p := \operatorname{rank}(A)$, then we obtain theoretically that after p steps the matrix

$$A^{(p)} = \left[\begin{array}{c|c} R & S \\ \hline 0 & 0 \end{array} \right]$$

with an invertible upper triangular matrix $R \in \mathrm{Mat}_p(\mathbf{R})$ and a matrix $S \in \mathrm{Mat}_{p,n-p}(\mathbf{R})$. Because of roundoff errors we obtain instead of this the following matrix

$$
A^{(p)} = \left[\begin{array}{c|c} R & S \\ \hline 0 & T^{(p+1)} \end{array} \right],
$$

where the elements of the rest matrix $T^{(p+1)} \in \mathrm{Mat}_{m-p,n-p}(\mathbf{R})$ are "very small". As the rank of the matrix is not generally known in advance we have to decide during the algorithm when to neglect the rest matrix. In the course of the QR factorization with column exchange the following criterion for the so called *rank decision* presents itself in a convenient way.   If we define the *numerical rank* p for a relative precision δ of the matrix A by the condition

$$
|r_{p+1,p+1}| < \delta\,|r_{11}| \le |r_{pp}|\,,
$$

then it follows directly that

$$
\|T^{(p+1)}\|_\square = |r_{p+1,p+1}| < \delta\,|r_{11}| = \delta\,\|A\|_\square\,,
$$

i.e. $T^{(p+1)}$ is below the error in A corresponding to the norm $\|\cdot\|_\square$. If $p = n$, then we can easily compute further the so called *subcondition number*

$$
\mathrm{sc}(A) := \frac{|r_{11}|}{|r_{nn}|}
$$

of DEUFLHARD and SAUTTER (1979) [25]. Analogously to the properties of the condition number $\kappa(A)$ we have

a) $\mathrm{sc}(A) \ge 1$

b) $\mathrm{sc}(\alpha A) = \mathrm{sc}(A)$

c) $A \ne 0$ singular $\Longleftrightarrow$ $\mathrm{sc}(A) = \infty$

d) $\mathrm{sc}(A) \le \kappa_2(A)$ (hence the name).

In harmony with the above definition of the numerical rank, A is called *almost singular* if we have

$$
\delta\mathrm{sc}(A) \ge 1 \quad \text{or equivalently} \quad \delta\,|r_{11}| \ge |r_{nn}|
$$

for a QR factorization with column exchange.

We have substantiated above that this concept makes sense. Each matrix which is almost singular according to this definition is also almost singular with respect to the condition number $\kappa_2(A)$, as shown by property d) (cf. Definition 2.32). The reverse is not true.

3.3 Generalized Inverses

According to Theorem 3.7 the solution x of the linear least squares problem $\|b - Ax\| = \min$, for $A \in \text{Mat}_{m,n}(\mathbf{R})$, $m \geq n$ and rank $A = n$, is uniquely determined. Clearly it depends linearly on b and it is formally denoted by $x = A^+b$. Under the above assumptions the normal equations imply that

$$A^+ = (A^T A)^{-1} A^T .$$

Because $A^+A = I$ is precisely the identity, A^+ is also called the *pseudo-inverse* of A. The definition of A^+ can be extended to arbitrary matrices $A \in \text{Mat}_{m,n}(\mathbf{R})$. In this case the solution of $\|b - Ax\| = \min$ is in general no longer uniquely determined. On the contrary, if we denote by

$$\bar{P} : \mathbf{R}^m \longrightarrow R(A) \subset \mathbf{R}^m$$

the orthogonal projection of $\mathbf{R}^m$ onto the image space $R(A)$, then according to Theorem 3.4 the solutions form an affine subspace

$$L(b) := \{ x \in \mathbf{R}^n \mid \|b - Ax\| = \min \} = \{ x \in \mathbf{R}^n \mid Ax = \bar{P}b \} .$$

Nevertheless, in order to enforce uniqueness we choose the smallest solution $x \in L(b)$ in the Euclidean norm $\| \cdot \|$, and we denote again $x = A^+b$. According to Remark 3.6, x is precisely the orthogonal projection of the origin $0 \in R^n$ onto the affine subspace $L(b)$ (see Figure 3.6). If $\bar{x} \in L(b)$ is an arbitrary solution of $\|b - Ax\| = \min$, then we obtain all the solutions by translating the nullspace $N(A)$ of A by $\bar{x}$, i.e.

$$L(b) = \bar{x} + N(A) .$$

Here the smallest solution x must be perpendicular onto the nullspace $N(A)$, in other words: x is the uniquely determined vector $x \in N(A)^\perp$ with $\|b - Ax\| = \min$.

Definition 3.15 The *pseudo inverse* of a matrix $A \in \text{Mat}_{m,n}(\mathbf{R})$ is a matrix $A^+ \in \text{Mat}_{n,m}(\mathbf{R})$, such that for all $b \in \mathbf{R}^m$ the vector $x = A^+b$ is the smallest solution $\|b - Ax\| = \min$, i.e.

$$A^+b \in N(A)^\perp \quad \text{and} \quad \|b - AA^+b\| = \min .$$

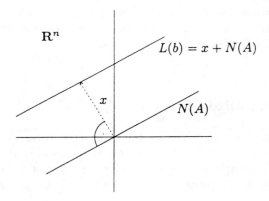

Figure 3.6: "Smallest" solution of the least squares problem as a projection of 0 onto $L(b)$

The situation can be most clearly represented by the following commutative diagram (where i denotes each time the inclusion operator).

$$
\begin{array}{ccc}
\mathbf{R}^n & \overset{A}{\underset{A^+}{\rightleftarrows}} & \mathbf{R}^m \\
P = A^+A \;\big\downarrow\big\uparrow\, i & & i \,\big\uparrow\big\downarrow\; \bar{P} = AA^+ \\
R(A^+) = N(A)^\perp & \cong & R(A)
\end{array}
$$

We can easily read that the projection $\bar{P}$ is precisely AA^+, while $P = A^+A$ describes the projection from $\mathbf{R}^n$ onto the orthogonal complement $N(A)^\perp$ of the nullspace. Furthermore, because of the projection property, we have obviously $A^+AA^+ = A^+$ and $AA^+A = A$. As seen in the following theorem the pseudo inverse is uniquely determined by these two properties and the symmetry of the orthogonal projections $P = A^+A$ and $\bar{P} = AA^+$.

Theorem 3.16 *The pseudo inverse $A^+ \in \mathrm{Mat}_{n,m}(\mathbf{R})$ of a matrix $A \in \mathrm{Mat}_{m,n}(\mathbf{R})$ is uniquely characterized by the following properties:*

    i) $(A^+A)^T = A^+A$

    ii) $(AA^+)^T = AA^+$

    iii) $A^+AA^+ = A^+$

iv) $AA^+A = A$.

The properties i) through iv) are also called the Penrose axioms.

Proof.   We have already seen that A^+ satisfies properties i) through iv), because A^+A and AA^+ are orthogonal projections onto $N(A)^\perp = R(A^+)$ and $R(A)$ respectively. Conversely i) through iv) imply that $P := A^+A$ and $\bar{P} := AA^+$ are orthogonal projections, because $P^T = P = P^2$ and $\bar{P}^T = \bar{P} = \bar{P}^2$. Analogously from iii) and $P = A^+A$ it follows that $N(P) = N(A)$. Thus the projections P and $\bar{P}$ are uniquely determined (independently of A^+) by properties i) through iv). From this it follows the uniqueness of A^+: If A_1^+ and A_2^+ satisfy conditions i) through iv), then $P = A_1^+A = A_2^+A$ and $\bar{P} = AA_1^+ = AA_2^+$ and therefore

$$A_1^+ \overset{iii)}{=} A_1^+AA_1^+ = A_2^+AA_2^+ \overset{iii)}{=} A_2^+ \ .$$

$\square$

Remark 3.17 If only part of the Penrose axioms hold, then we talk about *generalized inverses*. A detailed investigation is found e.g. in the book of NASHED [58].

Now we want to derive a way of computing the smallest solution $x = A^+b$ for an arbitrary matrix $A \in \mathrm{Mat}_{m,n}(\mathbf{R})$ and $b \in \mathbf{R}^m$ with the help of the QR factorization. Let $p := \mathrm{rank}\,A \leq \min(m,n)$ be the rank of the matrix A. In order to simplify notation we neglect permutations and bring A to upper triangular form by orthogonal transformations $Q \in \mathbf{O}(m)$ (e.g. Householder reflections)

$$QA = \left[\begin{array}{c|c} R & S \\ \hline 0 & 0 \end{array}\right] , \tag{3.9}$$

where $R \in \mathrm{Mat}_p(\mathbf{R})$ is an invertible upper triangular matrix and $S \in \mathrm{Mat}_{p,n-p}(\mathbf{R})$. We formally partition the vectors x and Qb in an analogous way as

$$x = \begin{pmatrix} x_1 \\ x_2 \end{pmatrix} \quad \text{with } x_1 \in \mathbf{R}^p \text{ and } x_2 \in \mathbf{R}^{n-p} , \tag{3.10}$$

$$Qb = \begin{pmatrix} b_1 \\ b_2 \end{pmatrix} \quad \text{with } b_1 \in \mathbf{R}^p \text{ and } b_2 \in \mathbf{R}^{m-p} . \tag{3.11}$$

Then we can characterize the solution of the least squares problem $\|b - Ax\| = \min$ as follows:

Lemma 3.18 *With the above notation x is a solution of $\|b - Ax\| = \min$, if and only if*

$$x_1 = R^{-1}b_1 - R^{-1}Sx_2 .$$

Proof. Because of the invariance of the Euclidean norm under orthogonal transformations we have

$$\|b - Ax\|^2 = \|Qb - QAx\|^2 = \|Rx_1 + Sx_2 - b_1\|^2 + \|b_2\|^2 .$$

The expression is minimal if and only if $Rx_1 + Sx_2 - b_1 = 0$.   □

The case $p = \operatorname{rank} A = n$ corresponds to the case of overdetermined full rank system that has already been treated. The matrix S vanishes and we get as in Theorem 3.13 the solution $x = x_1 = R^{-1}b_1$. In the underdetermined or rank deficient case $p < n$ the solution can be computed as follows:

Lemma 3.19 *Let $p < n$, $V := R^{-1}S \in \operatorname{Mat}_{p,n-p}(\mathbf{R})$ and $u := R^{-1}b_1 \in \mathbf{R}^p$. Then the smallest solution x of $\|b - Ax\| = \min$ is given by $x = (x_1, x_2) \in \mathbf{R}^p \times \mathbf{R}^{n-p}$ with*

$$(I + V^T V)x_2 = V^T u \quad and \quad x_1 = u - Vx_2 .$$

Proof. According to Lemma 3.18 the solutions of $\|b - Ax\| = \min$ are characterized by $x_1 = u - Vx_2$. By inserting into $\|x\|$ we obtain

$$
\begin{aligned}
\|x\|^2 &= \|x_1\|^2 + \|x_2\|^2 = \|u - Vx_2\|^2 + \|x_2\|^2 \\
&= \|u\|^2 - 2\langle u, Vx_2\rangle + \langle Vx_2, Vx_2\rangle + \langle x_2, x_2\rangle \\
&= \|u\|^2 + \langle x_2, (I + V^T V)x_2 - 2V^T u\rangle =: \varphi(x_2) .
\end{aligned}
$$

Here

$$\varphi'(x_2) = -2V^T u + 2(I + V^T V)x_2 \quad and \quad \varphi''(x_2) = 2(I + V^T V) .$$

Because $I + V^T V$ is a symmetric positive definite matrix, $\varphi(x_2)$ attains its minimum for x_2 and for this value we have $\varphi'(x_2) = 0$, i.e. $(I + V^T V)x_2 = V^T u$. This was exactly our claim.   □

Since $I + V^T V$ is an spd-matrix we can use the Cholesky factorization for computing x_2 . Altogether we obtain the following algorithm for computing the smallest solution $x = A^+ b$ of $\|b - Ax\| = \min$.

Algorithm 3.20 *Pseudo inverse via QR factorization.*
Let $A \in \mathrm{Mat}_{m,n}(\mathbf{R})$, $b \in \mathbf{R}^m$. Then $x = A^+ b$ is computed as follows

1. QR factorization (3.9) of A with $p = \mathrm{rank}\, A$, where $Q \in \mathbf{O}(m)$, $R \in \mathrm{Mat}_p(\mathbf{R})$ is an upper triangular matrix and $S \in \mathrm{Mat}_{p,n-p}(\mathbf{R})$.

2. Compute $V \in \mathrm{Mat}_{p,n-p}(\mathbf{R})$ from $RV = S$.

3. Cholesky factorization of $I + V^T V$

$$I + V^T V = LL^T \ ,$$

   where $L \in \mathrm{Mat}_{n-p}(\mathbf{R})$ is a lower triangular matrix.

4. $(b_1, b_2)^T := Qb$ with $b_1 \in \mathbf{R}^p$, $b_2 \in \mathbf{R}^{m-p}$.

5. Compute $u \in \mathbf{R}^p$ from $Ru = b_1$.

6. Compute $x_2 \in \mathbf{R}^{n-p}$ from $LL^T x_2 = V^T u$.

7. Set $x_1 := u - Vx_2$.

Then it follows that $x = (x_1, x_2)^T = A^+ b$.

Note that for different right hand sides b we have to perform steps 1. through 3. only once.

3.4 Exercises

Exercise 3.1 A Givens rotation

$$Q = \begin{bmatrix} c & s \\ -s & c \end{bmatrix}, \quad c^2 + s^2 = 1$$

can be stored, up to a sign, as a unique number ρ (naturally, the best storage location would be in the place of the eliminated matrix entry):

$$\rho := \begin{cases} 1 & \text{if } c = 0 \\ \mathrm{sgn}\,(c)\, s/2 & \text{if } |s| < |c| \\ 2\,\mathrm{sgn}\,(s)\,/c & \text{if } |s| \geq |c| \neq 0 \end{cases}$$

Give formulas which reconstruct up to a sign from ρ the Givens rotation $\pm Q$. Why is this representation meaningful although the sign is lost? Is this representation stable?

Exercise 3.2 Let the matrix $A \in \mathrm{Mat}_{m,n}(\mathbf{R})$, $m \geq n$, have full rank. Suppose that the solution $\hat{x}$ of a linear least squares problem $\|b - Ax\|_2 = \min$ computed in a stable way (by using a QR factorization) is not accurate enough. According to Björck (1967) the solution can be improved by residual correction. This is implemented on the linear system

$$\begin{bmatrix} I & A \\ A^T & 0 \end{bmatrix} \begin{bmatrix} r \\ x \end{bmatrix} = \begin{bmatrix} b \\ 0 \end{bmatrix}$$

where r is the residual $r = b - Ax$.

a) Show that the vector (r, x) is the solution of the above system of equations if and only if x is the solution of the least squares problem $\|b - Ax\|_2 = \min$, and r is the residual $r = b - Ax$.

b) Construct an algorithm for solving the above system that uses the available QR factorization of A . How much is the cost of one residual correction?

Exercise 3.3 In the special case rank $A = p = m = n - 1$ of an underdetermined system of codimension 1 the matrix S of the QR factorization (3.9) of A is a vector $s \in \mathbf{R}^m$. Show that Algorithm 3.20 for computing the pseudo inverse $x = A^+ b = (x_1, x_2)^T$ simplifies to

1. QR factorization $QA = [R, s]$.

2. $v := R^{-1}s \in \mathbf{R}^m$.

3. $b_1 := Qb \in \mathbf{R}^m$.

4. $u := R^{-1}b_1 \in \mathbf{R}^m$.

5. $x_2 := \langle v, u \rangle / (1 + \langle v, v \rangle) \in \mathbf{R}$.

6. $x_1 := u - x_2 v$.

Exercise 3.4 In chemistry one often measures the so called reaction rate constants K_i $(i = 1, \ldots, m)$ at temperature T_i with absolute precision (tolerance) δK_i . With the help of the law of Arrhenius

$$K_i = A \cdot \exp\left(-\frac{E}{RT_i}\right)$$

one determines in the sense of least squares both the factor A and the activising energy E, where the general gas constant R is given in advance. Formulate the above given nonlinear problem as a *linear* least squares problem. What simplifications are obtained for the following two special cases?

a) $\delta K_i = \varepsilon K_i$ (constant relative error),

b) $\delta K_i = \text{const}$ (constant absolute error).

Exercise 3.5 Program the Householder orthogonalization procedure without column interchange for (m, n)-matrices, $m \geq n$, and solve with it the linear least squares problem 3.4 for the data file of Table 3.1 $(\delta K_i = 1)$.

Table 3.1: Data file for the linear least squares problem

i	T_i	K_i
1	728.79	$7.4960 \cdot 10^{-6}$
2	728.61	$1.0062 \cdot 10^{-5}$
3	728.77	$9.0220 \cdot 10^{-6}$
4	728.84	$1.4217 \cdot 10^{-5}$
5	750.36	$3.6608 \cdot 10^{-5}$
6	750.31	$3.0642 \cdot 10^{-5}$
7	750.66	$3.4588 \cdot 10^{-5}$
8	750.79	$2.8875 \cdot 10^{-5}$
9	766.34	$6.2065 \cdot 10^{-5}$
10	766.53	$7.1908 \cdot 10^{-5}$
11	766.88	$7.6056 \cdot 10^{-5}$
12	764.88	$6.7110 \cdot 10^{-5}$
13	790.95	$3.1927 \cdot 10^{-4}$
14	790.23	$2.5538 \cdot 10^{-4}$
15	790.02	$2.7563 \cdot 10^{-4}$
16	790.02	$2.5474 \cdot 10^{-4}$
17	809.95	$1.0599 \cdot 10^{-3}$
18	810.36	$8.4354 \cdot 10^{-4}$
19	810.13	$8.9309 \cdot 10^{-4}$
20	810.36	$9.4770 \cdot 10^{-4}$
21	809.67	$8.3409 \cdot 10^{-4}$

Exercise 3.6 An experiment with m measurements leads to a linear least squares problem with $A \in \text{Mat}_{m,n}(\mathbf{R})$. Let a QR factorization of A, $A = QR$ be available. Then

a) a (first) measurement is added or,

b) the (first) measurement is omitted.

Give formulas for computing $\tilde{Q}\tilde{R} = \tilde{A}$ for

$$\tilde{A} = \begin{pmatrix} w^T \\ A \end{pmatrix} \quad \text{and} \quad A = \begin{pmatrix} z^T \\ \tilde{A} \end{pmatrix} \quad \text{respectively}$$

by using the QR factorization of A. How does the formula for modifying the k-th row of A read?

Exercise 3.7 Let $A = BC$. Show that $A^+ = C^+ B^+$ if and only if C or B is an orthogonal matrix (of appropriate dimension). Derive from it formally the solution of the linear least squares problem. Hint: In case of rank deficiency $(p < n)$ there exists an orthogonal transformation from the right such that only one more regular triangular matrix of dimension p is to be inverted. Consider such a transformation in detail.

Exercise 3.8 Let B^+ be the pseudo inverse of a matrix B and $A^- := LB^+R$ a generalized inverse, where L, R are regular matrices. Derive axioms for A^- corresponding to the Penrose axioms. What are the consequences for row or column scaling of linear least squares in the full rank case and the rank deficient case? Consider especially the influence on rank determination.

4 Nonlinear Systems and Least Squares Problems

So far we have been almost entirely concerned with linear problems. In this chapter we shall direct our attention to the solution of nonlinear problems. For this we should have a very clear picture of what is meant by a "solution" of an equation. Probably everybody knows the quadratic equation from high school

$$f(x) := x^2 - 2px + q = 0$$

and its analytic, closed form, solution

$$x_{1,2} = p \pm \sqrt{p^2 - q} \ .$$

(For a stable solution see Example 2.5.) In fact we have only transferred the problem of solving the quadratic equation to the problem of computing a square root, i.e., the solution of a simpler quadratic equation of the form

$$f(x) := x^2 - c = 0 \quad \text{with} \quad c = |p^2 - q| \ .$$

The question of how to determine this solution, i.e., how to solve such a problem *numerically*, still remains open.

4.1 Fixed Point Iterations

For the time being we continue with the scalar nonlinear equation

$$f(x) = 0$$

with an arbitrary function $f : \mathbf{R} \to \mathbf{R}$. The idea of fixed point iteration consists of transforming this equation into an equivalent *fixed point equation*

$$\phi(x) = x$$

and of constructing a sequence $\{x_0, x_1, \ldots\}$ with the help of the iterative scheme

$$x_{k+1} = \phi(x_k) \quad \text{with} \quad k = 0, 1, \ldots$$

for a given starting value x_0. We hope that the sequence $\{x_k\}$ defined in this way converges to a *fixed point* x^* with $\phi(x^*) = x^*$, which consequently is also a solution of the nonlinear equation, i.e. $f(x^*) = 0$.

Example 4.1 We consider the equation

$$f(x) := 2x - \tan x = 0 \ . \tag{4.1}$$

From Figure 4.1 we can read off the value $x^* \approx 1.2$ as an approximation for the solution of 4.1 in the interval $[0.5, 2]$. We choose $x_0 = 1.2$ as a

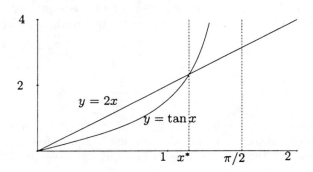

Figure 4.1: Graphical solution of $2x - \tan x = 0$

starting value for a fixed point iteration. The equation (4.1) can be easily transformed into a fixed point iteration, perhaps into

$$x = \frac{1}{2} \tan x =: \phi_1(x) \ \text{ or } \ x = \arctan(2x) =: \phi_2(x) \ .$$

If we try the two corresponding fixed point iterations with the starting value $x_0 = 1.2$, then we obtain the numerical values in Table 4.1. We see that the first sequence diverges ($\tan x$ has a pole at $\pi/2$ and $x_2 > \pi/2$), whereas the second one converges. The convergent sequence has the property that at about every second iteration there is another correct decimal.

Obviously not every naively constructed fixed point iteration converges. Therefore we consider now general sequences $\{x_k\}$, which are given by an *iteration mapping* ϕ

$$x_{k+1} = \phi(x_k).$$

If we want to estimate the difference of two consecutive terms

$$|x_{k+1} - x_k| = |\phi(x_k) - \phi(x_{k-1})|$$

Table 4.1: Comparison of the fixed point iterations ϕ_1 and ϕ_2

k	$x_{k+1} = \frac{1}{2}\tan x_k$	$x_{k+1} = \arctan(2x_k)$
0	1.2	1.2
1	1.2860	1.1760
2	$1.70\ldots > \pi/2$	1.1687
3		1.1665
4		1.1658
5		1.1656
6		1.1655
7		1.1655

by the difference of the previous terms $|x_k - x_{k-1}|$ (naturally we have the geometric series in mind), we are necessarily lead to the following theoretical characterization:

Definition 4.2 Let $I = [a,b] \subset \mathbf{R}$ be an interval and $\phi : I \to \mathbf{R}$ a mapping. ϕ is *contractive on* I if there is a $0 \le \theta < 1$ such that

$$|\phi(x) - \phi(y)| \le \theta|x - y| \quad \text{for all} \quad x, y \in I \ .$$

The *Lipschitz constant* θ can be easily computed if ϕ is continuously differentiable.

Lemma 4.3 *If $\phi : I \to \mathbf{R}$ is continuously differentiable, $\phi \in C^1(I)$, then*

$$\sup_{x,y \in I} \frac{|\phi(x) - \phi(y)|}{|x - y|} = \sup_{z \in I} |\phi'(z)| < \infty \ .$$

Proof. This is a simple application of the mean value theorem in $\mathbf{R}$: For all $x, y \in I$, $x < y$, there exists a $\xi \in [x, y]$, such that

$$\phi(x) - \phi(y) = \phi'(\xi)(x - y) \ .$$

$\square$

Theorem 4.4 *Let $I = [a,b] \subset \mathbf{R}$ be an interval and $\phi : I \to I$ a contractive mapping with Lipschitz constant $\theta < 1$. Then it follows that:*

a) *There exists a unique fixed point x^* of ϕ, $\phi(x^*) = x^*$.*

b) *For any starting point $x_0 \in I$, the fixed point iteration $x_{k+1} = \phi(x_k)$ converges to x^* such that*

$$|x_{k+1} - x_k| \le \theta |x_k - x_{k-1}| \quad \text{and} \quad |x^* - x_k| \le \frac{\theta^k}{1 - \theta} |x_1 - x_0| .$$

Proof. For all $x_0 \in I$ holds

$$|x_{k+1} - x_k| = |\phi(x_k) - \phi(x_{k-1})| \le \theta |x_k - x_{k-1}|$$

and therefore inductively

$$|x_{k+1} - x_k| \le \theta^k |x_1 - x_0| .$$

We want to show that $\{x_k\}$ is a Cauchy sequence and therefore we write

$$
\begin{aligned}
|x_{k+m} - x_k| &\le |x_{k+m} - x_{k+m-1}| + \cdots + |x_{k+1} - x_k| \\
&\le \underbrace{(\theta^{k+m-1} + \theta^{k+m-2} + \cdots + \theta^k)}_{= \theta^k (1 + \theta + \cdots + \theta^{m-1})} |x_1 - x_0| \\
&\le \frac{\theta^k}{1 - \theta} |x_1 - x_0|,
\end{aligned}
$$

where we have used the triangle inequality and the formula for the sum of geometric series $\sum_{k=0}^{\infty} \theta^k = 1/(1 - \theta)$. Thus $\{x_k\}$ is a Cauchy sequence in the complete metric space of all real numbers, and therefore it converges to a limit point

$$x^* := \lim_{k \to \infty} x_k$$

But then x^* is a fixed point of ϕ, because

$$
\begin{aligned}
|x^* - \phi(x^*)| &= |x^* - x_{k+1} + x_{k+1} - \phi(x^*)| \\
&= |x^* - x_{k+1} + \phi(x_k) - \phi(x^*)| \\
&\le |x^* - x_{k+1}| + |\phi(x_k) - \phi(x^*)| \\
&\le |x^* - x_{k+1}| + \theta |x^* - x_k| \quad \to 0 \quad \text{for} \quad k \to \infty .
\end{aligned}
$$

With this we have proved the second part of the theorem and the existence of a fixed point. If x^*, y^* are two fixed points, then

$$0 \le |x^* - y^*| = |\phi(x^*) - \phi(y^*)| \le \theta |x^* - y^*| .$$

Because $\theta < 1$ this is possible only if $|x^* - y^*| = 0$. This proves the uniqueness of the fixed point of ϕ. $\qquad\square$

Remark 4.5 Theorem 4.4 is a special case of the Banach fixed point theorem. The only properties used in the proof are the triangle inequality for the absolute value and the completeness of $\mathbf{R}$. Therefore the proof is valid in the much more general situation when $\mathbf{R}$ is replaced by a Banach space X, e.g. a function space, and the absolute value by the corresponding norm. Such theorems play a role not only in the theory but also in the numerics of differential and integral equations. In this introductory textbook we shall use only the extension to $X = \mathbf{R}^n$ with a norm $\|\cdot\|$ instead of the absolute value $|\cdot|$.

Remark 4.6 The algorithm of BRENT [7] has lately established itself as a standard code for solving scalar nonlinear equations, in the case when only a program for evaluating $f(x)$ and an interval enclosing the solution are available. It is based on a mixture of rather elementary techniques, such as bisection and inverse quadratic interpolation, which will not be further elaborated here. For a detailed description we refer the reader to [7]. If additional information regarding f, like convexity or differentiability, is available, then methods with faster convergence can be constructed, on which we will focus our attention in the following.

In order to assess the *speed of convergence* of a fixed point iteration we define the notion of *order of convergence* of a sequence $\{x_k\}$.

Definition 4.7 A sequence $\{x_k\}, x_k \in \mathbf{R}^n$, converges to x^*, *with order (at least)* $p \geq 1$, if there is a constant $C \geq 0$ such that

$$\|x_{k+1} - x^*\| \leq C\|x_k - x^*\|^p ,$$

where in case $p = 1$ we also require that $C < 1$. We use the term *linear convergence* in case $p = 1$, and *quadratic convergence* for $p = 2$. Furthermore we say that $\{x_k\}$ is *superlinearly* convergent if there exists a nonnegative null sequence $C_k \geq 0$ with $\lim_{k \to \infty} C_k = 0$ such that

$$\|x_{k+1} - x^*\| \leq C_k\|x_k - x^*\| .$$

Remark 4.8 Often, for reasons of simplicity, the convergence order p is alternatively defined by the analogue inequalities for the iterates, that is

$$\|x_{k+1} - x_k\| \leq C\|x_k - x_{k-1}\|^p$$

for convergence with order p and

$$\|x_{k+1} - x_k\| \leq C_k\|x_k - x_{k-1}\|$$

for superlinear convergence.

As we have seen above in the simple example $f(x) = 2x - \tan x$, in order to solve the nonlinear problem $f(x) = 0$ we must choose from many possible fixed point iterations a suitable one. In general this is not a simple task. Since

$$|x_{k+1} - x^*| = |\phi(x_k) - \phi(x^*)| \leq \theta |x_k - x^*|$$

and $0 \leq \theta < 1$, the fixed point iteration

$$x_{k+1} = \phi(x_k),$$

where ϕ is a contractive mapping, converges only linearly in the general case. We would expect that the convergence of a good iterative method be at least *superlinear* or *linear with a small constant* $C \ll 1$. Therefore in the next section we will turn our attention towards a quadratically convergent method.

4.2   Newton's Method for Nonlinear Systems

For the time being we consider again a *scalar nonlinear equation* $f(x) = 0$ and we are trying to find a zero x^* of f. As the function f is not given in global manner and we merely have the possibility of pointwise evaluation, we approximate it by its tangent $p(x)$ at the starting point x_0. Instead of the intersection point x^* of the graph of f with the x-axis, we compute the x-intercept of the tangent line (see Figure 4.2). The tangent line is represented

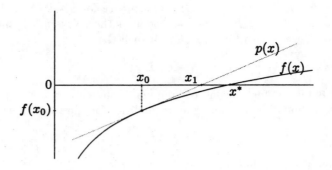

Figure 4.2: Idea of Newton's method in $\mathbf{R}^1$

by the first order polynomial

$$p(x) = f(x_0) + f'(x_0)(x - x_0) \ .$$

In case $f'(x_0) \neq 0$ the corresponding zero x_1 may be written as

$$x_1 = x_0 - \frac{f(x_0)}{f'(x_0)} \ .$$

The fundamental idea of *Newton's method* consists in repeated application of this rule:

$$x_{k+1} := x_k - \frac{f(x_k)}{f'(x_k)}, \quad k = 0, 1, 2, \dots \ .$$

This is obviously a specific fixed point iteration with iteration mapping

$$\phi(x) := x - \frac{f(x)}{f'(x)} \ .$$

Naturally, ϕ can only be constructed if f is differentiable and $f'(x)$ does not vanish at least in a neighborhood of the solution. The convergence properties of the method will be analyzed later in a general theoretical framework.

Example 4.9 *Computation of the square root.* We have to solve the equation

$$f(x) := x^2 - c = 0 \ .$$

In a computer the number c has the floating point representation

$$c = a2^p \quad \text{with} \ \ 0.5 < a \leq 1, p \in \mathbf{N}.$$

with mantissa a and exponent p. Therefore

$$\sqrt{c} = \begin{cases} 2^m \sqrt{a} & , \quad \text{if } p = 2m \\ 2^m \sqrt{a}\sqrt{0.5} & , \quad \text{if } p = 2m - 1 \end{cases} ,$$

where

$$\sqrt{0.5} < \sqrt{a} \leq 1 \ .$$

Once $\sqrt{0.5} = 1/\sqrt{2} \approx 0.71$ is computed and stored with the neccessary number of digits, then only the problem

$$f(x) := x^2 - a = 0 \ \ \text{for} \ \ a \in \,]0.5, 1]$$

remains to be solved. Because

$$f'(x) = 2x \neq 0 \ \ \text{for} \ \ x \in \,]0.5, 1] \ ,$$

Newton's method is applicable and the corresponding iteration mapping is
given by

$$\phi(x) = x - \frac{f(x)}{f'(x)} = x - \frac{x^2 - a}{2x} = x - \frac{x}{2} + \frac{a}{2x} = \frac{1}{2}\left(x + \frac{a}{x}\right),$$

Therefore the Newton iteration is defined as

$$x_{k+1} := \frac{1}{2}\left(x_k + \frac{a}{x_k}\right).$$

Division by 2 can be cheaply implemented by subtracting 1 from the exponent, so that only a division and an addition have to be carried out per iteration step. We have rendered the Newton iteration for $a = 0.81$ and $x_0 = 1$
in Table 4.2. It shows that the number of exact figures is approximately
doubled at each iteration step, a typical behavior of *quadratic* convergence.

Table 4.2: Newton's iteration for $a = 0.81$ and $x_0 = 1$

k	x_k
0	1.0000000000
1	0.9050000000
2	0.9000138122
3	0.90000000001

Newton's method can be easily extended to *nonlinear systems*

$$F(x) = 0$$

where $F : \mathbf{R}^n \to \mathbf{R}^n$ is a continuously differentialbe mapping satisfying
certain additional properties. The graphical derivation of the method is of
course no longer possible for dimention $n > 1$. In principle, however, we only
have to replace the nonlinear map by a linear one. The Taylor expansion of
F about a starting point x^0 yields

$$0 = F(x) = \underbrace{F(x^0) + F'(x^0)(x - x^0)}_{:= \bar{F}(x)} + o\left(|x - x^0|\right) \quad \text{for} \quad x \to x^0. \qquad (4.2)$$

The zero x^1 of the linear substitute map $\bar{F}$ is precisely

$$x^1 = x^0 - F'(x^0)^{-1}F(x^0),$$

as long as the Jacobian matrix $F'(x^0)$ at x^0 is invertible. This inspires the *Newton iteration* $(k = 0, 1, \ldots)$

$$F'(x^k)\Delta x^k = -F(x^k) \quad \text{with} \quad x^{k+1} = x^k + \Delta x^k .\qquad(4.3)$$

Of course, one does not compute the inverse $F'(x^k)^{-1}$ at each iteration step, but instead one determines the so called *Newton correction* Δx^k as the solution of the above linear system. Therefore we reduce the numerical solution of a system on nonlinear equations to the numerical solution of a *sequence of linear systems.*

Before turning to the analysis of the convergence properties of Newton's method we wish to point out an *invariance property.* Obviously, the problem of solving the equation $F(x) = 0$ is equivalent to solving

$$G(x) := AF(x) = 0,$$

where $A \in \mathrm{GL}(n)$ is an arbitrary invertible matrix. At the same time note that, for a given x^0, the Newton sequence $\{x^k\}$ is independent of A since

$$G'(x)^{-1}G(x) = (AF'(x))^{-1}AF(x) = F'(x)^{-1}A^{-1}AF(x) = F'(x)^{-1}F(x) .$$

The transformation $F \to G$ is an affine transformation (without the translation component). Therefore it has become common usage to say that the problem $F(x) = 0$ as well as Newton's method are *affine invariant.* Accordingly we require that the convergence properties of Newton's method be described by an affine invariant theory. Among many convergence theorems for Newton's method we select a relatively new one [24], because it yields particularly clear results and is nevertheless relatively easy to prove.

Theorem 4.10 *Let $D \subset \mathbf{R}^n$ be open and convex, and let $F : D \to \mathbf{R}^n$ be a continuously differentiable mapping, with an invertible Jacobian matrix $F'(x)$ for all $x \in D$. Suppose that for an $\omega \geq 0$ the following (affine invariant) Lipschitz condition holds:*

$$\|F'(x)^{-1}(F'(x + sv) - F'(x))v\| \leq s\omega\|v\|^2 \qquad(4.4)$$

for all $s \in [0, 1]$, $x \in D$ and $v \in \mathbf{R}^n$, so that $x + v \in D$. Furthermore let us assume that there exists a solution $x^* \in D$ and a starting point $x^0 \in D$ such that

$$\rho := \|x^* - x^0\| < \frac{2}{\omega} \quad \text{and} \quad B_\rho(x^*) \subseteq D .\qquad(4.5)$$

Then the sequence $\{x^k\}$, $k > 0$, defined by Newton's method stays in the open ball $B_\rho(x^*)$ and converges to x^*, i.e.

$$\|x^k - x^*\| < \rho \quad \text{for} \quad k > 0 \quad \text{and} \quad \lim_{k\to\infty} x^k = x^* .$$

The speed of convergence can be estimated by

$$\|x^{k+1} - x^*\| \leq \frac{\omega}{2}\|x^k - x^*\|^2 \quad for \quad k = 0, 1, \ldots$$

Moreover the solution x^* is unique in $B_{2/\omega}(x^*)$.

Proof.   First, we use the Lipschitz condition (4.4) to derive the following result for all $x, y \in D$:

$$\|F'(x)^{-1}(F(y) - F(x) - F'(x)(y - x))\| \leq \frac{\omega}{2}\|y - x\|^2 . \qquad (4.6)$$

Here we use the Lagrange form of the integral mean value theorem:

$$F(y) - F(x) - F'(x)(y - x) = \int_{s=0}^{1} (F'(x + s(y - x)) - F'(x))(y - x)ds .$$

The left hand side of (4.6) can thus be rewritten and estimated as

$$\left\| \int_{s=0}^{1} F'(x)^{-1}(F'(x + s(y - x)) - F'(x))(y - x)ds \right\|$$

$$\leq \int_{s=0}^{1} s\omega\|y - x\|^2 \, ds \;=\; \frac{\omega}{2}\|y - x\|^2 ,$$

which proves (4.6). After this preparation we can turn our attention to the question of the convergence of Newton's iteration. By using the iterative scheme (4.3) as well as the relation $F(x^*) = 0$ we get

$$\begin{aligned}
x^{k+1} - x^* \;&=\; x^k - F'(x^k)^{-1}F(x^k) - x^* \\
&=\; x^k - x^* - F'(x^k)^{-1}(F(x^k) - F(x^*)) \\
&=\; F'(x^k)^{-1}(F(x^*) - F(x^k) - F'(x^k)(x^* - x^k)) .
\end{aligned}$$

With the help of (4.6) this leads to the following estimate of the speed of convergence

$$\|x^{k+1} - x^*\| \leq \frac{\omega}{2}\|x^k - x^*\|^2 .$$

If $0 < \|x^k - x^*\| \leq \rho$, then

$$\|x^{k+1} - x^*\| \leq \underbrace{\frac{\omega}{2}\|x^k - x^*\|}_{\leq\, \rho\omega/2\, <\, 1}\|x^k - x^*\| < \|x^k - x^*\| .$$

Since $\|x^0 - x^*\| = \rho$, we have $\|x^k - x^*\| < \rho$ for all $k > 0$, and the sequence $\{x^k\}$ converges towards x^*.

In order to prove uniqueness in the ball $B_{2/\omega}(x^*)$ centered at x^* with radius $2/\omega$, we employ again inequality (4.6). Let $x^{**} \in B_{2/\omega}(x^*)$ be another solution so that $F(x^{**}) = 0$ and $\|x^* - x^{**}\| < 2/\omega$. By substituting in (4.6) we obtain

$$
\begin{aligned}
\|x^{**} - x^*\| &= \|F'(x^*)^{-1}(0 - 0 - F'(x^*)(x^{**} - x^*))\| \\
&\leq \underbrace{\frac{\omega}{2}\|x^{**} - x^*\|}_{< 1} \|x^{**} - x^*\| .
\end{aligned}
$$

This is possible only if $x^{**} = x^*$. $\qquad\qquad\square$

In short, the above theorem reads: *Newton's method has local quadratic convergence.*

Remark 4.11 Different variants of the above theorem prove the *existence* of a solution x^*, while we have assumed the existence of a solution. However the corresponding proofs are rather involved (see chapter 2 in [15]).

Because the solution x^* is not known a priori, the theoretical assumptions of Theorem 4.10 cannot be directly verified. On the other hand, of course, in order to save unnecessary iteration steps, we would like to know as early as possible whether the Newton iteration converges. Therefore we are looking for a *convergence criterion* that can be verified within the algorithm itself, and allows us to decide after one, or a few steps, if Newton's method converges. As in Section 2.4.3, when we considered the solution of linear systems, we will first look at the *residuals* $F(x^k)$. Obviously, solving the system of nonlinear equations is equivalent to minimizing the residual. Therefore one should expect that the residual is monotonically decreasing, i.e.

$$\|F(x^{k+1})\| \leq \bar\theta\,\|F(x^k)\| \quad \text{for} \quad k = 0, 1, \dots \quad \text{and a} \ \bar\theta < 1 . \tag{4.7}$$

However this so called *monotony test* is not affine invariant. Multiplication of F by an invertible matrix A may change completely the result of the monotony test (4.7). In trying to transform inequality (4.7) into an affine invariant and easily verifiable condition we come across the so called *natural monotony test*

$$\|F'(x^k)^{-1} F(x^{k+1})\| \leq \bar\theta\,\|F'(x^k)^{-1} F(x^k)\| \quad \text{for a} \ \bar\theta < 1 . \tag{4.8}$$

On the right-hand-side we rediscover the Newton correction Δx^k that has to be computed anyway. In order to estimate the left-hand-side we have to solve again a system of linear equations with the same matrix $F'(x^k)$, however this time for the function value $F(x^{k+1})$ taken at the next iteration point $x^{k+1} = x^k + \Delta x^k$. Even this is done with little additional work. Indeed, after applying an elimination method (requiring $O(n^3)$ operations for a full matrix) we only have to carry out one forward and one backward substitution (which means $O(n^2)$ additional operations). Therefore by defining the so called *simplified Newton correction* $\bar{\Delta} x^{k+1}$ as the solution of the linear system

$$F'(x^k)\bar{\Delta} x^{k+1} = -F(x^{k+1}) \ ,$$

the natural monotony test (4.8) can be written as

$$\|\bar{\Delta} x^{k+1}\| \le \bar{\theta} \|\Delta x^k\| \ . \tag{4.9}$$

From a theoretical analysis (see [15]) which is also justified by numerical practice the value $\bar{\theta} := 0.5$ turns out to be the correct choice for the constant $\bar{\theta}$. With this we have found our convergence criterion for Newton's method: If the natural monotony test (4.9) fails for some k, that is if

$$\|\bar{\Delta} x^{k+1}\| > \frac{1}{2} \|\Delta x^k\| \ ,$$

then the Newton iteration has to be stopped and we have no other option but to restart it with a different, hopefully better, starting point x^0.

Remark 4.12 The so called *damping* of the Newton correction Δx^k can be used to save the convergence of Newton's method even when the natural monotony test fails. Here we choose for the next iterate

$$x^{k+1} = x^{k+1}(\lambda_k) := x^k + \lambda_k \Delta x^k \ ,$$

where $0 < \lambda_k \le 1$ is the so called *damping factor*. In choosing it we may use again the natural monotony test. As a simple damping strategy we recommend choosing the damping factors λ_k such that the natural monotony test (4.9) is satisfied for $\bar{\theta} := 1 - \lambda_k/2$, i.e.

$$\|\bar{\Delta} x^{k+1}(\lambda_k)\| \le \left(1 - \frac{\lambda_k}{2}\right) \|\Delta x^k\| \ ,$$

where $\bar{\Delta} x^{k+1}(\lambda_k) = -F'(x^k)^{-1} F(x^k + \lambda_k \Delta x^k)$ is the simplified Newton correction of the damped method. Here we choose the λ_k's from a finite sequence $\{1, \frac{1}{2}, \frac{1}{4}, \ldots, \lambda_{\min}\}$ and we abort the iteration whenever the above monotony test would require $\lambda_k < \lambda_{\min}$. In order to avoid overflow (x^{k+1} larger than the largest number representable on the given computer), in critical examples one could experiment by starting with $\lambda_0 = \lambda_{\min}$. If λ_k is successful then in the next iteration we set $\lambda_{k+1} = \min(1, 2\lambda_k)$ in order to attain asymptotically the quadratic convergence of the pure Newton method (with $\lambda = 1$). In case the monotony test for λ_k is violated, we perform the monotony test for $\lambda_k/2$. A theoretically sounder, and in general substantially more efficient, damping strategy is presented in chapter 2 of the already often cited book [15].

Remark 4.13 In the implementation of Newton's method it is generally sufficient to replace the exact Jacobian matrix by a suitable approximation, which may be obtained for example by using finite differences instead of the corresponding partial derivatives. See also Exercise 4.7. Moreover, the convergence of the method is usually not impaired, if "nonessential" elements of the Jacobian matrix are omitted. This so called *sparsing* technique is particularly recommended for *large* nonlinear systems. However it requires a good insight into the underlying problem.

An alternative method of extending the convergence region requiring additional knowledge of the problem to be solved, is treated in Section 4.4.2.

4.3 Gauss-Newton Method for Nonlinear Least Squares Problems

In Section 3.1 we have dealt in detail with the general setting of the problem of Gaussian least squares computation. Our goal is to determine a parameter $x \in \mathbf{R}^n$, of a model function φ, such that it fits the measurements b in the least squares sense. If the parameter enters linearly in φ, then this leads to the linear least squares computation presented in Chapter 3. If the parameter enters nonlinearly in φ, then we have a *nonlinear least squares problem* of the form

$$g(x) := \|F(x)\|_2^2 = \min , \qquad (4.10)$$

where we assume that $F : D \subset \mathbf{R}^n \longrightarrow \mathbf{R}^m$ is a twice continuously differentiable function $F \in C^2(D)$ on an open set $D \subset \mathbf{R}^n$. In this section we treat only the overdetermined case $m > n$. Since in applications we have as a rule significantly more measurement data than parameters, i.e. $m \gg n$, we also use the term *data compression*. In what follows we omit consideration of the boundary ∂D and we are looking only for *interior local minima* $x^* \in D$ of g, that satisfy the sufficient conditions

$$g'(x^*) = 0 \quad \text{and} \quad g''(x^*) \text{ positive definite.} \qquad (4.11)$$

Since $g'(x) = 2F'(x)^T F(x)$, we therefore have to solve the following system of n nonlinear equations

$$G(x) := F'(x)^T F(x) = 0. \qquad (4.12)$$

The Newton iteration for this system of equations is

$$G'(x^k)\Delta x^k = -G(x^k) \quad \text{with} \quad k = 0, 1, \dots , \qquad (4.13)$$

where under the above assumptions the Jacobian matrix

$$G'(x) = F'(x)^T F'(x) + F''(x)^T F(x)$$

is positive definite in a neighborhood of x^* and hence invertible. When the model and data fully agree at x^*, i.e. when they are *compatible*, then we have

$$F(x^*) = 0 \quad \text{and} \quad G'(x^*) = F'(x^*)^T F'(x^*) .$$

The condition "$G'(x^*)$ is positive definite" is equivalent in this case with the condition that $F'(x^*)$ has full rank n. For compatible, or at least for "almost compatible" nonlinear least squares problems, we would like to save the effort of evaluating the tensor $F''(x)$. Therefore we substitute the Jacobian matrix $G'(x)$ in the Newton iteration (4.13) by $F'(x)^T F'(x)$ and we obtain the iterative scheme

$$F'(x^k)^T F'(x^k)\Delta x^k = -F'(x^k)^T F(x^k) .$$

Obviously these are the normal equations for the linear least squares problem

$$\|F'(x^k)\Delta x + F(x^k)\|_2 = \min .$$

By using the notation for the pseudo inverse from Section 3.3 we obtain the formal representation

$$\Delta x^k = -F'(x^k)^+ F(x^k) \quad \text{with} \quad x^{k+1} = x^k + \Delta x^k . \qquad (4.14)$$

In this way we have thus reduced the numerical solution of a nonlinear least squares problem to a *sequence of linear least squares problems*.

Remark 4.14 If the Jacobian matrix has full rank then

$$F'(x)^+ = (F'(x)^T F'(x))^{-1} F'(x)^T$$

and therefore the equation (4.12) to be solved for the nonlinear least squares problem is equivalent to

$$F'(x)^+ F(x) = 0 .$$

This characterization holds for the rank deficient case and the underdetermined case as well.

Similarly to Newton's method for nonlinear systems, we could have also deduced (4.14) directly from the original minimization problem by expanding in Taylor series and truncating after the linear term. Therefore (4.14) is also called the *Gauss–Newton method* for the nonlinear least squares problem $\|F(x)\|_2 = \min$. The convergence of the Gauss-Newton method is characterized in the following theorem (compare [23]), which is an immediate generalization of the corresponding convergence theorem 4.10 for Newton's method.

Theorem 4.15 *Let $D \subset \mathbf{R}^n$ be open and convex and $F : D \to \mathbf{R}^m$, $m \geq n$, a continuously differentiable mapping whose Jacobian matrix $F'(x)$ has full rank n for all $x \in D$. Suppose there is a solution $x^* \in D$ of the corresponding nonlinear least squares problem $\|F(x)\|_2 = \min$. Furthermore let $\omega > 0$ and $0 \leq \kappa_* < 1$, be two constants such that*

$$\|F'(x)^+(F'(x+sv) - F'(x))v\| \leq s\omega\|v\|^2 \qquad (4.15)$$

for all $s \in [0,1]$, $x \in D$ and $v \in \mathbf{R}^n$ with $x + v \in D$, and assume that

$$\|F'(x)^+F(x^*)\| \leq \kappa_*\|x - x^*\| \qquad (4.16)$$

for all $x \in D$. If for a given starting point $x^0 \in D$ we have

$$\rho := \|x^0 - x^*\| < 2(1 - \kappa_*)/\omega =: \sigma \qquad (4.17)$$

then the sequence $\{x^k\}$ defined by the Gauss–Newton method (4.14) stays in the open ball $B_\rho(x^*)$ and converges towards x^*, i.e.

$$\|x^k - x^*\| < \rho \quad for \quad k > 0 \quad and \quad \lim_{k \to \infty} x^k = x^* \, .$$

The speed of convergence can be estimated by

$$\|x^{k+1} - x^*\| \leq \frac{\omega}{2}\|x^k - x^*\|^2 + \kappa_*\|x^k - x^*\| \, . \qquad (4.18)$$

In particular quadratic convergence is obtained for compatible nonlinear least squares problems. Moreover the solution x^* is unique in $B_\sigma(x^*)$.

Proof. The proof follows directly the main steps of the proof of Theorem 4.10. From the Lipschitz condition (4.15) it follows immediately that

$$\|F'(x)^+(F(y) - F(x) - F'(x)(y - x))\| \leq \frac{\omega}{2}\|y - x\|^2 \qquad (4.19)$$

for all $x, y \in D$. In order to estimate the speed of convergence we use the definition (4.14) of the Gauss–Newton iteration as well as the property

$F'(x^*)^+ F(x^*) = 0$ of the solution x^* whose existence has been assumed. From the full rank Jacobian assumption it follows immediately, see Section 3.3, that

$$F'(x)^+ F'(x) = I_n \text{ for all } x \in D .$$

Therefore we obtain

$$x^{k+1} - x^* = x^k - x^* - F'(x^k)^+ F(x^k)$$
$$= F'(x^k)^+ (F(x^*) - F(x^k) - F'(x^k)^+ (x^* - x^k)) - F'(x^k)^+ F(x^*) .$$

By applying conditions (4.15) and (4.16) we get

$$\|x^{k+1} - x^*\| \leq \left(\frac{\omega}{2} \|x^k - x^*\| + \kappa_* \right) \|x^k - x^*\| .$$

Together with assumption (4.17) and induction on k this implies

$$\|x^{k+1} - x^*\| < \|x^k - x^*\| \leq \rho .$$

From here it follows immediately that the iterates remain in $B_\rho(x^*)$ and converge towards the solution x^*. For compatible least squares problems i.e. $F(x^*) = 0$, we can choose $\kappa_* = 0$ in (4.16) and hence obtain quadratic convergence. The uniqueness of the solution is obtained as in the proof of Theorem 4.10. $\square$

Remark 4.16 We note that in the above theorem the existence of a solution is assumed. A variant of the above theorem yields the *existence* of a solution x^* as well, and relaxes the full rank condition on the Jacobian (compare Chapter 3 in [15]). Only the Penrose axiom

$$F'(x)^+ F'(x) F'(x)^+ = F'(x)^+$$

is used. However the *uniqueness* requires, as in the particular case of a linear problem (see Section 3.3), a maximal rank assumption. Otherwise, in the nonlinear case there exists a solution manifold of dimension equal to the rank defect, and the Gauss–Newton method converges towards a point on this manifold. We will use this property in Section 4.4.2.

Finally we want to discuss in more detail the condition $\kappa_* < 1$. For $\kappa_* > 0$ the linear term $\kappa_* \|x^k - x^*\|$ dominates the speed of convergence estimate (4.18) as we approach the solution x^*. Therefore in this case the Gauss–Newton method converges *linearly* with asymptotic convergence factor κ_*. Obviously this enforces the condition $\kappa_* < 1$. The quantity κ_* reflects the omission of the tensor $F''(x)$ in the derivation of Newton's method (4.13).

Another interpretation comes from examining the influence of the statisti-cal measure of the error δb on the solution. In case the Jacobian matrix $F'(x^*)$ has full rank, the perturbation of the parameter induced by δb in the linearized error analysis is given by

$$\delta x^* = -F'(x^*)^+ \delta b.$$

A quantity of this type is given as a–posteriori error analysis by virtually all software packages in widespread use today in statistics. Obviously it does not reflect the possible effects of the nonlinearity of the model. A more accurate analysis of this problem was carried out by H.G. BOCK [5] who actually shows that one should perform the substitution

$$\|\delta x^*\| \longrightarrow \|\delta x^*\|/(1 - \kappa_*). \tag{4.20}$$

In the compatible case $F(x^*) = 0$ and $\kappa_* = 0$, while in the "almost compat-ible" case , $0 < \kappa_* \ll 1$, the linearized error theory is satisfactory. However, Bock shows that in case $\kappa_* \geq 1$ there are always statistical errors such that the solution "runs away unboundedly". Such models are therefore *statisti-cally ill posed* or *inadequate*. Conversely, a nonlinear least squares problem is called *statistically well posed* or *adequate* if $\kappa_* < 1$. Hence Theorem 4.15 can be restated in short form: *The ordinary Gauss-Newton converges lo-cally for adequate nonlinear least squares problems and the convergence is quadratic for compatible least squares problems.*

Intuitively it is clear that not every model and every set of measurements allow for determination of a unique suitable parameter. On the other hand only unique solutions permit a clear interpretation in connection with the basic theoretical model. The Gauss–Newton method presented here proves the uniqueness of the solution with the help of three criteria.

a) Checking the full rank condition of the corresponding Jacobians in the sense of a numerical rank determination, which can be done, for example, as in Section 3.2 on the basis of a QR-factorization.

b) Checking the statistical well posedness with the help of the condition $\kappa_* < 1$ by estimating

$$\kappa_* \doteq \frac{\|\Delta x^{k+1}\|}{\|\Delta x^k\|}$$

in the (linear) final phase of the Gauss–Newton iteration.

c) Checking the behaviour of the error by an a–posteriori error analysis as suggested by (4.20).

One should be aware of the fact that all three criteria are influenced by the choice of the measurement tolerances δb (cf. Secction 3.1) as well as by the scaling of the parameter x.

Remark 4.17 The convergence region of the Gauss-Newton method can be enlarged by a *damping strategy* similar to the one used for Newton's method. If we denote by Δx^k the ordinary Gauss–Newton correction then an iteration step reads as

$$x^{k+1} = x^k + \lambda_k \Delta x^k \quad \text{with} \quad 0 < \lambda_k \leq 1 \ .$$

There is also an effective strategy here with a strong theoretical foundation, which is implemented in a series of modern least squares software packages, see [15]. These programs check automatically whether the respective least squares problem is adequate. If this is not the case, as it rather infrequently happens, one should either improve the model or increase the precision of the measurements. Moreover these programs make sure automatically that the iteration is performed only until a relative precision matching the precision of the measurements is reached.

Example 4.18 *A biochemical reaction.* In order to illustrate the behavior of the damped Gauss-Newton method we give a nonlinear least squares problem from biochemistry, the Feulgen hydrolysis [61]. From an extensive series of measurements we choose a problem with $m = 30$ measurement data and $n = 3$ unknown parameters x_1, x_2 and x_3 (see Table 4.3).

Table 4.3: Measurement sequence (t_i, b_i), $i = 1, \ldots, 30$, for Feulgen Hydrolysis

t	b	t	b	t	b	t	b	t	b
6	24.19	42	57.39	78	52.99	114	49.64	150	46.72
12	35.34	48	59.56	84	53.83	120	57.81	156	40.68
18	43.43	54	55.60	90	59.37	126	54.79	162	35.14
24	42.63	60	51.91	96	62.35	132	50.38	168	45.47
30	49.92	66	58.27	102	61.84	138	43.85	174	42.40
36	51.53	72	62.99	108	61.62	144	45.16	180	55.21

Originally the model function was given in the form

$$\varphi(x; t) := \frac{x_1 x_2}{x_2 - x_3} \left(\exp(-x_3 t) - \exp(-x_2 t) \right),$$

where x_1 is the DNA concentration and x_2, x_3 the chemical reaction rate constants. For $x_2 = x_3$ we have obviously a limit of the form $0/0$, so that if $x_2 \approx x_3$ then numerical cancellation leads to difficulties in the behavior of the iteration. Therefore we perform a different parameterization that takes into account the inequalities $x_2 > x_3 \geq 0$ (coming from biochemistry) as well. Instead of x_1, x_2 and x_3 we consider the unknowns

$$\bar{x}_1 := x_1 x_2, \quad \bar{x}_2 := \sqrt{x_3} \quad \text{and} \quad \bar{x}_3 := \sqrt{(x_2 - x_3)/2}$$

and the transformed model function

$$\varphi(\bar{x};t) := \bar{x}_1 \exp\!\left(-(\bar{x}_2^2 + \bar{x}_3^2)t\right)\frac{\sinh(\bar{x}_3^2 t)}{\bar{x}_3^2} \; .$$

The property $\sinh(\bar{x}_3 t) = \bar{x}_3^2 t + o(|t|)$ for small arguments is surely established in every standard routine for calculating sinh, so that only the evaluation of φ for $\bar{x}_3 = 0$ has to be handled in a special way by the program. We choose as starting point $\bar{x}^0 = (80, 0.055, 0.21)$. The iterative conduct of $\varphi(\bar{x}^k; t)$ over the interval $t \in [0, 180]$ containing the measurements is represented in Figure 4.3. We have tol $= 0.142 \cdot 10^{-3}$ as "satisfactory" relative precision. At the solution x^* $\kappa_* = 0.156$ holds and the norm of the residual is $\|F(x^*)\|_2 \approx 3 \cdot 10^2$. Therefore in spite of a "large" residual, the problem is almost compatible according to the theoretically relevant characterization in terms of the constant κ_*.

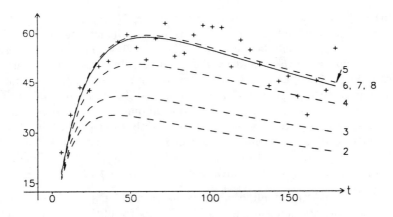

Figure 4.3: Measurements and iterated model function for Feulgen hydrolysis

Remark 4.19 Most software packages in statistics now contain yet another method with enlarged convergence region, the Levenberg-Marquardt method. This method is based on the idea that the local linearization

$$F(x) \doteq F(x^k) + F'(x^k)(x - x^k) \, ,$$

that leads to the (ordinary) Newton method is reasonable for iterates x^k that are "far away" from the solution x^* only in a neighbourhood of x^k.

Therefore instead of (4.2) one solves the following substitute problem

$$\|F(x^k) + F'(x^k)\Delta z\|_2 = \min \qquad (4.21)$$

under the *constraint*

$$\|\Delta z\|_2 \leq \delta , \qquad (4.22)$$

where the (locally fitted) parameter δ is to be discussed. The constraints (4.22) can be coupled with the help of a *Lagrange multiplier* $p \geq 0$ to the minimization problem (4.21) by

$$p(\|\Delta z\|_2^2 - \delta^2) \geq 0.$$

If $\|\Delta z\|_2 = \delta$, then we must have $p > 0$, while if $\|\Delta z\|_2 < \delta$, then $p = 0$. This leads to the formulation

$$\|F(x^k) + F'(x^k)\Delta z\|_2^2 + p\|\Delta z\|_2^2 = \min .$$

This quadratic function in Δz is minimized by a Δz^k satisfying the following equation:

$$\left(F'(x^k)^T F'(x^k) + pI_n\right)\Delta z^k = -F'(x^k)^T F(x^k) .$$

The fitting of the parameter δ is substituted here by the fitting of p. Obviously the symmetric matrix appearing here is positive definite for $p > 0$ even in case of a rank deficient Jacobian $F'(x^k)$, which gives the method a certain robustness. On the other hand we pay for this robustness with a series of disadvantages: The "solutions" are often not minima of $g(x)$, but merely saddle points with $g'(x) = 0$. Furthermore, the "covering up" of the rank of the Jacobian matrix makes unclear the uniqueness of a given solution, so that the numerical results are often incorrectly interpreted. Besides, the above linear system is a generalization of the normal equations having all the disadvantages discussed in Section 3.1.2 (cf. Exercise 5.6). Finally the above formulation is not affine invariant (cf. Exercise 4.12).

Altogether we see that nonlinear least squares problems represent, because of their statistical background, a considerably more subtle problem than the problem of solving systems of nonlinear equation.

4.4   Nonlinear Systems Depending on Parameters

In numerous applications in the natural sciences and engineering one has to solve not only "an" isolated nonlinear problem $F(x) = 0$, but rather a whole family of problems

$$F(x,\lambda) = 0 \ \text{ with } \ F : D \subset \mathbf{R}^n \times \mathbf{R}^p \to \mathbf{R}^n , \qquad (4.23)$$

that depend on one or several *parameters* $\lambda \in \mathbf{R}^p$. Here we limit ourselves to the case $p = 1$, i.e. to a scalar parameter $\lambda \in \mathbf{R}$, and begin with the analysis of the solution structure of the *parameterized nonlinear system*, whose main elements are illustrated by a simple example. With the insight gained there we construct in Section 4.4.2 a class of methods for solving parameterized systems, the so called *continuation methods*.

4.4.1 Structure of the solution

We consider a parameterized system of nonlinear equations

$$F : D \times [a, b] \to \mathbf{R}^n, \quad D \subset \mathbf{R}^n \text{ open}, \tag{4.24}$$

with a scalar parameter $\lambda \in [a, b] \in \mathbf{R}$, where we assume that F is continuously differentiable on $D \times [a, b]$. Our task is to determine eventually all solutions $(x, \lambda) \in D \times [a, b]$ of the equation $F(x, \lambda) = 0$, i.e. we are interested in the solution set

$$S := \{(x, \lambda) \mid (x, \lambda) \in D \times [a, b] \text{ with } F(x, \lambda) = 0\} .$$

In order to develop a feeling for the structure of the solution of a parameterized system, let us look first at the following simple example.

Example 4.20 We have to solve the scalar equation

$$F(x, \lambda) = x(x^3 - x - \lambda) = 0 ,$$

whose solution set is represented in Figure 4.4 for $\lambda \in [-1, 1]$. Because of the equivalence

$$F(x, \lambda) = 0 \iff x = 0 \text{ or } \lambda = x^3 - x$$

the solution set

$$S := \{(x, \lambda) \mid \lambda \in [-1, 1] \text{ and } F(x, \lambda) = 0\}$$

consists of two solution curves $S = S_0 \cup S_1$, the λ-axis

$$S_0 := \{(0, \lambda) \mid \lambda \in [-1, 1]\}$$

as *trivial solution* and the cubic parabola

$$S_1 := \{(x, \lambda) \mid \lambda \in [-1, 1] \text{ and } \lambda = x^3 - x\} .$$

Both solution curves intersect at $(0, 0)$. This is called a *bifurcation point*.

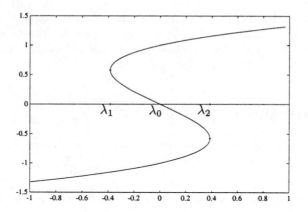

Figure 4.4: Solution of $x(x^3 - x - \lambda) = 0$ for $\lambda \in [-1, 1]$

At this point the Jacobian matrix of F vanishes since

$$F'(x, \lambda) = [F_x(x, \lambda), \ F_\lambda(x, \lambda)] = [4x^3 - 2x - \lambda, \ -x] .$$

A special role is played by the solutions at which only the derivative with respect to x vanishes, i.e.

$$F(x, \lambda) = 0, \quad F_x(x, \lambda) = 4x^3 - 2x - \lambda = 0 \ \text{ and } \ F_\lambda(x, \lambda) \neq 0 .$$

These so called *turning points* (x, λ) distinguish themselves by the fact that the solution cannot be expressed as a function of the parameter λ in any neighborhood (no matter how small) of (x, λ), while it may be expressed as a function of x. If we substitute the $\lambda = x^3 - x$ in $F_x(x, \lambda)$ and $F_\lambda(x, \lambda)$ it follows that the turning points of f are characterized by

$$x \neq 0, \quad \lambda = x^3 - x \ \text{ and } \ 3x^2 - 1 = 0.$$

Therefore in our example there are exactly two turning points, namely at $x_1 = 1/\sqrt{3}$ and $x_2 = -1/\sqrt{3}$. At these points the tangent to the solution curve is perpendicular to the λ-axis.

As a last property we want to note the *symmetry* of the equation

$$F(x, \lambda) = F(-x, -\lambda)$$

This is reflected in the point symmetry of the solution set: If (x, λ) is a solution of $F(x, \lambda) = 0$, then so is $(-x, -\lambda)$.

Unfortunately, we cannot go into all the phenomena observed in Example 4.20 within the frame of this introduction. We assume in what follows that the Jacobian $F'(x, \lambda)$ has maximal rank at each solution point $(x, \lambda) \in D \times [a, b]$, i.e.

$$\operatorname{rank} F'(x, \lambda) = n \quad \text{whenever} \quad F(x, \lambda) = 0 . \tag{4.25}$$

Thus we exclude bifurcation points because according to the implicit function theorem under the assumption (4.25) the solution set can be represented locally around a solution (x_0, λ_0) as the image of a differentiable curve, i.e. there is a neighborhood $U \subset D \times [a, b]$ of (x_0, λ_0) and continuously differentiable mappings

$$x :] - \varepsilon, \varepsilon [\longrightarrow D \quad \text{and} \quad \lambda :] - \varepsilon, \varepsilon [\longrightarrow [a, b] ,$$

such that $(x(0), \lambda(0)) = (x_0, \lambda_0)$ and the solutions of $F(x, \lambda) = 0$ in U are given exactly by

$$S \cap U = \big\{ (x(s), \lambda(s)) \mid s \in\,] - \varepsilon, \varepsilon [\big\}.$$

In many applications an even more special case is also interesting, where the partial derivative $F_x(x, \lambda) \in \operatorname{Mat}_n(\mathbf{R})$ with respect to x is invertible at every solution point, i.e.

$$F_x(x, \lambda) \quad \text{is invertible whenever} \quad F(x, \lambda) = 0 . \tag{4.26}$$

In this case both bifurcation points and turning points are excluded, and we can parameterize the solution curve with respect to λ , i.e. there are, as above, around each solution point (x_0, λ_0) a neighborhood $U \subset D \times [a, b]$ and a differentiable function $x :] - \varepsilon, \varepsilon [\to D$, such that $x(0) = x_0$ and

$$S \cap U = \big\{ (x(s), \lambda_0 + s) \mid s \in\,] - \varepsilon, \varepsilon [\big\} .$$

4.4.2 Continuation methods

Now we turn to the numerical computation of the solution of a parameterized system of equations. First we assume that the derivative $F_x(x, \lambda)$ is invertible for all $(x, \lambda) \in D \times [a, b]$. In the last section we have seen that in this case the solution set of the parameterized system is made up of differentiable curves that can be parameterized by λ. The idea of the *continuation methods* consists of computing successive points on such a solution curve. If we keep the parameter λ then

$$F(x, \lambda) = 0 \tag{4.27}$$

is a nonlinear system in x as treated in Section 4.2. Therefore we can try to compute a solution with the help of Newton's method

$$F_x(x^k, \lambda)\Delta x^k = -F(x^k, \lambda) \text{ and } x^{k+1} = x^k + \Delta x^k. \qquad (4.28)$$

Let us assume that we have found a solution (x_0, λ_0) and and we want to compute another solution (x_1, λ_1) on the solution curve $(x(s), \lambda_0 + s)$ through (x_0, λ_0). Then in selecting the *starting point* $x^0 := \hat{x}$ for the Newton Iteration (4.28), with fixed value of the parameter $\lambda = \lambda_1$ we can use the fact that both solutions (x_0, λ_0) and (x_1, λ_1) lie on the same solution curve. The simplest possibility is indicated in Figure 4.5. We take as starting point

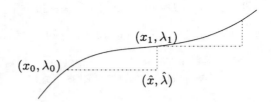

Figure 4.5: Idea of classical continuation method

the old solution and we set $\hat{x} := x_0$. This choice, originally suggested by Poincaré in his book on celestial mechanics [60] is called today *classical continuation*. A geometric view brings yet another choice: instead of going parallel to the λ-axis we can move along the *tangent* $(x'(0), 1)$ to the solution curve $(x(s), \lambda_0 + s)$ at the point (x_0, λ_0) and choose $\hat{x} := x_0 + (\lambda_1 - \lambda_0)x'(0)$ as starting point (see Figure 4.6). This is the *tangential continuation method*.

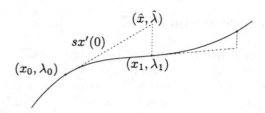

Figure 4.6: Tangential continuation method

If we differentiate the equation

$$F(x(s), \lambda_0 + s) = 0$$

with respect to s at $s = 0$, it follows that

$$F_x(x_0, \lambda_0)x'(0) = -F_\lambda(x_0, \lambda_0) \ ,$$

i.e. the slope $x'(0)$ is computed from a linear system similar to the Newton-corrector (4.28). Thus each continuation step contains two sub-steps: First the choice of a point $(\hat{x}, \lambda_1)$ as close as possible to the curve, and second the iteration from the starting point $\hat{x}$ back to a solution (x_1, λ_1) on the curve, where Newton's method appears to be the most appropriate because of its quadratic convergence. The first sub-step is frequently called *predictor*, the second sub-step *corrector*, and the whole process a *predictor-corrector method*. If we denote by $s := \lambda_1 - \lambda_0$ the *stepsize*, then the dependence of the starting point $\hat{x}$ on s for the both possibilities encountered so far can be expressed as

$$\hat{x}(s) = x_0$$

for the classical continuation and

$$\hat{x}(s) = x_0 + sx'(0)$$

for the tangential continuation. The most difficult problem in the construction of a continuation algorithm consists of choosing appropriately the steplength s in conjunction with the predictor-corrector strategy. The optimist who chooses a too large stepsize s must constantly reduce the steplength and therefore ends up with too many unsuccessful steps. The pessimist on the other hand chooses the stepsize too small and ends up with too many successful steps. Both variants waste computing time. In order to minimize cost, we want therefore to chose the steplength as large as possible while still ensuring the convergence of Newton's method.

Remark 4.21 In practice one should take care of a third criterion, namely not to leave the present solution curve and "jump" onto another solution curve without noticing it (see Figure 4.7). The problem of "jumping over" becomes important especially when considering bifurcations of solutions.

Naturally, the maximal *feasible stepsize* $s_{\max}$ for which Newton's method with starting point $x^0 := \hat{x}(s)$ and fixed parameter $\lambda = \lambda_0 + s$ converges depends on the quality of the predictor step. The better the curve is predicted the larger the stepsize. For example the point $\hat{x}(s)$ given by the tangential method appears graphically to be closer to the curve than the point given by the classical method. In order to describe more precisely this deviation of the predictor from the solution curve we introduce the *order of approximation* of two curves (see [18]):

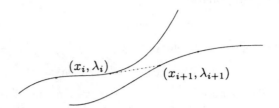

Figure 4.7: Unintentional abandoning of solution curve

Definition 4.22 Let x and $\hat{x}$ be two curves

$$x, \hat{x} : [-\varepsilon, \varepsilon] \to \mathbf{R}^n$$

in $\mathbf{R}^n$. We say that the curve $\hat{x}$ approximates the curve x with *order* $p \in \mathbf{N}$ at $s = 0$, if

$$\|x(s) - \hat{x}(s)\| = O(|s|^p) \quad \text{for} \quad s \to 0 \, ,$$

i.e, if there are constants $0 < s_0 \leq \varepsilon$ and $\eta > 0$ such that

$$\|x(s) - \hat{x}(s)\| \leq \eta |s|^p \quad \text{for all} \quad |s| < s_0 \, .$$

From the mean value theorem it follows immediately that for a sufficiently differentiable mapping F the classical continuation has order $p = 1$ while the tangential continuation has order $p = 2$. The constant η can be given explicitly. For the classical continuation we set $\hat{x}(s) = x(0)$.

Lemma 4.23 *For any continuously differentiable curve* $x : [-\varepsilon, \varepsilon] \to \mathbf{R}^n$ *holds*

$$\|x(s) - x(0)\| \leq \eta s \quad \text{with} \quad \eta := \max_{t \in [-\varepsilon, \varepsilon]} \|x'(t)\| \, .$$

Proof. According to the Lagrange form of the mean value theorem it follows that

$$\|x(s) - x(0)\| = \left\| s \int_{\tau=0}^{1} x'(\tau s) d\tau \right\| \leq s \max_{t \in [-\varepsilon, \varepsilon]} \|x'(t)\| \, .$$

$\square$

For the tangential continuation $\hat{x}(s) = x_0 + s x'(0)$ we obtain analogously the following statement:

Lemma 4.24 *Let* $x : [-\varepsilon, \varepsilon] \to \mathbf{R}^n$ *be a twice differentiable curve and* $\hat{x}(s) = x(0) + s x'(0)$. *Then*

$$\|x(s) - \hat{x}(s)\| \leq \eta s^2 \quad \text{with} \quad \eta := \frac{1}{2} \max_{t \in [-\varepsilon, \varepsilon]} \|x''(t)\| \, .$$

Proof. As in the proof of Lemma 4.23 we have

$$
\begin{aligned}
x(s) - \hat{x}(s) &= x(s) - x(0) - sx'(0) = \int_{\tau=0}^{1} sx'(\tau s) - sx'(0)d\tau \\
&= \int_{\sigma=0}^{1} \int_{\tau=0}^{1} x''(\tau \sigma s)s^2 \tau d\tau d\sigma
\end{aligned}
$$

and therefore

$$
\|x(s) - \hat{x}(s)\| \le \frac{1}{2}s^2 \max_{t\in[-\varepsilon,\varepsilon]} \|x''(t)\| .
$$

$\square$

The following theorem puts into context a continuation method of order p as predictor and Newton's method as corrector. It characterizes the maximal feasible stepsize $s_{\max}$, for which Newton's method applied to $x^0 := \hat{x}(s)$ with fixed parameter $\lambda_0 + s$ converges.

Theorem 4.25 *Let $D \subset \mathbf{R}^n$ be open and convex and let $F : D \times [a, b] \to \mathbf{R}^n$ be a continuously differentiable parameterized system, such that $F_x(x, \lambda)$ is invertible for all $(x, \lambda) \in D \times [a, b]$. Furthermore, let an $\omega > 0$ be given such that F satisfies the Lipschitz condition*

$$
\|F_x(x, \lambda)^{-1}(F_x(x + sv, \lambda) - F_x(x, \lambda))v\| \le s\omega\|v\|^2 \tag{4.29}
$$

Also let $(x(s), \lambda_0 + s)$, $s \in [-\varepsilon, \varepsilon]$ be a continuously differentiable solution curve around (x_0, λ_0), i.e.

$$
F(x(s), \lambda_0 + s) = 0 \quad and \quad x(0) = x_0 ,
$$

and $\hat{x}(s)$ a continuation method (predictor) of order p with

$$
\|x(s) - \hat{x}(s)\| \le \eta s^p \quad for \; all \; |s| \le \varepsilon . \tag{4.30}
$$

Then Newton's method (4.28) with starting point $x^0 = \hat{x}(s)$ converges towards the solution $x(s)$ of $F(x, \lambda_0 + s) = 0$, whenever

$$
s < s_{\max} := \max\left(\varepsilon, \sqrt[p]{\frac{2}{\omega\eta}}\right) . \tag{4.31}
$$

Proof. We must check the hypothesis of Theorem 4.10 for Newton's method (4.28) and the starting point $x^0 = \hat{x}(s)$. According to the condition (4.30) the following relation holds

$$
\rho(s) = \|x^* - x^0\| = \|x(s) - \hat{x}(s)\| \le \eta s^p .
$$

If we put this inequality into the convergence condition $\rho < 2/\omega$ of Theorem 4.10 we obtain the sufficient condition

$$\eta s^p < 2/\omega,$$

or, equivalently, (4.31).   □

This theorem guarantees that with the above described continuation methods, consisting of classical continuation (order $p = 1$) or tangential continuation (order $p = 2$) as predictor and Newton's method as corrector, will succeed in following a solution curve as long as the stepsize is chosen sufficiently small (depending on the problem). On the other hand the characterizing quantities ω and η are not known in general, and therefore the formula (4.31) for $s_{\max}$ cannot be used in practice. Hence we must develop a strategy for choosing the stepsize that uses exclusively information available from within the algorithm itself. Such a *stepsize control* consists first of an initial suggested s (most of the time the stepsize of the previous continuation step), and second, of a strategy for choosing a smaller stepsize in case Newton's method (as corrector) fails to converge for the starting point $\hat{x}(s)$. The convergence of Newton's method is assessed with the natural monotony test introduced in Section 4.2.

$$\|\bar{\Delta}x^{k+1}\| \leq \bar{\theta}\,\|\Delta x^k\| \quad \text{with} \quad \bar{\theta} := \frac{1}{2}\ . \tag{4.32}$$

Here Δx^k and $\bar{\Delta}x^{k+1}$ are the ordinary and the simplified Newton corrections of Newton's method (4.28), i.e. with $x^0 := \hat{x}(s)$ and $\hat{\lambda} := \lambda_0 + s$:

$$F_x(x^k, \hat{\lambda})\Delta x^k = -F(x^k, \hat{\lambda}) \quad \text{and} \quad F_x(x^k, \hat{\lambda})\bar{\Delta}x^{k+1} = -F(x^{k+1}, \hat{\lambda})\ .$$

If we establish with the help of the criterion (4.32) that Newton's method does not converge for the stepsize s, i.e. if

$$\|\bar{\Delta}x^{k+1}\| > \bar{\theta}\,\|\Delta x^k\|\ ,$$

then we reduce this stepsize by a factor $\beta < 1$ and we perform again the Newton iteration with the new stepsize

$$s' := \beta \cdot s\ ,$$

i.e. with the new starting point $x^0 = \hat{x}(s')$ and the new parameter $\hat{\lambda} := \lambda_0 + s'$. This process is repeated until either the convergence criterion (4.32) for Newton's method is satisfied or we get below a minimal stepsize $s_{\min}$. In the latter case we suspect that the assumptions on F are violated and we are for example in an immediate neighborhood of a turning point or a bifurcation point. On the other hand we can choose a larger stepsize for the next step if Newton's method converges "too fast". This can also be read from the two Newton corrections. If

$$\|\bar{\Delta}x^1\| \le \frac{\bar{\theta}}{4}\|\Delta x^0\| , \qquad (4.33)$$

then the method converges "too fast", and we can enlarge the stepsize for the next predictor step by a factor β, i.e. we suggest the stepsize

$$s' := s/\beta.$$

Here the choice

$$\beta := \sqrt[p]{\frac{1}{2}} ,$$

motivated by (4.31) is consistent with (4.32) and (4.33). The following algorithm describes the tangential continuation from a solution (x_0, a) up tp the right endpoint $\lambda = b$ of the parameter interval.

Algorithm 4.26 *Tangential Continuation.* The procedure *newton* $(\hat{x}, \hat{\lambda})$ contains the (ordinary) Newton method (4.28) for the starting point $x^0 = \hat{x}$ and fixed value of the parameter $\hat{\lambda}$. The Boolean variable *done* specifies whether the procedure has computed the solution accurately enough after at most $k_{\max}$ steps. Besides this information and (if necessary) the solution x the program will return the quotient

$$\theta = \frac{\|\bar{\Delta}x^1\|}{\|\Delta x^0\|}$$

of the norms of the simplified and ordinary Newton correctors. The procedure *continuation* realizes the continuation method with the "stepsize control described above". Beginning with a starting point $\hat{x}$ for the solution of $F(x, a) = 0$ at the left endpoint $\lambda = a$ of the parameter interval, the program tries to follow the solution curve up to the right endpoint $\lambda = b$. The program terminates if this is achieved or if the stepsize s becomes to small, or if the maximal number $i_{\max}$ of computed solution is exceeded.

> **function** [done,x, θ]=newton $(\hat{x}, \hat{\lambda})$
> $\quad x := \hat{x};$
> $\quad$ **for** $k = 0$ **to** $k_{\max}$ **do**
> $\qquad A := F_x(x, \hat{\lambda});$
> $\qquad$ solve $A\Delta x = -F(x, \hat{\lambda});$
> $\qquad x := x + \Delta x;$
> $\qquad$ solve $A\bar{\Delta}x = -F(x, \hat{\lambda});$ (use again the factorization of A)

```
        if k = 0 then
            θ := ‖Δ̄x‖/‖Δx‖; (for the next predicted stepsize)
        end
        if ‖Δx‖ < tol  then
            done:= true;
            break; (solution found)
        end
        if ‖Δ̄x‖ > θ̄‖Δx‖ then
            done:= false;
            break; (monotony violated)
        end
    end
    if k > k_max then
        done:= false; (too many iterations)
    end


function continuation (x̂)
    λ_0 := a;
    [done, x_0, θ] = newton (x̂, λ_0);
    if not done then
        poor starting point x̂ for F(x, a) = 0
    else
        s := s_0; (starting stepsize)
        for i = 0 to i_max do
            lvse F_x(x_i, λ_i)x' = −F_λ(x_i, λ_i);
            repeat
                x̂ := x_i + sx';
                λ_{i+1} := λ_i + s;
                [done, x_{i+1}, θ] = newton (x̂, λ_{i+1});
                if not done then
                    s = βs;
                elseif θ < θ̄/4 then
                    s = s/β;
                end
                s = min(s, b − λ_{i+1});
            until s < s_min or done
            if not done then
                break; (algorithm breaks down)
            elseif λ_{i+1} = b then
                break; (terminated, solution x_{i+1})
            end
        end
    end
```

Remark 4.27 There is a significantly more efficient stepsize control strategy which uses the fact that the quantities ω and η can be locally approximated by quantities accessible from the algorithm. That strategy is also well founded theoretically. However its description cannot be done within the frame of this introduction — it is presented in detail in [18].

We want to describe yet another variant of the tangential continuation because it fits well the context of Chapter 3 and Section 4.3. It allows at the same time dealing with *turning points* (x, λ) with

$$\operatorname{rank} F'(x, \lambda) = n \quad \text{and} \quad F_x(x, \lambda) \quad \text{singular} .$$

In the neighborhood of such a point the automatically chosen stepsizes s of the continuation method described above become arbitrarily small because the solution curve around (x, λ) cannot be parameterized anymore with respect to the parameter λ. We overcome this difficulty by giving up the "special role" of the parameter λ and consider instead directly the *underdetermined nonlinear system* in $y = (x, \lambda)$,

$$F(y) = 0 \quad \text{with} \quad F : \hat{D} \subset \mathbf{R}^{n+1} \longrightarrow \mathbf{R}^n .$$

We assume again that the Jacobian $F'(y)$ of this system is full rank for all $y \in \hat{D}$. Then for each solution $y_0 \in \hat{D}$ there is a neighborhood $U \subset \mathbf{R}^{n+1}$ and a differentiable curve $y :] - \varepsilon, \varepsilon[\to \hat{D}$, $S := \{y \in \hat{D} \mid F(y) = 0\}$ characterizing the curve around y_0, i.e.

$$S \cap U = \{y(s) \mid s \in] - \varepsilon, \varepsilon[\} .$$

If we differentiate the equation $F(y(s)) = 0$ with respect to s at $s = 0$, it follows that

$$F'(y(0))y'(0) = 0 , \tag{4.34}$$

i.e. the tangent $y'(0)$ to the solution curve spans exactly the nullspace of the Jacobian $F'(y_0)$. Since $F'(y_0)$ has maximal rank, the tangent through (4.34) is uniquely determined up to a scalar factor. Therefore we define for all $y \in \hat{D}$ the *normalized tangent* $t(y) \in \mathbf{R}^{n+1}$ by

$$F'(y)t(y) = 0 \quad \text{and} \quad \|t(y)\|_2 = 1 ,$$

which is uniquely determined up to its orientation (i.e. up to a factor ± 1). We choose the orientation of the tangent during the continuation process such that two successive tangents $t_0 = t(y_0)$ and $t_1 = t(y_1)$ form an acute angle, i.e.

$$\langle t_0, t_1 \rangle > 0 .$$

This guarantees that we are not going backward on the solution curve. With it we can also define tangential continuation for turning points by

$$\hat{y} = \hat{y}(s) := y_0 + s\, t(y_0).$$

Beginning with the starting vector $y^0 = \hat{y}$ we want to find $y(s)$ on the curve "as fast as possible". The vague expression "as fast as possible" can be interpreted geometrically as "almost orthogonal" to the tangent at a nearby point $y(s)$ on the curve. However, since the tangent $t(y(s))$ is at our disposal only *after* computing $y(s)$, we substitute $t(y(s))$ with the best approximation available at the present time $t(y^k)$. According to the geometric interpretation of the pseudo inverse (cf. Section 3.3) this leads to the iterative scheme

$$\Delta y^k := -F'(y^k)^+ F(y^k) \quad \text{and} \quad y^{k+1} := y^k + \Delta y^k . \qquad (4.35)$$

The iterative scheme (4.35) is obviously a Gauss-Newton method for the overdetermined system $F(y) = 0$. We mention without proof that if $F'(y)$ has maximal rank then this method is quadratically convergent in a neighborhood of $\bar{y}$, the same as the ordinary Newton method. The proof is found in Chapter 3 of the book [15].

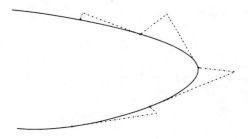

Figure 4.8: Tangential continuation through turning points .

Here we want to examine the computation of the correction Δy^k. We will drop the index k . The correction Δy in (4.35) is the shortest solution of the solution set $Z(y)$ of the overdetermined linear problem

$$Z(y) := \{ z \in \mathbf{R}^{n+1} | F'(y)z + F(y) = 0 \} .$$

By applying a Gauss elimination (with row pivoting and eventually column exchange, cf. Section 1.3) or a QR-factorization (with column exchange, cf. Section 3.2.2) we succeed relatively easily in computing *some* solution $z \in Z(y)$ as well as a nullspace vector $t(y)$ with

$$F'(y)t(y) = 0.$$

Then the following equation holds

$$\Delta y = -F'(y)^+ F(y) = F'(y)^+ F'(y)z .$$

As we have seen in Section 3.3, $P = F'(y)^+ F'(y)$ is the projection onto the orthogonal complement of $F'(y)$ and therefore

$$P = I - \frac{tt^T}{t^T t} .$$

For the correction Δy it follows that

$$\Delta y = \left(I - \frac{tt^T}{t^T t} \right) z = z - \frac{\langle t, z \rangle}{\langle t, t \rangle} t .$$

With this we have a simple computational scheme for the pseudo inverse (with rank defect 1) provided we only have at our disposal some solution z and nullspace vector t. The Gauss-Newton method given in (4.35) is also easily implementable in close interplay with tangential continuation.

For choosing the stepsize we grasp a similar strategy to that described in Algorithm 4.26. If the iterative method does not converge then we reduce the steplength s by a factor $\beta = 1/\sqrt{2}$. If the iterative method converges "too fast" we enlarge the stepsize for the next predictor step by a factor β^{-1}. This empirical continuation method is comparatively effective even in relatively complicated problems.

Remark 4.28 For this tangential continuation method there is also a theoretically based and essentially more effective stepsize control, whose description is found in [21]. Additionally, there one utilizes approximations of the Jacobian instead of $F'(y)$. Extremely effective programs for parameterized systems are working on this basis. (see Figure 4.9 and 4.10).

Remark 4.29 The description of the solutions of the parameterized system (4.23) is also called *parameter study*. On the other hand, parameterized systems are also used for *enlarging the convergence region* of a method for solving nonlinear systems. The idea here is to work our way, step by step, from a previously solved problem

$$G(x) = 0$$

to the actual problem

$$F(x) = 0.$$

For this we construct a parameterized problem

$$H(x, \lambda) = 0 \ , \quad \lambda \in [0, 1] \ ,$$

that connects the two problems:

$$H(x, 0) = G(x) \ \text{ and } \ H(x, 1) = F(x) \text{ for all } x.$$

Such a mapping H is called an *embedding* of the problem $F(x) = 0$, or a *homotopy*. The simplest example is the so-called *linear embedding*,

$$H(x, \lambda) := \lambda F(x) + (1 - \lambda)G(x) \ .$$

Problem specific embeddings are certainly preferable (see Example 4.30). If we apply a continuation method to this parameterized problem $H(x, \lambda) = 0$, where we start with a known solution x_0 of $G(x) = 0$, then we obtain a *homotopy method* for solving $F(x) = 0$.

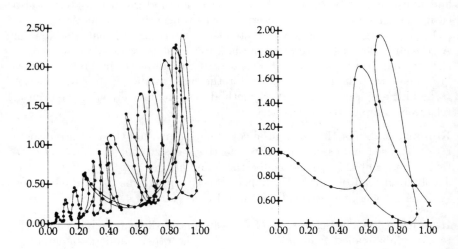

Figure 4.9: Continuation for the trivial embedding (left) and a problem specific embedding (right); represented here is x_9 with respect to λ.

Example 4.30 *Continuation for different embeddings.* The following problem is given in [36]:

$$F(x) \quad := \quad x - \phi(x) = 0, \qquad \text{where}$$

$$\phi_i(x) \quad := \quad \exp(\cos(i \cdot \sum_{j=1}^{10} x_j)), \quad i = 1, \ldots, 10.$$

The trivial embedding

$$H(x, \lambda) = \lambda F(x) + (1 - \lambda)x = x - \lambda\phi(x)$$

with starting point $x^0 = (0, \ldots, 0)$ at $\lambda = 0$ is suggested for it. The continuation with respect to λ leads indeed for $\lambda = 1$ to the solution (see Figure 4.9, left), but the problem specific embedding

$$\tilde{H}_i(x, \lambda) := x_i - \exp(\lambda \cdot \cos(i \cdot \sum_{j=1}^{10} x_j)), \quad i = 1, \ldots, 10$$

with starting point $x^0 = (0, \ldots, 0)$ at $\lambda = 0$ is clearly advantageous (see Figure 4.9, right). One should remark that there are no bifurcations in this example. The intersections of the solution curves appear only in the projection onto the coordinate plane (x_9, λ). The points on both solution branches mark the intermediate values computed automatically by the program: their number is a measure for the computing cost required to go from $\lambda = 0$ to $\lambda = 1$.

The above example has an illustrative character. It can be easily transformed into a purely scalar problem and solved as such (Exercise 4.6). Therefore we add another more interesting problem.

Example 4.31 *Brusselator.* In [63] a chemical reaction–diffusion equation is considered as a discrete model where two chemical substances with concentrations $z = (x, y)$ in several cells react with each other according to the rule

$$\dot{z} = \begin{pmatrix} \dot{x} \\ \dot{y} \end{pmatrix} = \begin{pmatrix} A - (B + 1)x + x^2 y \\ Bx - x^2 y \end{pmatrix} =: f(z). \tag{4.36}$$

Diffusion appears from coupling with neighboring cells. If one considers only solutions that are constant in time and diffusion is parameterized by λ, then the following nonlinear system is obtained

$$0 = f(z_i) + \frac{1}{\lambda^2} \sum_{(i,j)} D(z_j - z_i), \quad i = 1, \ldots, k.$$

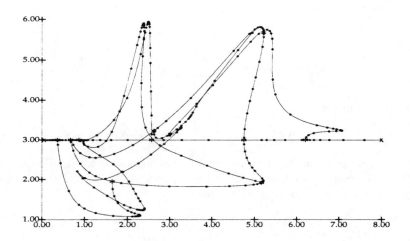

Figure 4.10: Brusselator with four cells in a linear chain $(A = 2, B = 6)$, represented here is x_8 with respect to λ.

Here $D = \mathrm{diag}(1, 10)$ is a $(2, 2)$ diagonal matrix. Because the equations reflect the symmetry of the geometrical arrangement of the cells, a rich set of bifurcations appears (see Figure 4.10), which is analyzed in [32] by using the symmetry of the system in interplay with methods from symbolic computation.

4.5 Exercises

Exercise 4.1 Explain the different convergence behaviors of the two fixed point iterations described in Section 4.1 for the solution of

$$f(x) = 2x - \tan x = 0.$$

Analyze the speed of convergence of the second method.

Exercise 4.2 In order to determine a fixed point x^* of a continuously differentiable mapping ϕ with $|\phi'(x)| \neq 1$ let us define the following iterative procedures for $k = 0, 1, \dots$:

(I) $x_{k+1} := \phi(x_k)$,

(II) $x_{k+1} := \phi^{-1}(x_k)$.

Show that at least one of the two iterations is locally convergent.

Exercise 4.3 Let $f \in C^1[a, b]$ be a function having a simple root $x^* \in [a, b]$, and let $p(x)$ be the uniquely determined quadratic interpolation polynomial through the three nodes

$$(a, f_a), \quad (c, f_c), \quad (b, f_b), \quad \text{with} \quad a < c < b, \quad f_a f_b < 0$$

a) Show that p has exactly one simple y zero in $[a, b]$.

b) Given a formal procedure

$$y = y(a, b, c, f_a, f_b, f_c),$$

that computes the zero of p in $[a, b]$, construct an algorithm for evaluating x^* with a prescribed precision eps.

Exercise 4.4 In order to accelerate the convergence of a linearly convergent fixed point method in $\mathbf{R}^1$

$$x_{i+1} := \phi(x_i), \quad x_0 \text{ given}, \quad x^* \text{ fixed point}$$

we can use the so-called Δ^2-method of Aitken. This consists of computing from the sequence $\{x_i\}$ a transformed sequence $\{\bar{x}_i\}$

$$\bar{x}_i := x_i - \frac{(\Delta x_i)^2}{\Delta^2 x_i},$$

where Δ is the difference operator $\Delta x_i := x_{i+1} - x_i$..

a) Show that: If the sequence $\{x_i\}$, with $x_i \neq x^*$, satisfies

$$x_{i+1} - x^* = (\kappa + \delta_i)(x_i - x^*) ,$$

where $|\kappa| < 1$ and $\{\delta_i\}$ is a null sequence, $\lim_{i \to \infty} \delta_i = 0$, then the sequence $\{\bar{x}_i\}$ is well defined for sufficiently large i and has the property that

$$\lim_{i \to \infty} \frac{\bar{x}_i - x^*}{x_i - x^*} = 0 .$$

b) For implementing the method one computes only x_0, x_1, x_2 and $\bar{x}_0$ and then one starts the iteration with the improved starting point $\bar{x}_0$ (Steffensen's method). Try this method on our trusted example

$$\phi_1(x) := (\tan x)/2 \quad \text{and} \quad \phi_1(x) := \arctan 2x$$

with starting point $x_0 = 1.2$.

Exercise 4.5 Compute the solution of the nonlinear least squares problem arising in Feulgen hydrolysis by the ordinary Gauss–Newton method (from a software package or written by yourself) for the data from Table 4.3 and the starting points given there. Hint: In this special case the ordinary Gauss–Newton method converges faster than the damped method (cf. Figure 4.3).

Exercise 4.6 Compute the solution of $F(x) = x - \phi(x) = 0$ with

$$\phi_i(x) := \exp(\cos(i \cdot \sum_{j=1}^{10} x_j)) \quad i = 1, \ldots, 10, \tag{4.37}$$

by first setting up an equation for $u = \sum_{j=1}^{10} x_j$ and then solving it.

Exercise 4.7 Let there be given a function $F : D \to \mathbf{R}^n$, $D \subset \mathbf{R}^n$, $F \in C^2(D)$. We consider the approximation of the Jacobian $J(x) = F'(x)$ by divided differences

$$\Delta_i F(x) \quad := \quad \frac{F(x + \eta_i e_i) - F(x)}{\eta_i} \ , \ \eta_i \neq 0$$

$$\hat{J}(x) \quad := \quad [\Delta_1 F(x), \ldots, \Delta_n F(x)] \ .$$

In order to obtain a sufficiently good approximation of the Jacobian we compute the quantity

$$\kappa(\eta_i) := \frac{\|F(x + \eta_i e_i) - F(x)\|}{\|F(x)\|}$$

and require that

$$\kappa(\hat{\eta}_i) \doteq \kappa_0 := \sqrt{10\text{eps}} \ ,$$

where eps is the relative machine precision. Show that

$$\kappa(\eta) \doteq c_1 \eta + c_2 \eta^2 \quad \text{for} \quad \eta \to 0 \ .$$

Specify a rule that provides an estimate for $\hat{\eta}$ in case $\kappa(\eta) \ll \kappa_0$. Why is a corresponding estimate for $\kappa(\eta) \gg \kappa_0$ not a useful result ?

Exercise 4.8 The zeros of $p_n(z) := z^n - 1$ for n even have to be determined with the (complex) NEWTON method .

$$z_{k+1} = \Phi(z_k) := z_k - \frac{p_n(z_k)}{p'_n(z_k)} \ , \quad k = 0, 1, \ldots \ .$$

Let us define:

$$L(s) := \{te^{i\frac{\pi}{n}s} | t \in \mathbf{R}\} \ , \quad s \in [0, n[\ .$$

a) Show that:
$$z_k \in L(s) \Rightarrow z_{k+1} \in L(s) \,.$$

b) Prepare a sketch describing the convergence behaviour. Compute
$$K(s) := L(s) \cap \{z \mid |\Phi'(z)| < 1\}$$

and all fixed points of Φ in $L(s)$ for
$$s = 0, 1, \ldots, n-1 \quad \text{and} \quad s = \frac{1}{2}, \frac{3}{3}, \ldots, n - \frac{1}{2} \,.$$

Exercise 4.9 Consider the system of $n = 10$ nonlinear equations ($s = \sum_{i=1}^{10} x_i$)

$$F(x, \lambda) := \begin{bmatrix} x_1 + x_4 - 3 \\ 2x_1 + x_2 + x_4 + x_7 + x_8 + x_9 + 2x_{10} - \lambda \\ 2x_2 + 2x_5 + x_6 + x_7 - 8 \\ 2x_3 + x_9 - 4\lambda \\ x_1 x_5 - 0.193 x_2 x_4 \\ x_6^2 x_1 - 0.67444 \cdot 10^{-5} x_2 x_4 s \\ x_7^2 x_4 - 0.1189 \cdot 10^{-4} x_1 x_2 s \\ x_8 x_4 - 0.1799 \cdot 10^{-4} x_1 s \\ (x_9 x_4)^2 - 0.4644 \cdot 10^{-7} x_1^2 x_3 s \\ x_{10} x_4^2 - 0.3846 \cdot 10^{-4} x_1^2 s \end{bmatrix} = 0 \,.$$

It describes a chemical equilibrium (for propane). All solutions of interest must be nonnegative because they are interpreted as chemical concentrations.

a) Show that we must have $\lambda \geq 3$ if $x_i \geq 0$, $i = 1, \ldots, n$. Compute (by hand) the special degenerate case $\lambda = 3$.

b) Write a program for a continuation method with ordinary Newton's method as local iterative procedure and empirical stepsize strategy.

c) Test the program on the above example with $\lambda > 3$.

Exercise 4.10 Prove the following theorem: Let $D \subseteq \mathbf{R}^n$ be open and convex and let $F : D \subseteq \mathbf{R}^n \rightarrow \mathbf{R}^n$ be differentiable. Suppose there exists a solution $x^* \in D$ such that $F'(x^*)$ is invertible. Assume further that the following (affine–invariant) Lipschitz condition is satisfied for all $x, y \in D$:

$$\|F'(x^*)^{-1}(F'(y) - F'(x))\| \leq \omega_* \|y - x\| .$$

Let $\rho := \|x^0 - x^*\| < 2/(3\omega_*)$ and $B_\rho(x^*) \subset D$. Then it follows that: The sequence $\{x^k\}$ defined by the ordinary Newton method stays in $B_\rho(x^*)$ and converges towards x^*. Moreover x^* is the unique solution in $B_\rho(x^*)$. Hint: Use the perturbation lemma, based on the Neumann series for the Jacobian $F'(x)$.

Exercise 4.11 Consider solving the quadratic equation

$$x^2 - 2px + q = 0 \quad \text{with} \quad p^2 - q \geq 0 \quad \text{and} \quad q = 0.123451234 .$$

Compute for $p \in \{1, 10, 10^2, \ldots\}$ the two solutions

$$x_1 = \hat{x}_1 = p + \sqrt{p^2 - r} ,$$
$$\hat{x}_2 = p - \sqrt{p^2 - q} , \quad x_2 = q/\hat{x}_1 .$$

Write down the results in a table and underline each time the correct figures.

Exercise 4.12 The principle used in deriving the Levenberg-Marquardt method

$$x^{k+1} = x^k + \Delta z^k , \quad k = 0, 1, \ldots$$

for solving nonlinear systems is not affine–invariant. This shortcoming is naturally inherited also by the method itself. An affine–invariant modification reads: Minimize

$$\|F'(x^k)^{-1}(F(x^k) + F'(x^k)\Delta z)\|_2$$

under the constraints

$$\|\Delta z\|_2 \leq \delta .$$

What is the resulting method?

5 Symmetric Eigenvalue Problems

The following chapter is dedicated to the study of the numerical eigenvalue problem of linear algebra. First we will perform a condition analysis for the general eigenvalue problem (Section 5.1). It will be shown that in general the eigenvalue problem is *well conditioned* only for *normal* matrices. Fortuitously this type of problem occurs most frequently in applications in the natural and technical sciences. Therefore we limit ourselves to the computation of eigenvalues and eigenvectors of symmetric matrices (Section 5.2 and 5.3). For general matrices, the problem of determining the singular value decomposition is well conditioned, and we present it finally in Section 5.4. This type of problem has also an enormous practical relevance.

5.1 Condition of General Eigenvalue Problems

We start with determining the *condition* of the eigenvalue problem

$$Ax = \lambda x$$

for an arbitrary complex matrix $A \in \mathrm{Mat}_n(\mathbf{C})$. For the sake of simplicity we assume that λ_0 is an (algebraically) simple eigenvalue of A , i.e. a simple zero of the characteristic polynomial $\chi_A(\lambda) = \det(A - \lambda I)$. Under these assumptions λ is differentiable in A, as will be seen in the following lemma.

Lemma 5.1 *Let $\lambda_0 \in \mathbf{C}$ be a simple eigenvalue of $A \in \mathrm{Mat}_n(\mathbf{C})$. Then there is a continuously differentiable mapping*

$$\lambda : V \subset \mathrm{Mat}_n(\mathbf{C}) \to \mathbf{C} , \quad B \mapsto \lambda(B)$$

from a neighborhood V of A in $\mathrm{Mat}_n(\mathbf{C})$, such that $\lambda(A) = \lambda_0$ and $\lambda(B)$ is a simple eigenvalue of B for all $B \in V$. If x_0 is an eigenvector of A for λ_0, and y_0 an (adjoint) eigenvector of $A^* := \bar{A}^T$ for the eigenvalue $\bar{\lambda}_0$, i.e.

$$Ax_0 = \lambda_0 x_0 \quad and \quad A^* y_0 = \bar{\lambda}_0 y_0 ,$$

then the derivative of λ at A satisfies

$$\lambda'(A)C = \frac{\langle Cx_0, y_0 \rangle}{\langle x_0, y_0 \rangle} \quad \text{for all } C \in \text{Mat}_n(\mathbf{C}) \ . \tag{5.1}$$

Proof. Let $C \in \text{Mat}_n(\mathbf{C})$ be an arbitrary complex matrix. Because λ_0 is a simple zero of the characteristic polynomial χ_A, we have

$$0 \neq \chi'_A(\lambda_0) = \frac{\partial}{\partial \lambda} \chi_{A+tC}(\lambda)\big|_{t=0}.$$

According to the implicit function theorem there is a neighborhood of the origin $\,]-\varepsilon, \varepsilon[\subset \mathbf{R}$ and a continuously differentiable mapping

$$\lambda :]-\varepsilon, \varepsilon[\to \mathbf{C} \ , \ t \mapsto \lambda(t) \ ,$$

such that $\lambda(0) = \lambda_0$ and $\lambda(t)$ is a simple eigenvalue of $A+tC$. Using again the fact that λ_0 is simple we deduce the existence of a continuously differentiable function

$$x :]-\varepsilon, \varepsilon[\to \mathbf{C}^n \ , \ t \mapsto x(t) \ ,$$

such that $x(0) = x_0$ and $x(t)$ is an eigenvector of $A + tC$ for the eigenvalue $\lambda(t)$ ($x(t)$ can be explicitly computed with adjoint determinants, see Exercise 5.2). If we differentiate the equation

$$(A + tC)x(t) = \lambda(t)x(t)$$

with respect to t at $t = 0$, then it follows that

$$Cx_0 + Ax'(0) = \lambda_0 x'(0) + \lambda'(0)x_0 \ .$$

If we multiply from the right by y_0 (in the sense of the scalar product), then we obtain

$$\langle Cx_0, y_0 \rangle + \langle Ax'(0), y_0 \rangle = \langle \lambda_0 x'(0), y_0 \rangle + \langle \lambda'(0)x_0, y_0 \rangle \ .$$

As $\langle \lambda'(0)x_0, y_0 \rangle = \lambda'(0)\langle x_0, y_0 \rangle$ and

$$\langle Ax'(0), y_0 \rangle = \langle x'(0), A^*y_0 \rangle = \lambda_0 \langle x'(0), y_0 \rangle = \langle \lambda_0 x'(0), y_0 \rangle \ ,$$

it follows that

$$\lambda'(0) = \frac{\langle Cx_0, y_0 \rangle}{\langle x_0, y_0 \rangle} \ .$$

Hence we have computed the derivative of λ in the direction of the matrix C . The continuous differentiability of the directional derivative implies the differentiability of λ with respect to A and

$$\lambda'(A)C = \lambda'(0) = \frac{\langle Cx_0, y_0 \rangle}{\langle x_0, y_0 \rangle}$$

for all $C \in \mathrm{Mat}_n(\mathbf{C})$. □

To compute the condition of the eigenvalue problem (λ, A) we must calculate the norm of the mapping $\lambda'(A)$ a as linear mapping,

$$\lambda'(A) : \mathrm{Mat}_n(\mathbf{C}) \to \mathbf{C} , \quad C \mapsto \frac{\langle Cx, y \rangle}{\langle x, y \rangle} ,$$

where x is an eigenvector for the simple eigenvalue λ_0 of A and y is an adjoint eigenvector for the eigenvalue $\bar{\lambda}_0$ of A^*. On $\mathrm{Mat}_n(\mathbf{C})$ we choose the matrix norm induced by the Euclidean vector norm, and on $\mathbf{C}$ the absolute value. For each matrix $C \in \mathrm{Mat}_n(\mathbf{C})$ we have (the Cauchy-Schwarz inequality)

$$|\langle Cx, y \rangle| \leq \|Cx\| \, \|y\| \leq \|C\| \, \|x\| \, \|y\| ,$$

where we have equality for $C = yx^*$, $x^* := \bar{x}^T$. Since $\|yx^*\| = \|x\| \, \|y\|$, it follows that

$$\|\lambda'(A)\| = \sup_{C \neq 0} \frac{|\langle Cx, y \rangle / \langle x, y \rangle|}{\|C\|} = \frac{\|x\| \, \|y\|}{|\langle x, y \rangle|} = \frac{1}{|\cos(\measuredangle \, (x, y))|} ,$$

where $\measuredangle \, (x, y)$ is the angle between the eigenvector x and the adjoint eigenvector y. For *normal* matrices each eigenvector x is also an adjoint eigenvector, i.e. $A^* x = \bar{\lambda}_0 x$, and therefore $\|\lambda'(A)\| = 1$. We can summarize our results as follows:

Theorem 5.2 *The absolute condition number of determining a simple eigenvalue λ_0 of a matrix $A \in \mathrm{Mat}_n(\mathbf{C})$ with respect to the 2-norm is*

$$\kappa_{\mathrm{abs}} = \|\lambda'(A)\| = \frac{\|x\| \, \|y\|}{|\langle x, y \rangle|} = \frac{1}{|\cos(\measuredangle \, (x, y))|}$$

and the relative condition number

$$\kappa_{\mathrm{rel}} = \frac{\|A\|}{|\lambda_0|} \|\lambda'(A)\| = \frac{\|A\|}{|\lambda_0 \cos(\measuredangle \, (x, y))|} ,$$

where x is an eigenvector of A for the eigenvalue λ_0, i.e. $Ax = \lambda_0 x$, and y an adjoint eigenvector, i.e. $A^* y = \bar{\lambda}_0 y$. In particular for normal matrices the eigenvalue problem is well conditioned with $\kappa_{\mathrm{abs}} = 1$.

Example 5.3 If A is not symmetric then the eigenvalue problem is in general not well conditioned anymore. As an example let us examine the matrices

$$A = \begin{pmatrix} 0 & 1 \\ 0 & 0 \end{pmatrix} \quad \text{and} \quad \tilde{A} = \begin{pmatrix} 0 & 1 \\ \delta & 0 \end{pmatrix},$$

with the eigenvalues $\lambda_1 = \lambda_2 = 0$ and $\tilde{\lambda}_{1,2} = \pm\sqrt{\delta}$. For the condition of the eigenvalue problem (A, λ_1) we have

$$\kappa_{\text{abs}} \geq \frac{|\tilde{\lambda}_1 - \lambda_1|}{\|\tilde{A} - A\|_2} = \frac{\sqrt{\delta}}{\delta} = \frac{1}{\sqrt{\delta}} \to \infty \quad \text{for} \quad \delta \to 0 \, .$$

The computation of the eigenvalue $\lambda = 0$ of A is therefore an ill posed problem (with respect to the absolute error).

A precise analysis of the behavior of eigenvalues and eigenvectors for non-symmetric matrices (and operators), as well as multiple eigenvalues, can be found in the book of KATO [50].

For the well conditioned real symmetric eigenvalue problem one could first think of setting up the characteristic polynomial and subsequently determining its zeros. Unfortunately the information on eigenvalues "disappears" once the characteristic polynomial is treated in coefficient representation. According to Section 2.2 the reverse problem is also ill conditioned.

Example 5.4 WILKINSON [80] has given the polynomial

$$P(\lambda) = (\lambda - 1) \cdots (\lambda - 20) \in \mathbf{P}_{20}$$

as a cautionary example. If we perform the multiplication in this root representation then the resulting coefficients have orders of magnitude between 1 (coefficient of λ^{20}) and 10^{20} (the constant term is e.g. 20!). Now let us perturb the coefficient of λ^{19} (which has an order of magnitude of 10^3) by the very small value $\varepsilon := 2^{-23} \approx 10^{-7}$. In Table 5.1 we have entered the

Table 5.1: Exact zeros of the polynomial $\tilde{P}(\lambda)$ for $\varepsilon := 2^{-23}$.

1.000 000 000	10.095 266 145 $\pm$ 0.643 500 904i
2.000 000 000	11.793 633 881 $\pm$ 1.652 329 728i
3.000 000 000	13.992 358 137 $\pm$ 2.518 830 070i
4.000 000 000	16.730 737 466 $\pm$ 2.812 624 894i
4.999 999 928	19.502 439 400 $\pm$ 1.940 330 347i
6.000 006 944	
6.999 697 234	
8.007 267 603	
8.917 250 249	
20.846 908 101	

exact zeros of the perturbed polynomial

$$\tilde{P}(\lambda) = P(\lambda) - \varepsilon \lambda^{19}$$

In spite of the extremely small perturbation the errors are considerable. In particular five pairs of zeros are complex.

5.2   Power Method

Computing the eigenvalues of a matrix $A \in \mathrm{Mat}_n(\mathbf{R})$ as zeros of the characteristic polynomial $\chi_A(\lambda) = \det(A - \lambda I)$, may be reasonable only for $n = 2$. Here we will develop more direct methods for determining eigenvalues and eigenvectors. The simplest direct possibility is the so called *power method*, and we will discuss in what follows both of its variations, the *direct* and the *inverse* power method.

The *direct power method* introduced by VON MISES is based on the following idea: we iterate the mapping given by the matrix $A \in \mathrm{Mat}_n(\mathbf{R})$ and define a sequence $\{x_k\}_{k=0,1,...}$ for an arbitrary starting point $x_0 \in \mathbf{R}^n$ by

$$x_{k+1} := A x_k \quad \text{for} \quad k = 0, 1, \dots \qquad (5.2)$$

If a simple eigenvalue λ of A is strictly grater in absolute value than all other eigenvalues of A, then we can suspect that λ "asserts" itself against all other eigenvalues during the iteration and x_k converges towards an eigenvector of A corresponding to the eigenvalue λ. This suspicion is confirmed by the following theorem. For the sake of simplicity we limit ourselves here to symmetric matrices, for which according to Theorem 5.2 the eigenvalue problem is well conditioned.

Theorem 5.5 *Let λ_1 be a simple eigenvalue of the symmetric matrix $A \in \mathrm{Mat}_n(\mathbf{R})$ that is strictly greater in absolute value than all other eigenvalues of A, i.e.*

$$|\lambda_1| > |\lambda_2| \geq \cdots \geq |\lambda_n| \, .$$

Furthermore let $x_0 \in \mathbf{R}^n$ be a vector that is not orthogonal to the eigenspace of λ_1. Then the sequence $y_k := x_k / \|x_k\|$ with $x_{k+1} = A x_k$ converges towards a normalized eigenvector of A corresponding to the eigenvalue λ_1.

Proof. Let $\eta_1, \ldots, \eta_n$ be an orthonormal basis of eigenvectors of A with $A\eta_i = \lambda_i \eta_i$. Then $x_0 = \sum_{i=1}^{n} \alpha_i \eta_i$ with $\alpha_1 = \langle x_0, \eta_1 \rangle \neq 0$. Consequently

$$x_k = A^k x_0 = \sum_{i=1}^{n} \alpha_i \lambda_i^k \eta_i = \alpha_1 \lambda_1^k \Big(\eta_1 + \underbrace{\sum_{i=2}^{n} \frac{\alpha_i}{\alpha_1} \Big(\frac{\lambda_i}{\lambda_1} \Big)^k \eta_i}_{=: \, z_k} \Big).$$

Because $|\lambda_i| < |\lambda_1|$ we have $\lim_{k \to \infty} z_k = \eta_1$ for all $i = 2, \ldots, n$, and therefore

$$y_k = \frac{x_k}{\|x_k\|} = \frac{z_k}{\|z_k\|} \to \pm \eta_1 \quad \text{for} \quad k \to \infty .$$

$\square$

The direct power method has however several disadvantages. First, we obtain only the eigenvector corresponding to the largest eigenvalue (in absolute value) λ_1 of A. Then the speed of convergence depends on the quotient $|\lambda_2 / \lambda_1|$. If the absolute values of the eigenvalues λ_1 and λ_2 are close then the direct power method converges very slowly.

The disadvantages of the direct power method described above are avoided by the *inverse power method* developed by WIELANDT (1945). Assuming that we had an estimated value $\bar{\lambda} \approx \lambda_i$ of an arbitrary eigenvalue λ_i of the matrix A at our disposal such that

$$|\bar{\lambda} - \lambda_i| < |\bar{\lambda} - \lambda_j| \quad \text{for all} \quad j \neq i . \tag{5.3}$$

Then $(\bar{\lambda} - \lambda_i)^{-1}$ is the largest eigenvalue of the matrix $(A - \bar{\lambda}I)^{-1}$. Consequently we apply the power method for this matrix. This idea delivers the iterative scheme

$$(A - \bar{\lambda}I)x_{k+1} = x_k \quad \text{for} \quad k = 0, 1, \ldots \tag{5.4}$$

This is called the *inverse power method*. One should be aware of the fact that at each iteration one has to solve the linear system (5.4) for different right hand sides x_k. Therefore one has to factor the matrix $A - \bar{\lambda}I$ only once (e.g. by LR-factorization). According to Theorem 5.5 the sequence $y_k := x_k / \|x_k\|$ converges under assumption (5.3) for $k \longrightarrow \infty$ towards a normalized eigenvector of A corresponding to the eigenvalue λ_i, provided the starting vector x_0 is not orthogonal to the eigenvector η_i of eigenvalue λ_i. Its *convergence factor* is

$$\max_{\cdot j \neq i} \left| \frac{\lambda_i - \bar{\lambda}}{\lambda_j - \bar{\lambda}} \right| < 1 .$$

If $\bar{\lambda}$ is a particularly good estimate of λ_i, then

$$\left| \frac{\lambda_i - \bar{\lambda}}{\lambda_j - \bar{\lambda}} \right| \ll 1 \quad \text{for all} \quad j \neq i \, ,$$

so that the method converges very rapidly in this case. Thus with appropriate choice of $\bar{\lambda}$ this method can be used with nearly arbitrary starting vector x_0 in order to pick out individual eigenvalues and eigenvectors. For an improvement of this method see Exercise 5.3.

Remark 5.6 Note that the matrix $A - \bar{\lambda}I$ is almost singular for "well chosen" $\bar{\lambda} \approx \lambda_i$. In the following this poses no numerical difficulties because we want to find only the *directions* of the eigenvectors whose calculation is well conditioned (cf. Example 2.33).

Example 5.7 Let us examine for example the 2×2–matrix

$$A := \begin{pmatrix} -1 & 3 \\ -2 & 4 \end{pmatrix} .$$

Its eigenvalues are $\lambda_1 = 1$ and $\lambda_2 = 2$. If we take as starting point an approximation $\bar{\lambda} = 1 - \varepsilon$ of λ_1 with $0 < \varepsilon \ll 1$, then the matrix

$$A - \bar{\lambda}I = \begin{pmatrix} -2 + \varepsilon & 3 \\ -2 & 3 + \varepsilon \end{pmatrix}$$

is almost singular and

$$(A - \bar{\lambda}I)^{-1} = \frac{1}{\varepsilon^2 + \varepsilon} \begin{pmatrix} 3 + \varepsilon & -3 \\ 2 & -2 + \varepsilon \end{pmatrix} .$$

Because the factor $1/(\varepsilon^2 + \varepsilon)$ simplifies through normalization, the computation of the direction of a solution x of $(A - \bar{\lambda}I)x = b$ is well conditioned. This can be also read from the component-wise relative condition number

$$\kappa_C = \frac{\| \, |(A - \bar{\lambda}I)^{-1}| \, |b| \, \|_\infty}{\|x\|_\infty}$$

corresponding to perturbations of the right hand side. For example if $b := (1, 0)^T$ we get

$$x = (A - \bar{\lambda}I)^{-1}b = |(A - \bar{\lambda}I)^{-1}| \, |b| = \frac{1}{\varepsilon(\varepsilon + 1)} \begin{pmatrix} 3 + \varepsilon \\ 2 \end{pmatrix}$$

and hence $\kappa_C\left((A - \bar{\lambda}I)^{-1},\, b\right) = 1$.

Actually in programs for a (genuinely) singular matrix $A - \bar{\lambda}I$ a pivot element $\varepsilon = 0$ is substituted by the relative machine precision eps and thus the inverse power method is performed only for nearly singular matrices. (cf. [81]).

5.3 QR-Algorithm for Symmetric Eigenvalue Problems

As described in Section 5.1 the eigenvalue problem for symmetric matrices is well conditioned. In the present section we are interested in the question of how to compute simultaneously *all* the eigenvalues of a real symmetric matrix $A \in \mathrm{Mat}_n(\mathbf{R})$ in an efficient way. We know that A has only real eigenvalues $\lambda_1, \ldots, \lambda_n \in \mathbf{R}$ and that there exists an orthonormal basis $\eta_1, \ldots, \eta_n \in \mathbf{R}^n$ of eigenvectors $A\eta_i = \lambda_i\eta_i$, i.e.

$$Q^T A Q = \Lambda = \mathrm{diag}\,(\lambda_1, \ldots, \lambda_n) \ \text{ with } \ Q = [\eta_1, \ldots, \eta_n] \in \mathbf{O}(n)\,. \qquad (5.5)$$

The first idea that comes to mind would be to determine Q in finitely many steps. Because the eigenvalues are zeros of the characteristic polynomial this would also give us a finite procedure for determining the zeros of polynomials of arbitrary degree (in case of symmetric matrices only with real roots). This would be in conflict with *Abel's theorem* which says that in general such a procedure (based on the operations $+, -, \cdot, /$ and radicals) does not exist.

The second idea, suggested by (5.5) is to bring A closer to diagonal form by a similarity transformation (conjugation), e.g. via orthogonal matrices, because the eigenvalues are invariant under similarity transformations. If we try to bring a symmetric matrix A to diagonal form by conjugation with Householder matrices, then we realize quickly that this is impossible.

$$
\begin{bmatrix} * & \cdots & * \\ \vdots & & \vdots \\ \vdots & & \vdots \\ * & \cdots & * \end{bmatrix}
\xrightarrow{Q_1\cdot}
\begin{bmatrix} * & * & \cdots & * \\ 0 & \vdots & & \vdots \\ \vdots & \vdots & & \vdots \\ 0 & * & \cdots & * \end{bmatrix}
\xrightarrow{\cdot Q_1^T}
\begin{bmatrix} * & * & \cdots & * \\ \vdots & * & \cdots & * \\ \vdots & \vdots & & \vdots \\ * & * & \cdots & * \end{bmatrix}
$$

What is done by multiplying with a Householder matrix from the left is undone by the multiplication from the right. Things look different if we only want to bring A to *tri-diagonal form*. Here the Householder transforms from the left and right do not disturb each other:

$$
\begin{bmatrix}
* & \cdots & * \\
\vdots & & \vdots \\
\vdots & & \vdots \\
\vdots & & \vdots \\
\vdots & & \vdots \\
* & \cdots & *
\end{bmatrix}
\xrightarrow{P_1\cdot}
\begin{bmatrix}
* & * & \cdots & * \\
 & * & & \vdots \\
0 & & & \vdots \\
\vdots & \vdots & & \vdots \\
0 & * & \cdots & *
\end{bmatrix}
\xrightarrow{\cdot P_1^T}
\begin{bmatrix}
* & * & 0 & \cdots & 0 \\
* & * & \cdots & \cdots & * \\
0 & \vdots & & & \vdots \\
\vdots & \vdots & & & \vdots \\
0 & * & \cdots & \cdots & *
\end{bmatrix}
\quad (5.6)
$$

We formulate this insight as a lemma.

Lemma 5.8 *Let $A \in \mathrm{Mat}_n(\mathbf{R})$ be symmetric. Then there is an orthogonal matrix $P \in \mathbf{O}(n)$, which is a product of $n-2$ Householder reflections such that PAP^T is tri-diagonal.*

Proof. We iterate the process shown in (5.6), and we obtain Householder reflections $P_1, \ldots, P_{n-2}$ such that

$$
\underbrace{P_{n-2} \cdots P_1}_{=\,P} A \underbrace{P_1^T \cdots P_{n-2}^T}_{=\,P^T} =
\begin{bmatrix}
* & * & & & \\
* & \ddots & \ddots & & \\
 & \ddots & \ddots & \ddots & * \\
 & & \ddots & \ddots & * \\
 & & & * & *
\end{bmatrix}
$$

□

With this we have transformed our problem to finding the eigenvalues of a symmetric tridiagonal matrix. Therefore we need an algorithm for this special case. The idea of the following algorithm goes back to Rutishauser. He has first tried to find out what happens when the factors of the LR factorization of a matrix $A = LR$ are interchanged, $A' = RL$, and if this process of factorization and interchange is iterated. It turned out that in many cases the matrices constructed in this way converged towards the diagonal matrix Λ of the eigenvalues. The QR algorithm that goes back to FRANCIS (1959) [31] and KUBLANOVSKAJA (1961) [52] employs the QR factorization instead of the of the LR factorization. This factorization always exists (no permutation is necessary) and above all it is inherently stable, as seen in Section 3.2.

Therefore we define a sequence $\{A_k\}_{k=1,2,\ldots}$ of matrices by

$$
\begin{array}{llll}
\text{a)} & A_1 & = & A \\
\text{b)} & A_k & = & Q_k R_k, \quad QR \text{ factorization} \\
\text{c)} & A_{k+1} & = & R_k Q_k \,.
\end{array}
\tag{5.7}
$$

Lemma 5.9 *The matrices A_k have the following properties:*

i) *The matrices A_k are all similar to A.*

ii) *If A is symmetric, then so are all A_k.*

iii) *If A is symmetric and tridiagonal, then so are all A_k.*

Proof. i) Let $A = QR$ and $A' = RQ$. Then

$$
QA'Q^T = QRQQ^T = QR = A \,.
$$

ii) The transformations of the form $A \to B^T A B, \quad B \in \mathrm{GL}(n)$, represent a change of basis for bilinear forms and therefore are symmetric. In particular this holds for orthogonal similarity transformations. This follows also directly from

$$
(A')^T = (A')^T Q^T Q = Q^T R^T Q^T Q = Q^T A^T Q = Q^T A Q = A' \,.
$$

iii) Let A be symmetric and tridiagonal. We realize Q with $n-1$ Givens rotations $\Omega_{12}, \ldots, \Omega_{n-1,n}$, so that $Q^T = \Omega_{n-1,n} \cdots \Omega_{12}$ ($\otimes$ to eliminate, $\oplus$ fill-in entry).

$$
A \qquad \longrightarrow \qquad R = Q^T A
$$

$$
\begin{bmatrix} * & * & \oplus & & & \\ & * & * & \oplus & & \\ & & \ddots & \ddots & \ddots & \\ & & & \ddots & \ddots & \oplus \\ & & & & \ddots & * \\ & & & & & * \end{bmatrix} \rightarrow \begin{bmatrix} * & * & \oplus & & & \\ * & * & * & \oplus & & \\ & \ddots & \ddots & \ddots & \ddots & \\ & & \ddots & \ddots & \ddots & \oplus \\ & & & \ddots & \ddots & * \\ & & & & * & * \end{bmatrix}
$$

$$R \qquad \rightarrow \qquad A' = RQ = Q^T A Q$$

According to ii) A' must be symmetric and therefore all $\oplus$ entries in A' vanish. Hence A' is also tridiagonal. □

We show the convergence properties only for the simple case when the absolute value of the eigenvalues of A are distinct.

Theorem 5.10 *Let $A \in \mathrm{Mat}_n(\mathbf{R})$ be symmetric with eigenvalues $\lambda_1, \ldots, \lambda_n$, such that*

$$|\lambda_1| > |\lambda_2| > \cdots > |\lambda_n| > 0 \tag{5.8}$$

and let A_k, Q_k, R_k be defined as in (5.7), with $A_k = (a_{ij}^{(k)})$. Then the following statements hold:

$$
\begin{aligned}
a) \quad & \lim_{k \to \infty} Q_k &=& \;\; I \\
b) \quad & \lim_{k \to \infty} R_k &=& \;\; \Lambda \\
c) \quad & a_{i,j}^{(k)} &=& \;\; O\left(\left|\frac{\lambda_i}{\lambda_j}\right|^k\right) \quad \text{for } i > j .
\end{aligned}
$$

Proof. The proof given here goes back to WILKINSON [79]. We show first that

$$A^k = \underbrace{Q_1 \cdots Q_k}_{=:\,P_k} \underbrace{R_k \cdots R_1}_{=:\,U_k} \quad \text{for } k = 1, 2, \ldots$$

The assertion is clear for $k = 1$ because $A = A_1 = Q_1 R_1$.

On the other hand from the construction of A_k it follows that

$$A_{k+1} = Q_{k+1} R_{k+1} = Q_k^T \cdots Q_1^T A Q_1 \cdots Q_k = P_k^{-1} A P_k$$

and from it the induction step

$$A^{k+1} = A A^k = A P_k U_k = P_k Q_{k+1} R_{k+1} U_k = P_{k+1} U_{k+1} .$$

Because $P_k \in \mathbf{O}(n)$ is orthogonal and U_k upper triangular, we can express the QR factorization $A^k = P_k U_k$ of A^k through the QR factorization of $A_1, \ldots, A_k$. Further

$$A^k = Q \Lambda^k Q^T, \quad \Lambda^k = \operatorname{diag}(\lambda_1^k, \ldots, \lambda_n^k) \ .$$

We assume for the sake of simplicity that Q has an LR factorization

$$Q = LR \ ,$$

where L is a unit lower triangular matrix and R an upper triangular matrix. We can always achieve this by conjugating A with appropriate permutations. With this we have

$$A^k = Q \Lambda^k LR = Q(\Lambda^k L \Lambda^{-k})(\Lambda^k R) \ . \tag{5.9}$$

The unit lower triangular matrix $\Lambda^k L \Lambda^{-k}$ satisfies

$$(\Lambda^k L \Lambda^{-k})_{ij} = l_{ij} \left(\frac{\lambda_i}{\lambda_j} \right)^k \ .$$

In particular all off-diagonal entries vanish for $k \to \infty$, i.e.

$$\Lambda^k L \Lambda^{-k} = I + E_k \quad \text{with} \quad E_k \to 0 \quad \text{for} \quad k \to \infty \ .$$

By substituting in (5.9) it follows that

$$A^k = Q(I + E_k) \Lambda^k R \ .$$

Now we (formally) apply a QR factorization

$$I + E_k = \tilde{Q}_k \tilde{R}_k$$

where all diagonal elements of $\tilde{R}_k$ are positive. Then from the uniqueness of the QR factorization and $\lim_{k \to \infty} E_k = 0$ it follows that

$$\tilde{Q}_k, \tilde{R}_k \to I \quad \text{for} \quad k \to \infty \ .$$

With this we have deduced a second QR factorization of A^k, because

$$A^k = (Q\tilde{Q}_k)(\tilde{R}_k \Lambda^k R) \ .$$

Therefore the following equality holds up to the sign of the diagonal elements:

$$P_k = Q\tilde{Q}_k, \quad U_k = \tilde{R}_k \Lambda^k R \ .$$

For $k \to \infty$ it follows that

$$
\begin{aligned}
Q_k &= P_{k-1}^T P_k = \tilde{Q}_{k-1}^T Q^T Q \tilde{Q}_k = \tilde{Q}_{k-1}^T \tilde{Q}_k \to I \\
R_k &= U_k U_{k-1}^{-1} = \tilde{R}_k \Lambda^k R R^{-1} \Lambda^{-(k-1)} \tilde{R}_{k-1}^{-1} = \tilde{R}_k \Lambda \tilde{R}_{k-1}^{-1} \to \Lambda
\end{aligned}
$$

and

$$
\lim_{k \to \infty} A_k = \lim_{k \to \infty} Q_k R_k = \lim_{k \to \infty} R_k = \Lambda .
$$

$\square$

Remark 5.11 A more precise analysis shows that the method converges also for multiple eigenvalues $\lambda_i = \cdots = \lambda_j$. However if $\lambda_i = -\lambda_{i+1}$ then the method does not converge. The 2×2 blocks are left as such.

If two eigenvalues λ_i, λ_{i+1} are very close in absolute value then the method converges very slowly. This can be improved with the so called *shift strategy*. In principle one tries to push both eigenvalues closer to the origin so as to reduce the quotients $|\lambda_{i+1}/\lambda_i|$. In order to do that one uses at each iteration step k a *shift parameter* σ_k and one defines the sequence $\{A_k\}$ by

$$
\begin{aligned}
\text{a)} \quad & A_1 && = && A \\
\text{b)} \quad & A_k - \sigma_k I && = && Q_k R_k, \quad QR\text{-factorization} \\
\text{c)} \quad & A_{k+1} && = && R_k Q_k + \sigma_k I .
\end{aligned}
$$

As above it follows that

1) $A_{k+1} = Q_k^T A_k Q_k \sim A_k$

2) $(A - \sigma_k I) \cdots (A - \sigma_1 I) = Q_1 \cdots Q_k R_k \cdots R_1.$

The sequence $\{A_k\}$ converges towards Λ with the speed

$$
a_{i,j}^{(k)} = O\left(\left| \frac{\lambda_i - \sigma_1}{\lambda_j - \sigma_1} \right| \cdots \left| \frac{\lambda_i - \sigma_{k-1}}{\lambda_j - \sigma_{k-1}} \right| \right) \quad \text{for} \;\; i > j .
$$

We have already met such a convergence behavior in Section 5.2 in the case of the inverse power method. In order to achieve a convergence acceleration the σ_k's have to be chosen as close as possible to the eigenvalues λ_i, λ_{i+1}. Willkinson has proposed the following shift strategy: we start with a symmetric tridiagonal matrix A. If the lower end of the tridiagonal matrix A_k is of the form

$$
\begin{pmatrix}
\ddots & \ddots & \\
\ddots & d_{n-1}^{(k)} & e_n^{(k)} \\
 & e_n^{(k)} & d_n^{(k)}
\end{pmatrix}
$$

then the 2×2 corner matrix has two eigenvalues and we choose as σ_k the one that is closer to $d_n^{(k)}$. Better than these *explicit* shift strategies, especially for badly scaled matrices, are the *implicit shift methods* for which we refer again to [39] and [72] . With these techniques one finally needs $O(n)$ arithmetic operations per computed eigenvalue, that is $O(n^2)$ for all eigenvalues.

Besides eigenvalues we are also interested in the *eigenvectors*, that can be computed as follows: If $Q \in \mathbf{O}(n)$ is an orthogonal matrix, then

$$A \approx Q^T \Lambda Q, \quad \Lambda = \mathrm{diag}(\lambda_1, \dots, \lambda_n),$$

then the columns of Q approximate the eigenvectors of A, i.e.

$$Q \approx [\eta_1, \dots, \eta_n] .$$

Together we obtain the following algorithm for determining all the eigenvalues and eigenvectors of a symmetric matrix.

Algorithm 5.12 *QR algorithm.*

a) Reduce the problem to tridiagonal form

$$A \to A_1 = PAP^T, \quad A_1 \text{ symmetric and tridiagonal}, \ P \in \mathbf{O}(n).$$

b) Approximate the eigenvalues with the QR algorithm with Givens rotations applied to A_1

$$\Omega A_1 \Omega^T \approx \Lambda, \quad \Omega \text{ Product of all Givens-rotations } \Omega_{ij}^{(k)} .$$

c) The columns of ΩP approximate the eigenvectors of A

$$\Omega P \approx [\eta_1, \dots, \eta_n] .$$

The cost amounts to

a) $\frac{4}{3}n^3$ multiplications for reduction to tridiagonal form,

b) $O(n^2)$ multiplications for the QR algorithm.

Hence for large n the cost of reduction to tridiagonal form dominates.

Remark 5.13 For non symmetric matrices an orthogonal conjugation reduces first the matrix to *Hessenberg form*. Finally the QR algorithm iteratively brings the matrix to the *Schur normal form* (complex upper triangular matrix). Details are found in the book of WILKINSON and REINSCH [81].

5.4   Singular Value Decomposition

A very useful tool for analyzing matrices is provided by the so called *singular value decomposition* of a matrix $A \in \mathrm{Mat}_{m,n}(\mathbf{R})$. First we prove the existence of such a decomposition and list some of its properties. Finally we will see how to compute the singular values by a variant of the QR algorithm described above.

Theorem 5.14 *Let $A \in \mathrm{Mat}_{m,n}(\mathbf{R})$ be an arbitrary real matrix. Then there are orthogonal matrices $U \in \mathbf{O}(m)$ and $V \in \mathbf{O}(n)$, such that*

$$U^T A V = \Sigma = \mathrm{diag}(\sigma_1, \ldots, \sigma_p) \in \mathrm{Mat}_{m,n}(\mathbf{R}) \ ,$$

where $p = \min(m, n)$ and $\sigma_1 \geq \sigma_2 \geq \ldots \geq \sigma_p \geq 0$.

Proof. It is sufficient to show that there are $U \in \mathbf{O}(m)$ and $V \in \mathbf{O}(n)$ such that

$$U^T A V = \begin{pmatrix} \sigma & 0 \\ 0 & B \end{pmatrix} .$$

The claim follows then by induction. Let $\sigma := \|A\|_2 = \max_{\|x\|=1} \|Ax\|$. Because the maximum is attained there are $v \in \mathbf{R}^n$, $u \in \mathbf{R}^m$, such that

$$Av = \sigma u \ \text{ and } \ \|u\|_2 = \|v\|_2 = 1 \ .$$

We can extend $\{v\}$ to an orthonormal basis $\{v = V_1, \ldots, V_n\}$ of $\mathbf{R}^n$ and $\{u\}$ to an orthonormal basis $\{u = U_1, \ldots, U_m\}$ of $\mathbf{R}^m$. Then

$$V := [V_1, \ldots, V_n] \ \text{ and } \ U := [U_1, \ldots, U_m]$$

are orthogonal matrices, $V \in \mathbf{O}(n)$, $U \in \mathbf{O}(m)$, and $U^T A V$ is of the form

$$A_1 := U^T A V = \begin{pmatrix} \sigma & w^T \\ 0 & B \end{pmatrix}$$

with $w \in \mathbf{R}^{n-1}$. Since

$$\left\| A_1 \begin{pmatrix} \sigma \\ w \end{pmatrix} \right\|_2^2 \geq (\sigma^2 + \|w\|_2^2)^2 \ \text{ and } \ \left\| \begin{pmatrix} \sigma \\ w \end{pmatrix} \right\|_2^2 = \sigma^2 + \|w\|_2^2 \ ,$$

we have $\sigma^2 = \|A\|_2^2 = \|A_1\|_2^2 \geq \sigma^2 + \|w\|_2^2$ and therefore $w = 0$, so that

$$U^T A V = \begin{pmatrix} \sigma & 0 \\ 0 & B \end{pmatrix} .$$

$\square$

Definition 5.15 The factorization $U^T A V = \Sigma$ is called the *singular value decomposition* of A, and the σ_i's are called the *singular values* of A.

With the singular value decomposition we have at our disposal the most important informations about the matrix. The following properties can be easily deduced from Theorem 5.14.

Corollary 5.16 *Let $U^T A V = \Sigma = \mathrm{diag}(\sigma_1, \ldots, \sigma_p)$ be the singular value decomposition of A with singular values $\sigma_1, \ldots, \sigma_p$, where $p = \min(m, n)$. Then:*

1. *If U_i and V_i are the columns of U and V respectively, then*

$$AV_i = \sigma_i U_i \quad \text{and} \quad A^T U_i = \sigma_i V_i \quad \text{for } i = 1, \ldots, p .$$

2. *If $\sigma_1 \geq \cdots \geq \sigma_r > \sigma_{r+1} = \cdots = \sigma_p = 0$, then $\mathrm{rank}\, A = r$,*

$$\ker A = \mathrm{span}\{V_{r+1}, \ldots, V_n\} \quad \text{and} \quad \mathrm{im}\, A = \mathrm{span}\{U_1, \ldots, U_r\} .$$

3. *The Euclidean norm of A is the largest singular value, i.e.*

$$\|A\|_2 = \sigma_1 .$$

4. *The Frobenius norm $\|A\|_F = (\sum_{i=1}^{n} \|A_i\|_2^2)^{1/2}$ is equal to*

$$\|A\|_F^2 = \sigma_1^2 + \ldots + \sigma_p^2 .$$

5. *The condition number of A relative to the Euclidean norm is equal to the quotient of the largest and the smallest singular values , i.e.*

$$\kappa_2(A) = \sigma_1 / \sigma_p .$$

6. *The squares of the singular values $\sigma_1^2, \ldots, \sigma_p^2$ are the eigenvalues of $A^T A$ and AA^T corresponding to the eigenvectors $V_1, \ldots, V_p$ and $U_1, \ldots, U_p$ respectively.*

Based on the invariance of the Euclidean norm $\|\cdot\|_2$ under the orthogonal transformations U and V we obtain from the singular value decomposition of A another representation of the pseudo inverse A^+ of A.

Corollary 5.17 *Let $U^T A V = \Sigma$ be the singular value decomposition of a matrix $A \in \mathrm{Mat}_{m,n}(\mathbf{R})$ with $p = \mathrm{rank}\, A$ and*

$$\Sigma = \mathrm{diag}(\sigma_1, \ldots, \sigma_p, 0, \ldots, 0) .$$

Then the pseudo inverse $A^+ \in \mathrm{Mat}_{n,m}(\mathbf{R})$ is given by

$$A^+ = V\Sigma^+ U^T \quad \text{with} \quad \Sigma^+ = \mathrm{diag}(\sigma_1^{-1}, \ldots, \sigma_p^{-1}, 0, \ldots, 0) .$$

Proof. We have to prove that the right hand side $B := V\Sigma^+U^T$ satisfies the Penrose axioms. The pseudo inverse of the diagonal matrix Σ is evidently Σ^+. Then the Moore-Penrose axioms for B follow immediately because $V^TV = I$ and $U^TU = I$. $\qquad\square$

Now we go over the problem of the *numerical computation* of the singular values. According to Corollary 5.16 the singular values of a matrix $A \in \mathrm{Mat}_n(\mathbf{R})$ are the square roots of the eigenvalues of A^TA,

$$\sigma_i(A) = \sqrt{\lambda_i(A^TA)}\,. \qquad (5.10)$$

The eigenvalue problem for the symmetric matrix A^TA is well conditioned and with it so is the singular value problem of A provided we can avoid the computation of A^TA. Relation (5.10) suggests a computational method for $\sigma_i(A)$. This detour is however inappropriate as will be easily seen from the following example.

Example 5.18 We compute with four significant figures (rounded).

$$A = A^T = \begin{pmatrix} 1.005 & 0.995 \\ 0.995 & 1.005 \end{pmatrix}, \quad \sigma_1 = \lambda_1(A) = 2, \quad \sigma_2 = \lambda_2(A) = 0.01\,.$$

For A^TA we obtain

$$\mathrm{fl}(A^TA) = \begin{pmatrix} 2.000 & 2.000 \\ 2.000 & 2.000 \end{pmatrix}, \quad \tilde{\sigma}_1^2 = 4, \quad \tilde{\sigma}_2^2 = 0.$$

As in the case of linear least squares we will search here also for a method operating only on the matrix A. For this we examine first the operations that leave the singular values invariant.

Lemma 5.19 *Let $A \in \mathrm{Mat}_{m,n}(\mathbf{R})$, and let $P \in \mathbf{O}(m)$, $Q \in \mathbf{O}(n)$ be orthogonal matrices. Then A and $B := PAQ$ have the same singular values*

Proof. Simple $\qquad\square$

Hence, we may pre- and post-multiply the matrix A with arbitrary orthogonal matrices, without changing the singular values. For the application of the QR algorithm it is desirable to transform the matrix A in such a way that A^TA is tridiagonal. The simplest way to accomplish this is by bringing A to *bidiagonal form*. The following Lemma shows that this goal is achievable by means of Householder transforms from right and left.

Lemma 5.20 *Let $A \in \mathrm{Mat}_{m,n}(\mathbf{R})$ be a general matrix and suppose, without loss of generality, that with $m \geq n$. Then there exist orthogonal matrices*

$P \in \mathbf{O}(m)$ and $Q \in \mathbf{O}(n)$, such that

$$
PAQ = \left[
\begin{array}{cccc|c}
* & * & & & \\
 & \ddots & & \ddots & \\
 & & \ddots & & * \\
 & & & & * \\
\hline
0 & \cdots & & \cdots & 0 \\
\vdots & & & & \vdots \\
0 & \cdots & & \cdots & 0
\end{array}
\right] = \left[\begin{array}{c} B \\ 0 \end{array} \right]
$$

where B is a (square) bidiagonal matrix.

Proof. We illustrate the construction of P and Q with Householder matrices:

$$
\begin{bmatrix}
* & \cdots & \cdots & * \\
\vdots & & & \vdots \\
\vdots & & & \vdots \\
\vdots & & & \vdots \\
* & \cdots & \cdots & *
\end{bmatrix}
\overset{P_1\cdot}{\rightarrow}
\begin{bmatrix}
* & * & \cdots & * \\
0 & & & \\
\vdots & \vdots & & \vdots \\
\vdots & \vdots & & \vdots \\
0 & * & \cdots & *
\end{bmatrix}
\overset{\cdot Q_1}{\rightarrow}
\begin{bmatrix}
* & * & 0 & \cdots & 0 \\
0 & * & & & * \\
\vdots & \vdots & & & \vdots \\
\vdots & \vdots & & & \vdots \\
0 & * & & \cdots & *
\end{bmatrix}
$$

$$
\overset{P_2\cdot}{\rightarrow}
\begin{bmatrix}
* & * & 0 & \cdots & 0 \\
0 & * & * & \cdots & * \\
\vdots & 0 & \vdots & & \vdots \\
\vdots & \vdots & \vdots & & \vdots \\
0 & 0 & * & \cdots & *
\end{bmatrix}
\rightarrow \cdots \overset{P_{n-1}\cdot}{\rightarrow}
\begin{bmatrix}
* & * & & & \\
 & * & * & & \\
 & & \ddots & \ddots & \\
 & & & \ddots & * \\
 & & & & *
\end{bmatrix}
$$

Therefore we have

$$
\binom{B}{0} = \underbrace{P_{n-1} \cdots \cdots P_1}_{=:\, P} A \underbrace{Q_1 \cdots \cdots Q_{n-2}}_{=:\, Q} \, .
$$

$\square$

In order to derive an effective algorithm we examine now the QR algorithm for the tridiagonal matrix $B^T B$. The goal is to find a simplified version that operates exclusively on B. If we perform the first Givens elimination steps of the QR algorithm for $A = B^T B$

$$A \longrightarrow \Omega_{12} B^T B \Omega_{12}^T = \underbrace{(B\Omega_{12}^T)^T}_{\tilde{B}^T} \underbrace{B\Omega_{12}^T}_{\tilde{B}} ,$$

then we obtain as $\tilde{B}$ the matrix

$$\tilde{B} = B\Omega_{12}^T = \begin{bmatrix} * & * & & & & \\ \oplus & * & * & & & \\ & & \ddots & \ddots & & \\ & & & \ddots & * \\ & & & & * \end{bmatrix} ,$$

where in position $\oplus$ a new fill-in element is generated. If we play back the QR algorithm for $B^T B$ in this way on B then it appears that the method corresponds to the following elimination process.

$$\begin{bmatrix} * & * & z_3 & & & & \\ z_2 & * & * & z_5 & & & \\ & z_4 & * & * & z_7 & & \\ & & \ddots & \ddots & \ddots & \ddots & \\ & & & z_{2n-6} & * & * & z_{2n-3} \\ & & & & z_{2n-4} & * & * \\ & & & & & z_{2n-2} & * \end{bmatrix} \qquad (5.11)$$

eliminate z_2 (Givens from left) $\rightarrow$ fill-in z_3

eliminate z_3 (Givens from right) $\rightarrow$ fill-in z_4

$\vdots$

eliminate z_{2n-3} (Givens from right) $\rightarrow$ fill-in z_{2n-2}

eliminate z_{2n-2} (Givens from left)

We "hunt down" fill-in elements alongside both diagonals and remove the newly generated fill-in entries with Givens rotations alternating from left and right. Whence the name *chasing* given to this process. In the end the matrix has bidiagonal form and we have performed one iteration of the QR method for $B^T B$ operating only on B. According to Theorem 5.10 we have

$$B_k^T B_k \to \Lambda = \mathrm{diag}(\sigma_1^2, \ldots, \sigma_n^2) = \Sigma^2 \text{ for } k \to \infty .$$

Therefore the sequence B_k converges towards the diagonal matrix of the singular values of B, i.e.

$$B_k \to \Sigma \text{ for } k \to \infty .$$

Summing up we obtain the following algorithm for determining the singular values of A (for details we refer to [39]):

Algorithm 5.21 *QR algorithm for singular values.*

a) Bring $A \in \mathrm{Mat}_{m,n}(\mathbf{R})$ to bidiagonal form via orthogonal transforms, $P \in \mathbf{O}(m)$ and $Q \in \mathbf{O}(n)$ (i.e. Householder reflections).

$$PAQ = \begin{pmatrix} B \\ 0 \end{pmatrix}, \quad B \in \mathrm{Mat}_n(\mathbf{R}) \text{ upper bidiagonal matrix}$$

b) Perform the QR algorithm for $B^T B$ by the "chasing" method (5.11) on B and obtain a sequence of bidiagonal matrices $\{B_k\}$ that converges towards a diagonal matrix Σ of the singular values.

For the cost in case $m = n$ we count

a) $\sim \frac{4}{3}n^3$ multiplications for reduction to bidiagonal form,

b) $O(n^2)$ multiplications for the modified QR algorithm.

5.5   Exercises

Exercise 5.1 Determine the eigenvalues, eigenvectors and the determinant of a Householder matrix

$$Q = I - 2\frac{vv^T}{v^T v}.$$

Exercise 5.2 Give a formula (in terms of determinants) for an eigenvector $x \in \mathbf{C}^n$ corresponding to a simple eigenvalue $\lambda \in \mathbf{C}$ of a matrix $A \in \mathrm{Mat}_n(\mathbf{C})$.

Exercise 5.3 The computation of an eigenvector η_j corresponding to an eigenvalue λ_j of a given matrix A can be done, according to Wielandt, by the inverse iteration

$$A z_i - \hat{\lambda}_j z_i = z_{i-1}$$

with an approximation $\hat{\lambda}_j$ to the eigenvalue λ_j. Deduce from the relation

$$r(\delta) := A z_i - (\hat{\lambda}_j + \delta) z_i = z_{i-1} - \delta z_i$$

a correction δ for the approximation $\hat{\lambda}_j$ such that $\|r(\delta)\|_2$ is minimal.

Exercise 5.4 Let there be given a so called arrow matrix Z of the form

$$Z = \begin{bmatrix} A & B \\ B^T & D \end{bmatrix},$$

where $A = A^T \in \mathrm{Mat}_n(\mathbf{R})$ is symmetric, $B \in \mathrm{Mat}_{n,m}$ and D is a diagonal matrix, $D = \mathrm{diag}(d_1, \ldots, d_m)$. For $m \gg n$ it is recommended to use the sparsity structure of Z.

a) Show that

$$Z - \lambda I = L^T(\lambda)(Y(\lambda) - \lambda I)L(\lambda) \quad \text{for } \lambda \neq d_i, i = 1, \ldots, m,$$

where

$$L(\lambda) := \begin{bmatrix} I_n & 0 \\ (D - \lambda I_m)^{-1} B^T & I_m \end{bmatrix} \quad \text{and } Y(\lambda) := \begin{bmatrix} M(\lambda) & 0 \\ 0 & D \end{bmatrix}$$

and $M(\lambda) := A - B(D - \lambda I_m)^{-1} B^T$.

b) Modify the method handled in Exercise 5.3 in such a way that one operates essentially only on (n, n) matrices.

Exercise 5.5 Prove the properties of the singular value decomposition from Corollary 5.16.

Exercise 5.6 Let there be given an (m, n)-matrix $A, m \geq n$, and an m-vector b. The following linear system is to be solved for different values of $p \geq 0$ (Levenberg-Marquardt method, compare Section 4.3):

$$(A^T A + p I_n) x = A^T b . \tag{5.12}$$

a) Show, that the matrix $A^T A + p I_n$ is invertible for rank $A < n$ and $p > 0$.

b) Let A have the singular values $\sigma_1 \geq \sigma_2 \geq \cdots \geq \sigma_n \geq 0$. Show that: If $\sigma_n \geq \sigma_1 \sqrt{\text{eps}}$ then:

$$\kappa_2(A^T A + p I_n) \leq \frac{1}{\text{eps}} \quad \text{for} \quad p \geq 0 .$$

If $\sigma_n < \sigma_1 \sqrt{\text{eps}}$, then there exists a $\overline{p} \geq 0$ such that:

$$\kappa_2(A^T A + p I_n) \leq \frac{1}{\text{eps}} \quad \text{for} \quad p \geq \overline{p} .$$

Determine $\overline{p}$.

c) Develop an efficient algorithm for solving (5.12) by using the singular value decomposition of A.

Exercise 5.7 Determine the eigenvalues $\lambda_i(\epsilon)$ and the eigenvectors $\eta_i(\epsilon)$ of the matrix

$$A(\epsilon) := \begin{bmatrix} 1 + \epsilon \cos(2/\epsilon) & -\epsilon \sin(2/\epsilon) \\ -\epsilon \sin(2/\epsilon) & 1 - \epsilon \cos(2/\epsilon) \end{bmatrix} .$$

How do $A(\epsilon)$, $\lambda_i(\epsilon)$ and $\eta_i(\epsilon)$ behave for $\epsilon \to 0$?

6 Three-Term Recurrence Relations

There are many problems in mathematics and science, where a solution function can be represented in terms of *special functions*. These functions are distinguished by special properties which make them particularly suitable for the problem under consideration, and which often allow for simple construction. The study and use of special functions is an old branch of mathematics to which many outstanding mathematicians have contributed. Recently this area has experienced a resurgance because of new discoveries and extended computational capabilities (e.g. through symbolic computation). As examples of classical special functions, let us mention here the Chebyshev, Legendre, Jacobi, Laguerre and Hermite polynomials and the Bessel functions. In the next section, we shall use some of these polynomials and derive the pertinent and important properties.

Here we want to consider the aspect of evaluating linear combinations

$$f(x) = \sum_{k=0}^{N} \alpha_k P_k(x) \tag{6.1}$$

of special functions $P_k(x)$, where we consider the coefficients α_k as given. The computation or even approximation of these coefficients may be very difficult. We shall address this question in Section 7.2 when we discuss the discrete Fourier transform.

One property which is common to all special functions is their *orthogonality*. So far, we only considered orthogonality in connection with a scalar product on a finite dimensional vector space. Many of the familiar structures carry over to (infinite dimensional) function spaces. Here the scalar product is usually an integral. To illustrate this, we consider the following example.

Example 6.1 Define a scalar product

$$(f, g) \quad = \quad \int_{-\pi}^{\pi} f(x)g(x)dx$$

for functions $f, g : [-\pi, \pi] \to \mathbf{R}$. It is easy to convince oneself that the special functions $P_{2k}(x) = \cos kx$ and $P_{2k+1}(x) = \sin kx$ for $k = 0, 1, \ldots$ are orthogonal with respect to this scalar product, i.e.

$$(P_k, P_l) = \delta_{kl}(P_k, P_k) \quad \text{for all} \quad k,l=0,1,\ldots$$

As in the finite dimensional case, this scalar product induces a norm

$$\|f\| = \sqrt{(f,f)} = \left(\int_{-\pi}^{\pi} |f(x)|^2 dx \right)^{\frac{1}{2}} .$$

The functions, for which this norm is well defined and finite, can be approximated arbitrarily well with respect to this norm by the partial sums of the *Fourier series*

$$f_N(x) = \sum_{k=0}^{2N} \alpha_k P_k(x) = \alpha_0 + \sum_{k=1}^{N} (\alpha_{2k} \cos kx + \alpha_{2k-1} \sin kx) , \qquad (6.2)$$

if N is large enough.

Here we can compute the functions $\cos kx$ and $\sin kx$ via the three-term recurrence relation

$$T_k(x) = 2 \cos x \cdot T_{k-1}(x) - T_{k-2}(x) \quad \text{for} \quad k = 2, 3, \ldots \qquad (6.3)$$

as in Example 2.27.

It is not by accident that we can compute the trigonometric functions $\cos kx$ and $\sin kx$ by a three-term recurrence relation in k, since the existence of a three-term recurrence relation for special functions is connected with their orthogonality.

First we shall study this connection, and we shall in particular be concerned with orthogonal polynomials. The theoretical investigation of three-term recurrence relations as difference equations is the central part of Section 6.1.2. A detailed numerical example will show that the obvious and naive idea of using the three-term recurrence relation as an algorithm may not lead to useful results. In Section 6.2.1 we shall therefore analyze the conditioning of the three-term recurrence relation and thus obtain a classification of the solutions. This will finally enable us to give stable algorithms for the computation of special functions and linear combinations of the form (6.1).

6.1   Theoretical Foundations

Three-term recurrence relations, as e.g. the trigonometric recurrence relation (6.3), are of central importance in the computation of special functions.

In the following section, we shall study the general connection between orthogonality and three-term recurrence relations. Subsequently, we shall be concerned with the theory of homogeneous and nonhomogeneous three-term recurrence relations.

6.1.1 Orthogonality and three-term recurrence relations

As a generalization of the scalar product in Example 6.1, we consider the scalar product

$$(f,g) := \int_a^b \omega(t)f(t)g(t)\,dt \tag{6.4}$$

with an additional *positive weight function* $\omega :]a,b[\to \mathbf{R}$, $\omega(t) > 0$. We assume that the induced norm

$$\|P\| = \sqrt{(P,P)} = \left(\int_a^b \omega(t)P(t)^2\,dt\right)^{\frac{1}{2}} < \infty$$

is well defined and finite for all of the polynomials $P \in \mathbf{P}_k$ and all $k \in \mathbf{N}$. In particular, under this assumption all *moments*

$$\mu_k := \int_a^b t^k \omega(t)\,dt$$

exist, because $1, t^k \in \mathbf{P}_k$ implies that

$$|\mu_k| = \left|\int_a^b t^k \omega(t)\,dt\right| = |(1,t^k)| \leq \|1\|\,\|t^k\| < \infty$$

by the Cauchy-Schwarz inequality. Suppose that $\{P_k(t)\}$ is a sequence of pairwise orthogonal polynomials $P_k \in \mathbf{P}_k$ of degree k, i.e. with non-vanishing leading coefficient and

$$(P_i, P_j) = \int_a^b \omega(t)P_i(t)P_j(t)\,dt = \delta_{ij}\gamma_i \quad \text{where } \gamma_i := \|P_i\|^2 > 0\,,$$

then the P_k are called *orthogonal polynomials* on $[a,b]$ with respect to the weight function $\omega(t)$. In order to define the orthogonal polynomials uniquely, we require an additional normalization condition, e.g. by assuming that $P_k(0) = 1$ or that the leading coefficient is one, i.e.

$$\mathbf{P}_k(t) = t^k + \cdots$$

The existence and uniqueness of a system of orthogonal polynomials with respect to the scalar product (6.4) will now be shown by employing the three-term recurrence relation.

Theorem 6.2 *For each weighted scalar product (6.4), there exist uniquely determined orthogonal polynomials $P_k \in \mathbf{P}_k$ with leading coefficient one. These satisfy the three-term recurrence relation*

$$P_k(t) = (t + a_k)P_{k-1}(t) + b_k P_{k-2}(t) \quad for \quad k = 2, 3, \ldots$$

with starting values $P_0 := 1$, $P_1(t) = t + a_1$ and coefficients

$$a_k = -\frac{(tP_{k-1}, P_{k-1})}{(P_{k-1}, P_{k-1})} \ , \quad b_k = -\frac{(P_{k-1}, P_{k-1})}{(P_{k-2}, P_{k-2})} \ .$$

Proof. The only polynomial of degree 0 with leading coefficient one is $P_0 \equiv 1 \in \mathbf{P}_0$. Suppose that $P_0, \ldots, P_{k-1}$ have already been constructed as pairwise orthogonal polynomials $P_j \in \mathbf{P}_j$ of degree j with leading coefficient one. If $P_k \in \mathbf{P}_k$ is an arbitrary normalized polynomial of degree k, then $P_k - tP_{k-1}$ is a polynomial of degree $\leq k - 1$. On the other hand, the $P_0, \ldots, P_{k-1}$ form an orthogonal basis of $\mathbf{P}_{k-1}$ with respect to the weighted scalar product $(\cdot, \cdot)$, so that

$$P_k - tP_{k-1} = \sum_{j=0}^{k-1} c_j P_j \quad \text{with} \quad c_j = \frac{(P_k - tP_{k-1}, P_j)}{(P_j, P_j)} \ .$$

If P_k is to be orthogonal to $P_0, \ldots, P_{k-1}$, then

$$c_j = -\frac{(tP_{k-1}, P_j)}{(P_j, P_j)} = -\frac{(P_{k-1}, tP_j)}{(P_j, P_j)} \ .$$

This implies that $c_0 = \cdots = c_{k-3} = 0$,

$$c_{k-1} = -\frac{(tP_{k-1}, P_{k-1})}{(P_{k-1}, P_{k-1})}, \quad c_{k-2} = -\frac{(P_{k-1}, tP_{k-2})}{(P_{k-2}, P_{k-2})} = -\frac{(P_{k-1}, P_{k-1})}{(P_{k-2}, P_{k-2})} \ .$$

We therefore obtain the next orthogonal polynomial from the formula

$$P_k = (t + c_{k-1})P_{k-1} + c_{k-2}P_{k-2} = (t + a_k)P_{k-1} + b_k P_{k-2},$$

and the statement follows by induction. □

Example 6.3 By putting $\cos \alpha = x$ and viewing $\cos k\alpha$ as a function of x, one is led to the *Chebyshev polynomials*

$$T_k(x) = \cos(k \arccos x) \quad \text{for} \quad x \in [-1, 1] \ .$$

The three-term recurrence relation for $\cos kx$ implies that the T_k are in fact polynomials which satisfy the recurrence relation

$$T_k(x) = 2xT_{k-1}(x) - T_{k-2}(x) \quad \text{for} \quad k \geq 2$$

with the starting values $T_0(x) = 1$ and $T_1(x) = x$. Using this, we can define $T_k(x)$ for all $x \in \mathbf{R}$. From the variable substitution $x = \cos \alpha$, i.e. $dx = -\sin \alpha \, d\alpha$, we can see that the Chebyshev polynomials are indeed the orthogonal polynomials on $[-1, 1]$ with respect to the weight function $\omega(x) = 1/\sqrt{1 - x^2}$, i.e.

$$\int_0^1 \frac{1}{\sqrt{1-x^2}} T_n(x) T_m(x) dx = \begin{cases} 0 & \text{, if} \quad n \neq m \\ \pi & \text{, if} \quad n = m = 0 \\ \pi/2 & \text{, if} \quad n = m \neq 0 \end{cases}.$$

The Chebyshev polynomials are particularly important in approximation theory. We shall encounter them several times in the next chapters.

By carefully analyzing the proof of Theorem 6.2, we can understand the connection between orthogonality and three-term recurrence relations in greater generality. We shall encounter this structure again in Section 8.3 when studying the method of conjugate gradients.

Theorem 6.4 *Let $V_1 \subset V_2 \subset \ldots \subset X$ be an increasing chain of subspaces of dimension $\dim V_k = k$ in a vector space X, and let $A : X \to X$ be a self-adjoint linear mapping with respect to a scalar product $(\cdot, \cdot)$ on X, i.e.*

$$(Au, v) = (u, Av) \quad \text{for all } u, v \in X ,$$

such that

$$A(V_k) \subset V_{k+1} \quad \text{and} \quad A(V_k) \not\subset V_k .$$

Then for each $p_1 \in V_1$, there exists a unique extension to an orthogonal system $\{p_k\}$ with $p_k \in V_k$ for all k and

$$(p_k, p_k) = (Ap_{k-1}, p_k) \quad \text{for all } k \geq 2 .$$

The family $\{p_k\}$ satisfies the three-term recurrence relation

$$p_k = (A + a_k)p_{k-1} + b_k p_{k-2} \quad \text{for } k = 2, 3, \ldots \tag{6.5}$$

with $p_0 := 0$ and

$$a_k := -\frac{(Ap_{k-1}, p_{k-1})}{(p_{k-1}, p_{k-1})} \ , \quad b_k := -\frac{(p_{k-1}, p_{k-1})}{(p_{k-2}, p_{k-2})} \ .$$

Proof. Completely analogous to theorem 6.2, where the self-adjoint operator $A : X \to X$ was multiplication by t,

$$t : \mathbf{P}_k \to \mathbf{P}_{k+1} \ , \quad P(t) \mapsto tP(t) \ .$$

The self-adjointness is used in the proof of Theorem 6.2 in the transition:

$$(tP_{k-1}, P_j) = (P_{k-1}, tP_j) \ .$$

□

A remarkable property of orthogonal polynomials is that they possess only real and simple roots, which lie in the interval $]a, b[$.

Theorem 6.5 *The orthogonal polynomials $P_k(t) \in \mathbf{P}_k$ possess exactly k simple roots in $]a, b[$.*

Proof. Let $t_1, \ldots, t_m$ be the m distinct points $t_i \in]a, b[$, at which P_k changes sign. The polynomial

$$Q(t) := (t - t_1)(t - t_2) \cdots (t - t_m)$$

then changes sign at the same points, so that the function $\omega(t)Q(t)P_k(t)$ does not change sign in $]a, b[$, and therefore

$$(Q, P_k) = \int_a^b \omega(t)Q(t)P_k(t) \, dt \neq 0 \ .$$

Since P_k is orthogonal to all polynomials $P \in \mathbf{P}_{k-1}$, it follows that $\deg Q = m \geq k$ as required. □

6.1.2 Homogeneous and non-homogeneous recurrence relations

Because of their importance, which became clear in the last section, we shall now study real three-term recurrence relations of the form

$$p_k = a_k p_{k-1} + b_k p_{k-2} + c_k \quad \text{for} \ \ k = 2, 3, \ldots \tag{6.6}$$

for values $p_k \in \mathbf{R}$ with coefficients $a_k, b_k, c_k \in \mathbf{R}$. We assume that $b_k \neq 0$ for all k, so that this actually is a three-term recurrence relation. Under this assumption, we can perform the recurrence relation backwards, i.e.

$$p_{k-2} = -\frac{a_k}{b_k} p_{k-1} + \frac{1}{b_k} p_k - \frac{c_k}{b_k} \quad \text{for} \ \ k = N, N-1, \ldots, 2 \ . \tag{6.7}$$

As in the trigonometric or the Bessel recurrence relation, it is often the case that $b_k = -1$ for all k, so that the three-term recurrence relation (6.7) can be obtained from the original one by interchanging p_k and p_{k-2}. We shall call such a three-term recurrence relation *symmetric*. If all c_k vanish, then the three-term recurrence relation is called *homogeneous*, and otherwise *non-homogeneous*. So far all of our examples were homogeneous and symmetric.

For each pair p_j, p_{j+1} of starting values, the three-term recurrence relation (6.6) determines exactly one sequence $p = (p_0, p_1, \ldots) \in \mathbf{R}^\mathbf{N}$. The solutions $p = (p_k)$ of the homogeneous three-term recurrence relation

$$p_k = a_k p_{k-1} + b_k p_{k-2} \quad \text{for} \quad k = 2, 3, \ldots \tag{6.8}$$

depend linearly on the starting values p_j, p_{j+1}, and they therefore form a two-dimensional subspace

$$\mathcal{L} := \{p \in \mathbf{R}^\mathbf{N} \mid p_k = a_k p_{k-1} + b_k p_{k-2} \quad \text{for} \quad k = 2, 3, \ldots\}$$

of $\mathbf{R}^\mathbf{N}$. Two solutions $p, q \in \mathcal{L}$ are *linearly independent*, if and only if the *Casorati determinants* of p and q,

$$D(k, k+1) := p_k q_{k+1} - q_k p_{k+1} , \tag{6.9}$$

do not vanish. It is easy to compute that

$$D(k, k+1) = -b_{k+1} D(k-1, k) ,$$

and, because $b_k \neq 0$, then either all $D(k, k+1)$ vanish or none do. In particular, for all symmetric recurrence relations, i.e. $b_k = -1$ for all k, we have

$$D(k, k+1) = D(0, 1) \quad \text{for all } k.$$

Example 6.6 For $x \notin \mathbf{Z}\pi$, the *trigonometric recurrence relation*

$$p_k = a_k p_{k-1} + b_k p_{k-2} , \quad a_k = 2 \cos x , \quad b_k = -1 ,$$

has the linearly independent solutions $\cos kx$ and $\sin kx$, since

$$D(0, 1) = \cos 0 \sin x - \sin 0 \cos x = \sin x \neq 0 .$$

If $x = l\pi$ with $l \in \mathbf{Z}$, then $D(0, 1) = 0$; and $\cos kx$ and $\sin kx$ would not be linearly independent. Instead

$$p_k = \cos kx = (-1)^{lk} , \quad q_k = k(-1)^{lk}$$

are two linearly independent solutions with $D(0, 1) = 1$. Note that this value of the Casorati determinant can obviously not be obtained by passing to the limit $x \to l\pi$, a theoretical weakness. In the following, we shall learn about a different characteristic quantity which does satisfy the required limiting property (compare Exercise 6.8).

We shall now try to put together the solution of the general *non-homogeneous* recurrence relation (6.6) from solutions of non-homogeneous recurrence relations, which are as simple as possible. In order to do this, we study how a single non-homogeneity $c_k = \delta_{jk}$ propagates at the position j.

Definition 6.7 Let $g^+(j,k)$ and $g^-(j,k)$ be the solutions of the non-homogeneous three-term recurrence relation

$$
\begin{aligned}
g^-(j,k) - a_k g^-(j,k-1) - b_k g^-(j,k-2) &= \delta_{jk} \\
g^+(j,k) - a_k g^+(j,k-1) - b_k g^+(j,k-2) &= -b_k \delta_{j,k-2}
\end{aligned}
$$

for $j, k \in \mathbf{N}$ and $k \geq 2$ with the starting values

$$
\begin{aligned}
g^-(j,j-2) &= g^-(j,j-1) = 0 \quad \text{resp.} \\
g^+(j,j+2) &= g^+(j,j+1) = 0.
\end{aligned}
$$

Then the *discrete Green's function* $g(j,k)$ of the three-term recurrence relation (6.6) is defined by

$$
g(j,k) := \begin{cases} g^-(j,k) & \text{if } k \geq j \\ g^+(j,k) & \text{if } k \leq j \end{cases}.
$$

Here note that $g^-(j,j) = g^+(j,j) = 1$. The solutions of the non-homogeneous recurrence relation (6.6) with the starting values $p_0 = c_0$ and $p_1 = c_1$ can now be obtained by superposition according to

$$
p_k = \sum_{j=0}^{k} c_j g(j,k) = \sum_{j=0}^{k} c_j g^-(j,k) \quad \text{for } k = 0, 1, \ldots \tag{6.10}
$$

(proof as an exercise). Conversely, for the backward recurrence relation (6.7), it follows that

$$
p_k = \sum_{j=k}^{N+1} c_j g(j,k) = \sum_{j=k}^{N+1} c_j g^+(j,k) \quad \text{for } k = 0, \ldots, N+1
$$

is the solution for the starting values $p_N = c_N$ and $p_{N+1} = c_{N+1}$.

Remark 6.8 Readers who are well versed in the theory of ordinary differential equations, may find the above method for difference equations familiar. In fact, the name 'discrete Green's function' is chosen analogous to the terminology used in differential equations. Similarly, the Casorati determinant corresponds to the *Wronski determinant*, and the special starting values, of the non-homogeneous differential equation which are defined by δ_{ij}, correspond to the δ distribution.

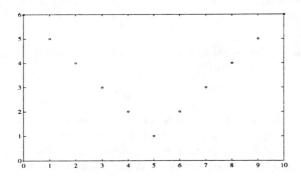

Figure 6.1: Discrete Green's function $g(5, k)$ over $k = 0, \ldots, 10$ for $a_k = 2$ and $b_k = -1$

6.2 Numerical Aspects

The mathematical structure of the three-term recurrence relation suggests a direct translation into an algorithm (simple loop). In Example 2.27, we have already seen that this way of computing special functions has to be treated with special care. At least in that case, it was possible to stabilize the trigonometric three-term recurrence relation numerically. The following example shows that this is not always possible.

Example 6.9 *Bessel's maze.* The *Bessel functions,* $J_k = J_k(x)$, satisfy the three-term recurrence relation

$$J_{k+1} = \frac{2k}{x} J_k - J_{k-1} \quad \text{for} \quad k \geq 1 . \tag{6.11}$$

We start, for example, with $x = 2.13$ and the values

$$
\begin{aligned}
J_0 &= 0.14960677044884 \\
J_1 &= 0.56499698056413 ,
\end{aligned}
$$

which can be taken from a table (e.g. [67]). At the end of the chapter we shall be able to confirm these values (see Exercise 6.7). We can now try to compute the values $J_2, \ldots, J_{23}$ by employing the three-term recurrence relation in forward mode. In order to 'verify' (see below) the results $\hat{J}_2, \ldots, \hat{J}_{23}$, we solve the recurrence relation (6.11) with respect to J_{k-1}, and insert $\hat{J}_{23}$ and $\hat{J}_{22}$ into the recurrence relation in backward mode. This way we get $\bar{J}_{21}, \ldots, \bar{J}_0$

back and actually expect that $\bar{J}_0$ coincides approximately with the starting value J_0. However, with a relative machine precision of eps $= 10^{-16}$, we obtain

$$\bar{J}_0/J_0 \approx 10^9 \ .$$

A comparison of the computed value $\hat{J}_{23}$ with the actual value J_{23} reveals that it is much worse, namely

$$\hat{J}_{23}/J_{23} \approx 10^{27} \ ,$$

i.e. the result misses by several orders of magnitude! In Figure 6.2, we have plotted the repetition of this procedure, i.e. the renewed start with $\bar{J}_0$ etc.: Numerically, one does not find the way back to the starting value, hence this phenomenon is called *Bessel's maze*. What happened? A first analysis of

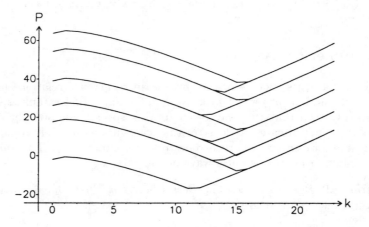

Figure 6.2: Bessel's maze for $x = 2.13$, $\ln(|\hat{J}_k(x)|)$ is plotted over k for 5 loops until $k = 23$

the behavior of the rounding errors shows that

$$\frac{2k}{x} J_k \approx J_{k-1} \quad \text{for } k > x$$

(compare Table 6.1). Thus cancellation occurs in the forward recurrence relation every time when J_{k+1} is computed (see Exercise 6.9). Moreover, besides the Bessel functions J_k, the *Neumann functions* Y_k also satisfy the same recurrence relation (Bessel and Neumann functions are called cylinder functions). However these possess an opposite growth behavior. The

Table 6.1: Cancellation in the three-term recurrence relation for the Bessel functions $J_k = J_k(x)$, $x = 2.13$

k	J_{k-1}	$\frac{2k}{x} J_k$
1	$1.496 \cdot 10^{-1}$	$5.305 \cdot 10^{-1}$
2	$5.649 \cdot 10^{-1}$	$7.153 \cdot 10^{-1}$
3	$3.809 \cdot 10^{-1}$	$4.234 \cdot 10^{-1}$
4	$1.503 \cdot 10^{-1}$	$1.597 \cdot 10^{-1}$
5	$4.253 \cdot 10^{-2}$	$4.425 \cdot 10^{-2}$
6	$9.425 \cdot 10^{-3}$	$9.693 \cdot 10^{-3}$
7	$1.720 \cdot 10^{-3}$	$1.756 \cdot 10^{-3}$
8	$2.672 \cdot 10^{-4}$	$2.716 \cdot 10^{-4}$
9	$3.615 \cdot 10^{-5}$	$3.662 \cdot 10^{-5}$
10	$4.333 \cdot 10^{-6}$	$4.379 \cdot 10^{-6}$

Bessel functions decrease when k increases, whereas the Neumann functions increase rapidly. It is through the input error for J_0 and J_1 (in the order of magnitude of machine precision),

$$\tilde{J}_0 = J_0 + \varepsilon_0 Y_0 , \quad \tilde{J}_1 = J_1 + \varepsilon_1 Y_1 ,$$

that the input $\tilde{J}_0, \tilde{J}_1$ always contains a portion of the Neumann function Y_k, which at first is very small, but which in the course of the recurrence increasingly overruns the Bessel function. Conversely in the backward direction, the Bessel functions superimpose the Neumann functions.

In the following section we shall try to understand the observed numerical phenomena.

6.2.1   Condition numbers

We view the three-term recurrence relation (6.6) as a mapping which relates the starting values p_0, p_1 and the a_k, b_k as input quantities to the values $p_2, p_3, \ldots$ as resulting quantities. Only two multiplications and one addition have to be carried out in each step, and we have verified the stability of these operations in Lemma 2.19. The execution of the three-term recurrence relation in floating point arithmetic is therefore *stable*. Thus only the *condition number* of the three-term recurrence relation determines whether it is numerically useful. In order to analyze the numerical usefulness, we prescribe perturbed starting values

$$\tilde{p}_0 = p_0(1 + \theta_0) , \quad \tilde{p}_1 = p_1(1 + \theta_1)$$

and perturbed coefficients

$$\tilde{a}_k = a_k(1 + \alpha_k) , \quad \tilde{b}_k = b_k(1 + \beta_k) \quad \text{for} \quad k \geq 2 ,$$

whose errors are bounded by $\delta > 0$,

$$|\theta_0|, |\theta_1|, |\alpha_k|, |\beta_k| \leq \delta ,$$

and we compute the error

$$\Delta p_k := \tilde{p}_k - p_k ,$$

where $\tilde{p}$ is the solution of the perturbed three-term recurrence relation. By employing the recursion for p and $\tilde{p}$, it turns out that Δp satisfies the *non-homogeneous* recurrence relation

$$\Delta p_k = a_k \Delta p_{k-1} + b_k \Delta p_{k-2} + E_k \quad \text{for} \quad k \geq 2 \qquad (6.12)$$

with the starting values $\Delta p_0 = E_0 := \theta_0 p_0$, $\Delta p_1 = E_1 := \theta_1 p_1$ and coefficients

$$E_k = \alpha_k a_k \tilde{p}_{k-1} + \beta_k b_k \tilde{p}_{k-2} \doteq \alpha_k a_k p_{k-1} + \beta_k b_k p_{k-2} \quad \text{for} \quad \delta \to 0 .$$

By utilizing the Green's function as in (6.10), we obtain

$$\Delta p_k = \sum_{j=0}^{k} E_j g(j, k) .$$

The Green's function thus characterizes the *absolute condition* of the three-term recurrence relations. Similarly, it follows that the relative error

$$\theta_k := \frac{\tilde{p}_k - p_k}{p_k} = \frac{\Delta p_k}{p_k} , \quad p_k \neq 0 ,$$

is the solution of the non-homogeneous recurrence relation

$$\theta_k = \frac{a_k p_{k-1}}{p_k} \theta_{k-1} + \frac{b_k p_{k-2}}{p_k} \theta_{k-2} + \varepsilon_k \quad \text{for} \quad l \geq 2$$

with the starting values $\varepsilon_0 := \theta_0$, $\varepsilon_1 := \theta_1$, where

$$\varepsilon_k := \frac{E_k}{p_k} \doteq \alpha_k \frac{a_k p_{k-1}}{p_k} + \beta_k \frac{b_k p_{k-2}}{p_k} ,$$

and we therefore have

$$\theta_k = \sum_{j=0}^{k} \varepsilon_j r(j,k) \quad \text{with} \quad r(j,k) := \frac{p_j}{p_k} g(j,k). \tag{6.13}$$

The functions $r(j,k)$ obviously describe the propagation of the relative errors and characterize therefore the *relative condition* of the three-term recurrence relations. Motivated by the Bessel and Neumann functions, we distinguish between two types of solutions in order to judge $r(j,k)$.

Definition 6.10 *A solution $p \in \mathcal{L}$ is called* recessive *or a* minimal solution, *if for each solution $q \in \mathcal{L}$ which is linearly independent of p we have*

$$\lim_{k \to \infty} \frac{p_k}{q_k} = 0 \ .$$

The solutions q which are linearly independent of p are called dominant.

It is clear that the minimal solution is only uniquely determined up to a scalar factor. In many cases the free factor is determined by a normalization condition

$$G_\infty := \sum_{k=0}^{\infty} m_k p_k = 1 \tag{6.14}$$

with the weights m_k. Conversely, such relations generally hint that the corresponding solutions p_k are minimal. If they exist, then the minimal solutions form a one dimensional subspace of $\mathcal{L}$. The existence can be guaranteed by imposing certain assumptions on the coefficients a_k and b_k.

Theorem 6.11 *Suppose that the three-term recurrence relation is symmetric, i.e. $b_k = -1$ for all k, and that there exists a $k_0 \in \mathbf{N}$ such that*

$$|a_k| \geq 2 \quad \text{for all} \quad k > k_0 \ .$$

Then there is a minimal solution p with the properties

$$|p_k| \leq \frac{1}{|a_{k+1}| - 1} |p_{k-1}| \quad \text{and} \quad p_{k+1}(x) \neq 0 \tag{6.15}$$

for all $k > k_0$. Furthermore, for each dominant solution q, there is an index $k_1 \geq k_0$, such that

$$|q_{k+1}| > (|a_{k+1}| - 1)|q_k| \quad \text{for} \quad k > k_1 \ . \tag{6.16}$$

Proof. The proof is by *continued fractions* and can be found in MEIXNER and SCHÄFFKE [56]. □

Example 6.12 The three-term recurrence relations of the trigonometric functions
$\cos kx, \sin kx$ satisfies $b_k = -1$ and

$$|a_k| = 2|\cos x| \geq 2 \iff x = l\pi \quad \text{with} \quad l \in \mathbf{Z} .$$

If $x = l\pi \in \mathbf{Z}\pi$, then $p_k = (-1)^{lk}$ is a minimal solution, and the sequences

$$q_k = \beta k(-1)^{lk} + \alpha p_k \quad \text{with} \quad \beta \neq 0$$

are dominant solutions.

Example 6.13 For the recurrence relations of the cylinder functions, we have $b_k = -1$ and

$$|a_k| = 2\frac{k-1}{|x|} \geq 2 \Leftrightarrow k > k_0 := [|x|] .$$

The minimal solution is the Bessel function J_k, whereas the Neumann function Y_k is dominant. This can be proved by invoking the asymptotic approximations for J_k respectively Y_k for $k \to \infty$, because

$$J_k(x) \sim \frac{1}{\sqrt{2\pi k}} \left(\frac{ex}{2k}\right)^k \quad , \quad Y_k(x) \sim -\sqrt{\frac{2}{\pi k}} \left(\frac{ex}{2k}\right)^{-k} .$$

The Bessel functions $J_k(x)$ satisfy the normalization condition (see e.g. [1])

$$G_\infty := J_0 + 2\sum_{k=1}^{\infty} J_{2k} = 1 .$$

Under the assumptions of Theorem 6.11, it can be shown that

$$|g(j,k)| \geq |k - j + 1| \quad \text{for all} \quad k \geq j > k_0 .$$

So the discrete Green's functions $g(j,k)$ are themselves dominant solutions and increase beyond any bounds. On the other hand, because of (6.15), a *minimal solution* p satisfies $|p_j/p_k| \geq 1$, and therefore

$$|r(j,k)| = \left|\frac{p_j}{p_k}g(j,k)\right| \geq |g(j,k)| \geq |k - j + 1|$$

for all $k \geq j > k_0$. Beginning with the index k_0, the three-term recurrence relation is thus *ill-conditioned* for the computation of a minimal solution. For dominant solutions, the growth of the discrete Green's functions can be compensated by the growth of the solution itself, so that the relative error amplification, which is expressed by $r(j,k)$, stays moderate, and the three-term recurrence relation is well-conditioned. Thus the three-term recurrence relation (in forward direction) is ill-conditioned for the Bessel functions as minimal solution, but well-conditioned for the Neumann functions.

Example 6.14 *Spherical harmonics.* We now harvest what was presented above by giving a more complicated example, which plays an important role in many applications, for example in theoretical physics or geodesy. In general, one has to compute expansions with respect to spherical harmonics, as well as entire sets of spherical harmonics. They are usually denoted by $Y_k^l(\theta, \varphi)$, where the Euler angles, θ and φ, are variables on the sphere subject to

$$0 \le \theta \le \pi \ \text{ and } \ 0 \le \varphi \le 2\pi \ .$$

Among the numerous representations of spherical harmonics, we choose the complex representation

$$Y_k^l(\theta, \varphi) = e^{il\varphi} P_k^l(\cos\theta) = C_k^l(\theta, \varphi) + i S_k^l(\theta, \varphi) \ ,$$

where $P_k^l(x)$ denotes the so-called *associated Legendre functions of the first kind* for $|x| \le 1$. They can be given explicitly as follows:

$$P_k^l(x) := \frac{(-1)^{k+l}}{(k+l)! \, k! \, 2^k}(1-x^2)^{\frac{l}{2}}\frac{d^{k+l}}{dx^{k+l}}(1-x^2)^k \ . \tag{6.17}$$

Among the numerous normalizations of these functions, which appear in the literature, we have chosen the one according to [34]. Using the relations

$$P_k^l(x) \equiv 0 \ \text{ for } \ l > k \ge 0 \ \text{ and } \ l < -k \le 0 \tag{6.18}$$

and

$$P_k^{-l}(x) = (-1)^l P_k^l(x) \text{ for } \ l > 0, \tag{6.19}$$

it is sufficient to compute the real spherical harmonics

$$\begin{aligned} C_k^l(\theta,\varphi) &= P_k^l(\cos\theta)\cos(l\varphi) \ \text{ for } \ 0 \le l \le k \\ S_k^l(\theta,\varphi) &= P_k^l(\cos\theta)\sin(l\varphi) \ \text{ for } \ 0 < l \le k. \end{aligned}$$

We have earlier discussed the three-term recurrence relations for the trigonometric functions in great detail. We therefore draw our attention to the Legendre functions for the argument $x = \cos\theta$. All three-term recurrence relations, which are valid for these doubly indexed Legendre functions of the first kind (see e.g. [38]), are also valid for the Legendre functions of the second kind, which, in contrast to those of the first kind, have singularities of order l at $x = 1$ and $x = -1$. This property carries directly over to the corresponding discrete Green's functions (compare Exercise 6.8). Thus recurrence relations with variable l would be ill-conditioned for the P_k^l. Consequently, among the many three-term recurrence relations, we choose those with *constant* l. This leads to the recurrence relation

$$P_k^l = \frac{(2k-1)x P_{k-1}^l - P_{k-2}^l}{(k-l)(k+l)} \ , \tag{6.20}$$

with the only running index k. It is well-conditioned for the P_k^l in forward direction with respect to k (see e.g. the survey-like paper by GAUTSCHI [34]). Still missing is a well-conditioned interlink for different l. From Definition (6.17), we obtain for $k = l$ the representation

$$P_l^l(x) = \frac{(-1)^l}{2^l \cdot l!}(1 - x^2)^{\frac{l}{2}} \, ,$$

which leads immediately to the two-term recurrence relation

$$P_l^l = -\frac{(1 - x^2)^{\frac{1}{2}}}{2l} P_{l-1}^{l-1}, \tag{6.21}$$

which as such is well-conditioned. In order to start the recurrence relations, the value $P_0^0 = 1$ is used for (6.21), and for (6.20) we use the recurrence relation for $k = l + 1$, which, because of $P_{l-1}^l \equiv 0$ and according to (6.18), also degenerates into a two-term recurrence relation:

$$P_{l+1}^l = xP_l^l \, . \tag{6.22}$$

If we replace the argument x by $\cos\theta$, then we expect, by the results of Section 2.3, that the corresponding algorithm is *numerically unstable*. As in the stabilization of the trigonometric recurrence relation, we here try again to replace the argument $\cos\theta$ by $1 - \cos\theta = -2\sin^2(\theta/2)$ for $\theta \to 0$. Unfortunately, the stabilization is not as easily accomplished as for the trigonometric functions. We therefore seek solutions of the form:

$$P_k^l = q_k^l \bar{P}_k^l \quad \text{and} \quad \bar{P}_k^l = r_k^l \bar{P}_{k-1}^l + \Delta P_k^l$$

with suitably chosen transformations q_k^l and r_k^l. Observe that the relative condition numbers in (6.13) are the same for P_k^l and $\bar{P}_k^l$, independent of the choice of the transformations q_k^l. Insertion of P_k^l and P_{k-2}^l into (6.20) then gives

$$q_k^l \Delta P_k^l = \sigma_k(\theta)\bar{P}_{k-1}^l + \frac{q_{k-2}^l}{(k - l)(k + l)r_{k-1}^l} \cdot \Delta P_{k-1}^l \, ,$$

where

$$\sigma_k(\theta) := \frac{(2k - 1)\cos\theta}{(k - l)(k + l)} \cdot q_{k-1}^l - q_k^l r_k^l - \frac{q_{k-2}^l}{(k - l)(k + l)\, r_{k-1}^l} \, .$$

In order for the expression $(1 - \cos\theta)$ to be a factor in $\sigma_k(\theta)$, $\theta = 0$ obviously has to be a root of σ_k, i.e. $\sigma_k(0) = 0$. Because of (6.22), we require in addition that $q_{l+1}^l r_{l+1}^l = 1$. These two requirements regarding the transformations q_k^l and r_k^l are satisfied by the choice

$$q_k^l = \frac{1}{r_k^l} = k - l + 1 \, .$$

With this transformation, and by starting with $P_0^0 = \bar{P}_0^0 = 1$, one obtains the following numerically stable recurrence relation representation:

Algorithm 6.15 Computation of the spherical harmonics $P_k^l(\cos\theta)$ for $l = 0, \ldots, L$ and $k = l, \ldots, K$.

$\quad P_0^0 := \bar{P}_0^0 := 1;$

$\quad$ **for** $l := 0$ **to** L **do**

$$P_{l+1}^{l+1} := \bar{P}_{l+1}^{l+1} := -\frac{\sin\theta}{2(l+1)}P_l^l;$$

$$\Delta P_l^l := -\sin^2(\theta/2)P_l^l;$$

$\qquad$ **for** $k := l+1$ **to** K **do**

$$\Delta P_k^l := \frac{(k-l-1)\Delta P_{k-1}^l - 2(2k-1)\sin^2(\theta/2)\bar{P}_{k-1}^l}{(k+l)(k-l+1)};$$

$$\bar{P}_k^l := \frac{1}{(k-l+1)}\bar{P}_{k-1}^l + \Delta P_k^l;$$

$$P_k^l := (k-l+1)\bar{P}_k^l;$$

$\qquad$ **end for**

$\quad$ **end for**

Remark 6.16 For the successful computation of orthogonal polynomials, one obviously needs a kind of 'look-up table of condition numbers' for as many orthogonal polynomials as possible. A first step in this direction is the paper [34]. However, the numerically necessary information is in many cases more hidden than published. Moreover, the literature often does not clearly distinguish between the notions 'stability' and 'condition'.

6.2.2   Idea of the Miller algorithm

Do we have to abandon the three-term recurrence relation for the computation of a minimal solution because of the above error analysis? This is not the case, as we shall show here. The remedy, which is due to MILLER [57], is based on two ideas. The first consists in analyzing the three-term recurrence relation in backward mode with the starting values p_n, p_{n+1} with respect to its condition. By carrying the above considerations over to this case (see Exercise 6.5), it can be shown that the three-term recurrence relation is well-conditioned for a minimal solution in backward mode. The second idea consists of utilizing the *normalization condition* (6.14). Since the minimal solutions $p_k(x)$ become arbitrarily small in absolute value for $k \to \infty$, G_∞ can be approximated by the finite partial sums

$$G_n := \sum_{k=0}^{n} m_k p_k .$$

By computing an arbitrary solution $\hat{p}_k$ of the three-term recurrence relation in backward mode, e.g. with the starting values $\hat{p}_{n+1} = 0$ and $\hat{p}_n = 1$, and normalizing these with the help of G_n, one obtains for increasing n increasingly better approximations of the minimal solution. These considerations motivate the following algorithm for the computation of p_N with a relative precision ε.

Algorithm 6.17 MILLER *algorithm for the computation of a minimal solution* p_N.

1. Choose a break-off index $n > N$ and put

$$\hat{p}_{n+1}^{(n)} := 0 , \quad \hat{p}_n^{(n)} := 1 .$$

2. Compute $\hat{p}_{n-1}^{(n)}, \ldots, \hat{p}_0^{(n)}$ from

$$\hat{p}_{k-2}^{(n)} = \frac{1}{b_k}(\hat{p}_k^{(n)} - a_k \hat{p}_{k-1}^{(n)}) \quad \text{for} \quad k = n+1, \ldots, 2 .$$

3. Compute

$$\hat{G}_n := \sum_{k=0}^{n} m_k \hat{p}_k .$$

4. Normalize according to

$$p_k^{(n)} := \hat{p}_k^{(n)}/\hat{G}_n .$$

5. Repeat steps 1. to 4. for increasing $n = n_1, n_2, \ldots$, and while doing this, test the accuracy by comparing $p_N^{(n_i)}$ and $p_N^{(n_{i-1})}$. If

$$|p_N^{(n_i)} - p_N^{(n_{i-1})}| \leq \varepsilon \, p_N^{(n_i)} ,$$

then $p_N^{(n_i)}$ is a sufficiently accurate approximation of p_N.

In the following theorem it is shown that this algorithm converges indeed .

Theorem 6.18 *Let $p \in \mathcal{L}$ be a minimal solution of a homogeneous three-term recurrence relation, which satisfies the normalization condition*

$$\sum_{k=0}^{\infty} m_k p_k = 1 .$$

In addition, it is assumed that there is a dominant solution $q \in \mathcal{L}$ such that

$$\lim_{n \to \infty} \frac{p_{n+1}}{q_{n+1}} \sum_{k=0}^{n} m_k q_k = 0 .$$

Then the sequence of Miller approximations $p_N^{(n)}$ converges to p_N,

$$\lim_{n \to \infty} p_N^{(n)} = p_N .$$

Proof.   The solution $\hat{p}_k^{(n)}$ of the three-term recurrence relation with the starting values $\hat{p}_n^{(n)} := 1$ and $\hat{p}_{n+1}^{(n)} := 0$ can be represented as a linear combination of p_k and q_k, because

$$\hat{p}_k^{(n)} = \frac{p_k q_{n+1} - q_k p_{n+1}}{p_n q_{n+1} - q_n p_{n+1}} = \frac{q_{n+1}}{p_n q_{n+1} - q_n p_{n+1}} p_k - \frac{p_{n+1}}{p_n q_{n+1} - q_n p_{n+1}} q_k .$$

This implies

$$\hat{G}_n := \sum_{k=0}^{n} m_k \hat{p}_k^{(n)} = \frac{q_{n+1}}{p_n q_{n+1} - q_n p_{n+1}} \left(\sum_{k=0}^{n} m_k p_k - \frac{p_{n+1}}{q_{n+1}} \sum_{k=0}^{n} m_k q_k \right) .$$

For the Miller approximations $p_N^{(n)} = \hat{p}_N^{(n)} / \hat{G}_n$, this yields

$$p_N^{(n)} = \Big(p_N - \underbrace{\frac{p_{n+1}}{q_{n+1}}}_{\to 0} q_N \Big) \Big(\underbrace{\sum_{k=0}^{n} m_k p_k}_{\to 1} - \underbrace{\frac{p_{n+1}}{q_{n+1}} \sum_{k=0}^{n} m_k q_k}_{\to 0} \Big) \to p_N$$

for $n \to \infty$. $\square$

This algorithm, which was developed by Miller in 1952, is now out-dated, since it uses too much storage space and computation cost. The idea however enters into a more effective algorithm, which we shall present in the next chapter.

6.3   Adjoint Summation

We turn our attention again towards the original problem, namely to evaluate linear combinations of the form

$$f(x) = S_N = \sum_{k=0}^{N} \alpha_k p_k, \qquad (6.23)$$

where $p_k := p_k(x)$ satisfies a homogeneous three-term recurrence relation (6.8) with starting values p_0, p_1 and given coefficients α_k. The subsequent presentation essentially follows the lines of [16] and [17]. As an illustration we start with a two-term recurrence relation.

Example 6.19 The evaluation of a polynomial

$$S_N := p(x) = \alpha_0 + \alpha_1 x + \cdots + \alpha_N x^N$$

can be considered as a computation of a linear combination (6.23), where $p_k := x^k$ satisfies the two-term recurrence relation

$$p_0 := 1, \quad p_k := x p_{k-1} \quad \text{for} \quad k \geq 1 \,.$$

This sum can be computed in two different ways. The direct way is the *forward recurrence relations*

$$p_k := x p_{k-1} \quad \text{and} \quad S_k := S_{k-1} + \alpha_k p_k \qquad (6.24)$$

for $k = 1, \ldots, N$ with starting values $S_0 := \alpha_0$ and $p_0 := 1$. It corresponds to the naive evaluation of a polynomial. But one can factor cleverly

$$S_N = \alpha_0 + x(\alpha_1 + x(\cdots (\alpha_{N-1} + x \underbrace{\alpha_N}_{u_N}) \cdots))$$

$$\underbrace{\phantom{\alpha_{N-1} + x \alpha_N}}_{u_{N-1}}$$

and compute the sum by the *backward recurrence relation*

$$
\begin{aligned}
u_{N+1} &:= 0 \\
u_k &:= x u_{k+1} + \alpha_k \quad \text{for} \quad k = N, N-1, \ldots, 0 \qquad (6.25) \\
S_N &:= u_0.
\end{aligned}
$$

This is the *Horner algorithm*. Compared with the first algorithm (6.25), it saves N multiplications and is therefore approximately twice as fast.

6.3.1 Summation of dominant solutions

An obvious way of computing the sum (6.23) consists of computing the values p_k in each step by the three-term recurrence relation

$$p_k = a_k p_{k-1} + b_k p_{k-2}, \tag{6.26}$$

multiplying them by the coefficients α_k, and adding the result up. The resulting algorithm corresponds to the forward recurrence relation (6.24):

$$p_k := a_k p_{k-1} + b_k p_{k-2} \quad \text{and} \quad S_k := S_{k-1} + \alpha_k p_k \quad \text{for} \quad k = 2, \ldots, N$$

with the starting value $S_1 := \alpha_0 p_0 + \alpha_1 p_1$. We wonder, if the procedure that we used in the derivation of the Horner algorithm, carries over to the case at hand. In order to construct an analogue to "factoring", we extend the three-term recurrence relation (6.8) by the two trivial equations $p_0 = p_0$ and $p_1 = p_1$. For the computation of the values $p = (p_0, \ldots, p_N)$, we thus obtain the triangular system

$$\underbrace{\begin{bmatrix} 1 & & & & \\ & 1 & & & \\ -b_2 & -a_2 & 1 & & \\ & \ddots & \ddots & \ddots & \\ & & -b_N & -a_N & 1 \end{bmatrix}}_{=:\, L} \underbrace{\begin{pmatrix} p_0 \\ \vdots \\ \vdots \\ \vdots \\ p_N \end{pmatrix}}_{=:\, p} = \underbrace{\begin{pmatrix} p_0 \\ p_1 \\ 0 \\ \vdots \\ 0 \end{pmatrix}}_{=:\, r}$$

with a unipotent lower triangular matrix $L \in \mathrm{Mat}_{N+1}(\mathbf{R})$ and the trivial right hand side r. The linear combination S_N is just the (Euclidean) scalar product

$$S_N = \sum_{k=0}^{N} \alpha_k p_k = \langle \alpha, p \rangle \quad \text{with} \quad Lp = r$$

of p and $\alpha = (\alpha_0, \ldots, \alpha_N)^T$. Therefore

$$S_N = \langle \alpha, L^{-1} r \rangle = \langle L^{-T} \alpha, r \rangle . \tag{6.27}$$

If u denotes the solution of the (adjoint) triangular system $L^T u = \alpha$, i.e. $u := L^{-T} \alpha$, then it follows that

$$S_N = \langle u, r \rangle = u_0 p_0 + u_1 p_1 .$$

Explicitly, u is the solution of

$$
\begin{bmatrix}
1 & -b_2 & & & \\
& 1 & -a_2 & \ddots & \\
& & 1 & \ddots & -b_N \\
& & & \ddots & -a_N \\
& & & & 1
\end{bmatrix}
\begin{pmatrix}
u_0 \\ \vdots \\ \vdots \\ \vdots \\ u_N
\end{pmatrix}
=
\begin{pmatrix}
\alpha_0 \\ \vdots \\ \vdots \\ \vdots \\ \alpha_N
\end{pmatrix}.
$$

By solving this triangular system of equations, we obtain the desired analogue of the algorithm (6.25):

$$
\begin{aligned}
u_k &:= a_{k+1}u_{k+1} + b_{k+2}u_{k+2} + \alpha_k \quad \text{for} \quad k = N, N-1, \ldots, 1 \\
u_0 &:= b_2 u_2 + \alpha_0 \\
S_N &:= u_0 p_0 + u_1 p_1 ,
\end{aligned}
\tag{6.28}
$$

where we set $u_{N+1} = u_{N+2} := 0$. The homogeneous three-term recurrence relation, which is defined by $L^T u = 0$, is called the *recurrence relation* adjoint to (6.26). Similarly, because of the relation (6.27), we call (6.28) the *adjoint summation* for $S_N = \sum_{k=0}^{N} \alpha_k p_k$. Compared with the first algorithm, we here save again N multiplications.

Example 6.20 The adjoint summation, as applied to the partial Fourier sums

$$
S_N := \sum_{k=1}^{N} \alpha_k \sin kx \quad \text{resp.} \quad C_N := \sum_{k=0}^{N} \alpha_k \cos kx
$$

with the trigonometric three-term recurrence relation

$$
p_k = 2\cos x \cdot p_{k-1} - p_{k-2}
$$

for $s_k = \sin kx$ and $c_k = \cos kx$ and the starting values

$$
s_0 = 0 , \quad s_1 = \sin x \quad \text{resp.} \quad c_0 = 1 , \quad c_1 = \cos x ,
$$

leads to the recurrence relation

$$
\begin{aligned}
u_{N+2} &= u_{N+1} = 0 \\
u_k &= 2\cos x \cdot u_{k+1} - u_{k+2} + \alpha_k \quad \text{for} \quad k = N, \ldots, 1
\end{aligned}
\tag{6.29}
$$

and to the results

$$
S_N = u_1 \sin x \quad \text{resp.} \quad C_N = \alpha_0 + u_1 \cos x - u_2 .
$$

This algorithm, which GOERTZEL [38] devised in 1958, is unstable for $x \to l\pi$, $l \in \mathbf{Z}$, like the three-term recurrence relation on which it is based (as an algorithm for the mapping $x \mapsto \cos kx$, compare Example 2.27). However, the recurrence relation (6.29) can be stabilized by introducing the differences $\Delta u_k = u_k - u_{k+1}$ for $\cos x \geq 0$, and by changing to a system of two-term recurrence relation. As in Example 2.27, we obtain the following stable form of the recurrence relation for $k = N, N-1, \ldots, 1$, the algorithm of Goertzel and Reinsch:

$$\begin{aligned} \Delta u_k &= -4\sin^2(x/2) \cdot u_{k+1} + \Delta u_{k+1} + \alpha_k \\ u_k &= u_{k+1} + \Delta u_k \end{aligned}$$

with the starting values $u_{N+1} = \Delta u_{N+1} = 0$. For the sums we obtain

$$S_N = u_1 \sin x \quad \text{and} \quad C_N = \alpha_0 - 2\sin^2(x/2)u_1 + \Delta u_1 \ .$$

As in the error analysis of the three-term recurrence relation, the execution of the additions and multiplications of the adjoint summation is stable; the resulting errors can be interpreted as modifications of the input errors in the coefficients a_k, b_k and α_k. Only the condition determines the numerical usefulness; however, note that the condition is independent of the algorithmic realization. It is therefore true that if the original recurrence relation (6.26) is well-conditioned in forward mode, then this is also the case for the adjoint three-term recurrence relation (6.28), which of course runs backward. The algorithm (6.28) is thus only suitable for the summation of *dominant* solutions.

Example 6.21 *Adjoint summation of spherical harmonics.* To illustrate this, we consider the evaluation of expansions of the form

$$C(K, L; \theta, \varphi) := \sum_{l=0}^{L} \cos(l\varphi) \sum_{k=l}^{K} A_k^l P_k^l(\cos\theta)$$

$$S(K, L; \theta, \varphi) := \sum_{l=1}^{L} \sin(l\varphi) \sum_{k=l}^{K} A_k^l P_k^l(\cos\theta) \ ,$$

where $P_k^l(x)$ are the Legendre functions of the first kind as introduced in Example 6.14. Negative indices l are omitted here because of $P_k^{-l} = (-1)^l P_k^l$. A set of well-conditioned recurrence relations was already given in Example 6.14. Also in this situation, the application of similar stabilization techniques yields an economical numerically stable version. After a few intermediate calculations, one obtains the following algorithm for $\cos\theta \geq 0$ and $\cos\varphi \geq 0$:

Algorithm 6.22 *Computation of the sums $C(K, L; \theta, \varphi)$ and $S(K, L; \theta, \varphi)$.*

$V := 0;\ \Delta V := 0;$

for $l := L$ **to** 1 **step** -1 **do**

   $U := 0;\ \Delta U := 0;$

   for $k := K$ **to** $l+1$ **step** -1 **do**

      $\Delta U := (A_k^l - 2(2k+1)\sin^2(\theta/2)\,U + \Delta U)/(k-l);$

      $U\quad := (U + \Delta U)/(k+l);$

   end for

   $U_l\quad := A_l^l - 2(2l+1)\sin^2(\theta/2)\,U + \Delta U;$

   if $l > 0$ **then**

      $\Delta V\quad := U_l - \dfrac{\sin\theta}{2(l+1)}\left(-4\sin^2(\varphi/2)V + \Delta V\right);$

      $V\quad := -\dfrac{\sin\theta}{2(l+1)}V + \Delta V;$

   end

end for

$C(K,L;\theta,\varphi) := U_0 - \dfrac{\sin\theta}{2}\left(-2\sin^2(\varphi/2)V + \Delta V\right);$

$S(K,L;\theta,\varphi) := -\dfrac{1}{2}\sin\theta\sin\varphi \cdot V;$

6.3.2   Summation of minimal solutions

For the summation of minimal solutions, like e.g. the Bessel functions, we go back to the idea of the Miller algorithm. We again assume that the minimal solution $p \in \mathcal{L}$, which we have to compute, satisfies the normalization condition (6.14). In this case one can, in principle, derive from the MILLER algorithm 6.17 a method to compute the approximations of the sum S_N:

$$S_N^{(n)} = \sum_{k=0}^{N} \alpha_k \hat{p}_k^{(n)}/\hat{G}_n^{(n)} \quad \text{with} \quad \hat{G}_n^{(n)} := \sum_{k=0}^{n} m_k \hat{p}_k^{(n)} \ .$$

Under the assumptions of Theorem 6.18, we then have

$$\lim_{n\to\infty} S_N^{(n)} = S_N \ .$$

However, the cost of computing the $S_N^{(n)}$ would be quite high, since for each new n all values have to be computed again. Can that be avoided by employing some kind of adjoint summation? In order to answer this question, we proceed as in the derivation of the previous section, and we describe for given $n > N$ one step of the Miller algorithm through a linear system $M_n p^{(n)} = r^{(n)}$:

$$
\underbrace{\begin{bmatrix}
b_2 & a_2 & -1 & & & \\
& \ddots & \ddots & \ddots & & \\
& & b_n & a_n & -1 & \\
& & & b_{n+1} & a_{n+1} & \\
m_0 & \cdots & \cdots & \cdots & m_n
\end{bmatrix}}_{=:\, M_n \,\in\, \mathrm{Mat}_{n+1}(\mathbf{R})}
\underbrace{\begin{pmatrix}
p_0^{(n)} \\ \vdots \\ \vdots \\ \vdots \\ p_n^{(n)}
\end{pmatrix}}_{=:\, p^{(n)}}
=
\underbrace{\begin{pmatrix}
0 \\ \vdots \\ \vdots \\ 0 \\ 1
\end{pmatrix}}_{=:\, r^{(n)}}.
$$

With $\alpha^{(n)} := (\alpha_0, \ldots, \alpha_N, 0, \ldots, 0)^T \in \mathbf{R}^{n+1}$, the sum $S_N^{(n)}$ can again be written as a scalar product

$$
S_N^{(n)} = \sum_{k=0}^N \alpha_k p_k^{(n)} = \langle \alpha^{(n)}, p^{(n)} \rangle \quad \text{with} \quad M_n p^{(n)} = r^{(n)} .
$$

If we assume that M_n is invertible (otherwise $\hat{G}_n^{(n)} = 0$ and the normalization cannot be carried out), then it follows that

$$
S_N^{(n)} = \langle \alpha^{(n)}, M_n^{-1} r^{(n)} \rangle = \langle M_n^{-T} \alpha^{(n)}, r^{(n)} \rangle . \tag{6.30}
$$

By setting $u^{(n)} := M_n^{-T} \alpha^{(n)}$, we have

$$
S_N^{(n)} = \langle u^{(n)}, r^{(n)} \rangle = u_n^{(n)} ,
$$

where $u^{(n)}$ is the solution of the system $M_n^T u^{(n)} = \alpha^{(n)}$. More explicitly:

$$
\begin{bmatrix}
b_2 & & & & m_0 \\
a_2 & \ddots & & & \vdots \\
-1 & \ddots & b_n & & \vdots \\
& \ddots & a_n & b_{n+1} & \vdots \\
& & -1 & a_{n+1} & m_n
\end{bmatrix}
\begin{pmatrix}
u_0^{(n)} \\ \vdots \\ \vdots \\ \vdots \\ u_n^{(n)}
\end{pmatrix}
=
\begin{pmatrix}
\alpha_0^{(n)} \\ \vdots \\ \vdots \\ \vdots \\ \alpha_n^{(n)}
\end{pmatrix}.
$$

We solve this system by the Gaussian elimination method. The arising computations and results are listed in the following theorem:

Theorem 6.23 *Define* $e^{(n)} = (e_0, \ldots, e_n)$ *and* $f^{(n)} = (f_0, \ldots, f_n)$ *by*

a) $e_{-1} := 0$, $e_0 := \alpha_0/b_2$ *and*

$$e_{k+1} := \frac{1}{b_{k+3}}(\alpha_{k+1} + e_{k-1} - a_{k+2}e_k) \quad for \quad k = 0, \ldots, n-1, \quad (6.31)$$

b) $f_{-1} := 0$, $f_0 := m_0/b_2$ *and*

$$f_{k+1} := \frac{1}{b_{k+3}}(m_{k+1} + f_{k-1} - a_{k+2}f_k) \quad for \quad k = 0, \ldots, n-1, \quad (6.32)$$

where $\alpha_k := 0$ *for* $k > N$. *Then, under the assumption* $f_n \neq 0$, *we have*

$$S_N^{(n)} = \sum_{k=0}^{N} \alpha_k p_k^{(n)} = \frac{e_n}{f_n}.$$

Proof.   The values $f_0, \ldots, f_n$ are computed by *LU-factorization* $M_n^T = L_n U_n$ of M_n^T, where

$$L_n := \begin{bmatrix} b_2 & & & & \\ a_2 & \ddots & & & \\ -1 & \ddots & \ddots & & \\ & \ddots & \ddots & \ddots & \\ & & & -1 & a_{n+1} & b_{n+2} \end{bmatrix}, \quad U_n := \begin{bmatrix} 1 & & & & f_0 \\ & \ddots & & & \vdots \\ & & \ddots & & \vdots \\ & & & 1 & f_{n-1} \\ & & & & f_n \end{bmatrix}$$

and therefore $L_n f^{(n)} = m^{(n)}$. This is equivalent to the recurrence relation b) for $f_0, \ldots, f_n$. The recurrence relation a) for $e_0, \ldots, e_n$ is equivalent to $L_n e^{(n)} = \alpha^{(n)}$. By inserting the factorization $M^T = L_n U_n$ into the system $M^T u^{(n)} = \alpha^{(n)}$, it follows that

$$U_n u^{(n)} = e^{(n)},$$

and therefore

$$S_N^{(n)} = u_n^{(n)} = \frac{e_n}{f_n}.$$

$\square$

With the recurrence relations (6.31) and (6.32), we need $O(1)$ operations in order to compute the next approximation $S_N^{(n+1)}$ from $S_N^{(n)}$, as opposed to

$O(n)$ operations with the method which is directly derived from the Miller algorithm. In addition, we need less memory not depending on n but only on N (if the coefficients $\{\alpha_k\}$ are given as a field). Because of (6.30), we call the method developed in Theorem 6.23 the *adjoint summation of minimal solutions*.

We now want to illustrate, how this method can be employed to obtain a useful algorithm from the theoretical description of Theorem 6.23. First we replace the three-term recurrence relation (6.31) for e_k by a system of two-term recurrence relations for

$$u_k := u_k^{(k)} = \frac{e_k}{f_k} \quad \text{and} \quad \Delta u_k := u_k - u_{k-1} ,$$

because we are interested in precisely these two values (u_k as solution, Δu_k to scrutinize the precision). Furthermore, one has to consider that the f_n and e_k get very large, and they may fall outside the domain of numbers which can be represented on the computer. Instead of the f_k, we therefore use the new quantities

$$g_k := \frac{f_{k-1}}{f_k} \quad \text{and} \quad \bar{f}_k := \frac{1}{f_k} .$$

In the transformation of the recurrence relations (6.31) and (6.32) to the new quantities $u_k, \Delta u_k, g_k$ and $\bar{f}_k$, it turns out to be appropriate to introduce the notation

$$\bar{g}_k := \frac{b_{k+2}}{g_k} = b_{k+2}\frac{f_k}{f_{k-1}} \quad \text{and} \quad \bar{m}_k := m_k\bar{f}_{k-1} = \frac{m_k}{f_{k-1}} .$$

From (6.32), it thus follows that (multiplication by b_{k+2}/f_{k-1})

$$\bar{g}_k = \bar{m}_k - a_{k+1} + g_{k-1} \quad \text{for} \quad k \geq 1 , \tag{6.33}$$

and from (6.31) (multiplication with b_{k+2}/f_{k-1} and insertion of (6.33)) that

$$\bar{g}_k \Delta u_k = \bar{f}_{k-1}\alpha_k - g_{k-1}\Delta u_{k-1} - \bar{m}_k u_{k-1} .$$

By arranging the operations so that as little storage space as possible is used and by omitting the no longer required indices, we then obtain the following numerically useful algorithm.

Algorithm 6.24 *Computation of* $S_N = \sum_{k=0}^{N} \alpha_k p_k$ *for a minimal solution* (p_k) *with relative precision* ε.

$$g := \Delta u := 0; \quad \bar{f} := b_2/m_0; \quad u := \alpha_0/m_0; \quad k := 1$$

repeat

$\bar{m} \;:= m\bar{f};$

$\Delta u := \bar{f}\alpha_k - g\Delta u - \bar{m}u;$

$g \;\;\; := \bar{m} - a_{k+1} + g;$

$\Delta u := \Delta u / g;$

$u \;\;\; := u + \Delta u;$

if $(k > N$ **and** $|\Delta u| \leq |u| \cdot \varepsilon)$ **then exit**; (Solution $S_N \approx u$)

$g \;\;\; := b_{k+2}/g;$

$\bar{f} \;\;\; := \bar{f}g;$

$k \;\;\; := k + 1;$

until $(k > n_{\max})$

6.4 Exercises

Exercise 6.1 On a computer, calculate the value of the Chebyshev polynomial $T_{31}(x)$ for $x = 0.923$,

a) by using the Horner scheme (by computing the coefficients of the monomial representation of T_{31} on a computer, or by looking them up in a table),

b) by using the three-term recurrence relation.

Compare the results with the value

$$T_{31}(x) = 0.948715916161 \;,$$

which is precise up to 12 digits, and explain the error.

Exercise 6.2 Consider the three-term recurrence relation

$$T_k = a_k T_{k-1} + b_k T_{k-2} \;.$$

For the relative error $\theta_k = (\tilde{T}_k - T_k)/T_k$ of T_k, there is a non-homogeneous three-term recurrence relation, which is of the form

$$\theta_k = a_k \frac{T_{k-1}}{T_k}\theta_{k-1} + b_k \frac{T_{k-2}}{T_k}\theta_{k-2} + \varepsilon_k \;.$$

Consider the case $a_k \geq 0$, $b_k > 0$, $T_0, T_1 > 0$, and verify that:

a) $|\varepsilon_k| \leq 3\mathrm{eps}$

b) $|\theta_k| \leq (3k - 2)\mathrm{eps}$, $k \geq 1$.

Exercise 6.3 The *Legendre polynomials* are defined through the recurrence relation

$$kP_k(x) = (2k-1)xP_{k-1}(x) - (k-1)P_{k-2}(x) , \qquad (6.34)$$

with the starting values $P_0(x) = 1$ und $P_1(x) = x$. (6.34) is well conditioned in forward mode. Show that the computation of $S_k(\theta) := P_k(\cos\theta)$ according to (6.34) is numerically unstable for $\theta \to 0$. For $\cos\theta > 0$, find an economical, stable version of (6.34) for the computation of $S_k(\theta)$.
Hint: Define $D_k = \alpha_k(S_k - S_{k-1})$, and determine a suitable α_k.

Exercise 6.4 Consider the three-term recurrence relation

$$T_{k-1} - 2\alpha T_k + T_{k-1} = 0 . \qquad (6.35)$$

a) Find the general solution of (6.35) by seeking solutions of the form $T_k = \omega^k$ (distinguish cases!).

b) Show the existence of a minimal solution, if $|\alpha| \geq 1$.

Exercise 6.5 Under the assumptions of Theorem 6.11, analyze the condition of the adjoint three-term recurrence relation for a minimal solution.

Exercise 6.6 Show that the symmetry of three-term recurrence relation carries over to the discrete Green's function, i.e. for symmetric three-term recurrence relation, with $b_k = -1$ for all k, it is true that

$$g(j,k) = g(k,j) \text{ for all } j,k \in \mathbf{N}.$$

Exercise 6.7 Compute the Bessel functions $J_0(x)$ and $J_1(x)$ for $x = 2.13$, and compute $J_k(x)$ for $x = 1024$ and $k = 0, \dots, 1024$,

a) by employing the Miller algorithm,

b) by specializing the adjoint summation.

Compare both algorithms with respect to storage space and computational cost.

Exercise 6.8 Starting from two arbitrary linearly independent solutions $\{P_k\}, \{Q_k\}$, show that the discrete Green's functions can be written in the form

$$g^-(j,k) = \frac{D(j-1,k)}{D(j-1,j)} \text{ and } g^+(j,k) = \frac{D(j+1,k)}{D(j+1,j)} ,$$

where the

$$D(l,m) := P_lQ_m - Q_lP_m$$

denote the generalized Casorati determinants. For the special case of the trigonometric recurrence relation, find a closed formula for $g(j,k)$, and carry out the limiting process $x \to l\pi$, $l \in \mathbf{Z}$. Sketch $g(j,k)$ for selected j,k and x.

Exercise 6.9 Consider the three-term recurrence relation for the cylinder functions

$$T_{k+1} = \frac{2k}{x}T_k - T_{k-1} . \tag{6.36}$$

Starting with the asymptotic representations for the Bessel functions

$$J_k(x) \doteq \frac{1}{\sqrt{2\pi k}} \left(\frac{ex}{2k}\right)^k \quad \text{for} \quad k \to \infty ,$$

and the Neumann functions

$$Y_k(x) \doteq -\sqrt{\frac{2}{\pi k}} \left(\frac{ex}{2k}\right)^{-k} \quad \text{for} \quad k \to \infty ,$$

show that there is cancellation in (6.36) for J_{k+1} in forward mode, and cancellation for Y_{k-1} in backward mode.

Exercise 6.10 Consider a three-term recurrence relation of the form

$$p_k = a_k p_{k-1} + b_k p_{k-2} . \tag{6.37}$$

a) Transform p_k by $p_k = c_k \bar{p}_k$, such that the $\bar{p}_k$ satisfy the symmetric three-term recurrence relation

$$\bar{p}_k = \bar{a}_k \bar{p}_{k-1} - \bar{p}_{k-2} . \tag{6.38}$$

b) Using the assumption

$$|b_k| + 1 \leq |a_k| ,$$

prove a theorem about the existence of minimal solutions, which is similar to Theorem 6.11. Compare the application of Theorem 6.11 to the recurrence relation (6.38) with the application of a) to (6.37).

Exercise 6.11 Consider a symmetric tri-diagonal matrix

$$T_n := \begin{bmatrix} d_1 & e_2 & & & \\ e_2 & \ddots & \ddots & & \\ & \ddots & \ddots & e_n \\ & & e_n & d_n \end{bmatrix}$$

Show:

a) The polynomials $p_i(\lambda) := \det(T_i - \lambda I_i)$ satisfy a three-term recurrence relation.

b) Under the assumption $\prod_{i=2}^{n} e_i \neq 0$ it is true that:
For $i \geq 1$, p_i has only real simple roots. (The roots of p_i separate those of p_{i+1}.)

c) If T_n possesses an eigenvalue of multiplicity k, then at least $k - 1$ non-diagonal elements e_i vanish.

Hint: $\{p_i\}$ is called a *Sturm sequence*.

7 Interpolation and Approximation

In numerical analysis, one often encounters the situation that instead of a function $f : \mathbf{R} \to \mathbf{R}$, only a few discrete function values $f(t_i)$ are given, and maybe derivatives $f^{(j)}(t_i)$ at finitely many points t_i. This is the case, for example, when the function f is given in the form of experimental data. Also, most methods for solving differential equations calculate a solution $f(t)$ (including its derivative) only at finitely many positions. Historically, this problem occurred in the computation of additional function values between tabulated ones. Nowadays, one of the most important applications occurs in computer graphics, known under the abbreviations CAD (*Computer Aided Design*) and CAGD (*Computer Aided Geometric Design*).

If one is interested in total behavior of the function, then, from the given data

$$f^{(j)}(t_i) \quad \text{for} \quad i = 0, \ldots, n \quad \text{and} \quad j = 0, \ldots, c_i \,,$$

one should construct a function φ, which differs as little as possible from the original function f. In addition, the function φ should be simple to evaluate, like, for example, (piecewise) polynomials, trigonometric, exponential

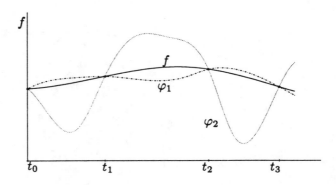

Figure 7.1: Various interpolating functions for f

or rational functions. A first obvious requirement regarding the function φ is the *interpolation property*: φ should coincide with the function f at the *knots* or *nodes* t_i,

$$\varphi^{(j)}(t_i) = f^{(j)}(t_i) \quad \text{for all} \quad i, j .$$

The values $f^{(j)}(t_i)$ are called *node values*. If we compare the two functions φ_1 and φ_2 in Figure 7.1, then both obviously satisfy the interpolation condition at given values $f(t_i)$. Nevertheless, we would prefer φ_1. In addition to the interpolation property, we therefore require also the *approximation property*: φ should differ as little as possible from f with respect to a norm $\| \cdot \|$ in a suitable function space,

$$\|\varphi - f\| \quad \text{'small'} .$$

7.1 Classical Polynomial Interpolation

We start with the simplest case, when only the values

$$f_i := f(t_i) \quad \text{for} \quad i = 0, \ldots, n$$

are given at the pairwise distinct knots $t_0, \ldots, t_n$. We now seek a polynomial $P \in \mathbf{P}_n$ of degree $\deg P \leq n$,

$$P(t) = a_n t^n + a_{n-1} t^{n-1} + \cdots + a_1 t + a_0 \quad \text{with} \quad a_0, \ldots, a_n \in \mathbf{R} ,$$

which interpolates f at the $n+1$ knots $t_0, \ldots, t_n$, i.e.

$$P(t_i) = f_i \quad \text{for} \quad i = 0, \ldots, n . \tag{7.1}$$

7.1.1 Uniqueness and condition number

The following argument shows that the $n+1$ unknown coefficients $a_0, \ldots, a_n$ are uniquely determined by the condition (7.1): If $P, Q \in \mathbf{P}_n$ are two interpolating polynomials with $P(t_i) = Q(t_i)$ for $i = 0, \ldots, n$, then $P - Q$ is a polynomial of at most n-th degree with the $n+1$ roots $t_0, \ldots, t_n$, and is therefore the null-polynomial. But the rule

$$\mathbf{P}_n \to \mathbf{R}^{n+1}, \quad P \mapsto (P(t_0), \ldots, P(t_n))$$

is also a linear mapping between the two $(n+1)$-dimensional real vector spaces $\mathbf{P}_n$ and $\mathbf{R}^{n+1}$, so that injectivity already implies surjectivity. We have therefore proven the following theorem.

Theorem 7.1 *Suppose $n + 1$ nodes (t_i, f_i) for $i = 0, \ldots, n$ are given with pairwise distinct knots $t_0, \ldots, t_n$, then there exists a unique interpolating polynomial $P \in \mathbf{P}_n$, i.e. $P(t_i) = f_i$ for $i = 0, \ldots, n$.*

The unique polynomial P, which is given by theorem 7.1, is called *interpolating polynomial* of f for the pairwise distinct knots $t_0, \ldots, t_n$, and it is denoted by

$$P = P(f \,|\, t_0, \ldots, t_n) \ .$$

In order to actually compute the interpolating polynomial, we have to choose a basis of the space of polynomials $\mathbf{P}_n$. In order to illustrate this, we first give two classical representations. If we write P as above in *coefficient representation*

$$P(t) = a_n t^n + a_{n-1} t^{n-1} + \cdots + a_1 t + a_0 \ ,$$

i.e. with respect to the *monomial basis* $\{1, t, \ldots, t^n\}$ of $\mathbf{P}_n$, then the interpolation conditions $P(t_i) = f_i$ can be formulated as a linear system

$$\underbrace{\begin{bmatrix} 1 & t_0 & t_0^2 & \cdots & t_0^n \\ \vdots & \vdots & \vdots & & \vdots \\ 1 & t_n & t_n^2 & \cdots & t_n^n \end{bmatrix}}_{=: \, V_n} \begin{pmatrix} a_0 \\ \vdots \\ a_n \end{pmatrix} = \begin{pmatrix} f_0 \\ \vdots \\ f_n \end{pmatrix} \ .$$

The matrix V_n is called *Vandermonde matrix*. For the determinant of V_n, we have

$$\det V_n = \prod_{i=0}^{n} \prod_{j=i+1}^{n} (t_i - t_j) \ ,$$

which is proven in virtually every linear algebra textbook (see e.g. [8]). It is different from zero, exactly when the knots $t_0, \ldots, t_n$ are pairwise distinct (which is agreement with Theorem 7.1). However, the solution of the system requires an excessive amount of computational effort when compared with the methods which will be discussed below.

In addition, the Vandermonde matrices are almost singular in higher dimensions n. Gaussian elimination *without* pivoting is recommended for its solution, because pivoting strategies may perturb the structure of the matrix (compare [48]). For special knots, the above Vandermonde matrix can easily be inverted analytically. In Section 7.2 we shall encounter an example of this.

An alternative basis for the representation of the interpolation polynomial is formed by the so-called *Lagrange polynomials* $L_0, \ldots, L_n$.   They are

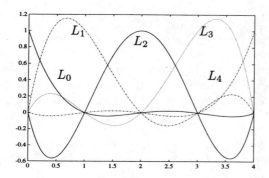

Figure 7.2: Lagrange polynomials L_i for $n = 4$ and equidistant knots t_i

defined as the uniquely determined interpolation polynomials $L_i \in \mathbf{P}_n$ with

$$L_i(t_j) = \delta_{ij} \ .$$

The corresponding explicit form is

$$L_i(t) = \prod_{\substack{j=0 \\ j \neq i}}^{n} \frac{t - t_j}{t_i - t_j} \ .$$

The interpolating polynomial P for arbitrary nodes $f_0, \ldots, f_n$ can easily be built from the Lagrange polynomials by superposition: With

$$P(t) := \sum_{i=0}^{n} f_i L_i(t) \ , \tag{7.2}$$

we obviously have

$$P(t_j) = \sum_{i=0}^{n} f_i L_i(t_j) = \sum_{i=0}^{n} f_i \delta_{ij} = f_j \ .$$

Remark 7.2 The above statement can also be phrased as follows: The Lagrange polynomials form an orthogonal basis of $\mathbf{P}_n$ with respect to the scalar product

$$\langle P, Q \rangle := \sum_{i=0}^{n} P(t_i) Q(t_i)$$

for $P, Q \in \mathbf{P}_n$. Let $\langle P, L_i \rangle = P(t_i)$, then, obviously,

$$P = \sum_{i=0}^{n} \langle P, L_i \rangle L_i = \sum_{i=0}^{n} P(t_i) L_i = \sum_{i=0}^{n} f_i L_i .$$

For practical purposes, the Lagrange representation (7.2) is computationally too costly, however, in many theoretical questions it is advantageous. An example of this is the determination of the condition number of the interpolation problem.

Theorem 7.3 *Let $a \leq t_0 < \ldots < t_n \leq b$ be pairwise distinct knots, and let L_{in} be the corresponding Lagrange polynomials. Then the absolute condition number κ_{abs} of the polynomial interpolation*

$$\phi = P(\cdot \,|\, t_0, \ldots, t_n) : C[a, b] \longrightarrow \mathbf{P}_n$$

with respect to the supremum norm is the so-called Lebesgue constant

$$\kappa_{\mathrm{abs}} = \Lambda_n := \max_{t \in [a,b]} \sum_{i=0}^{n} |L_{in}(t)| .$$

for the knots $t_0, \ldots, t_n$.

Proof. The polynomial interpolation is linear, i.e. $\phi'(f)(g) = \phi(g)$. We have to show that $\|\phi'\| = \Lambda_n$. For every continuous function $f \in C[a, b]$, we have

$$
\begin{aligned}
|\phi(f)(t)| &= |\sum_{i=0}^{n} f(t_i) L_{in}(t)| \leq \sum_{i=0}^{n} |f(t_i)| \, |L_{in}(t)| \\
&= \|f\|_\infty \max_{t \in [a,b]} \sum_{i=0}^{n} |L_{in}(t)| ,
\end{aligned}
$$

and thus $\kappa_{\mathrm{abs}} \leq \Lambda_n$. For the opposite direction, we construct a function $g \in C[a, b]$, such that

$$|\phi(g)(\tau)| = \|g\|_\infty \max_{t \in [a,b]} \sum_{i=0}^{n} |L_{in}(t)|$$

for a $\tau \in [a, b]$. For this let $\tau \in [a, b]$ be the place where the maximum is attained, i.e.

$$\sum_{i=0}^{n} |L_{in}(\tau)| = \max_{t \in [a,b]} \sum_{i=0}^{n} |L_{in}(t)| ,$$

and let $g \in C[a,b]$ be a function with $\|g\|_\infty = 1$ and $g(t_i) = \operatorname{sgn} L_i(\tau)$, e.g. the piecewise linear interpolation function corresponding to the points $(t_i, \operatorname{sgn} L_i(\tau))$. Then, as desired

$$|\phi(g)(\tau)| = \sum_{i=0}^n |L_{in}(\tau)| = \|g\|_\infty \max_{t \in [a,b]} \sum_{i=0}^n |L_{in}(t)| \,,$$

and thus $\kappa_{\text{abs}} \geq \Lambda_n$, and altogether $\kappa_{\text{abs}} = \Lambda_n$. $\qquad\square$

One easily computes that the Lebesgue constant Λ_n is invariant under affine transformations (see Exercise 7.1), and it therefore depends only on the *relative position* of the knots t_i with respect to each other. In Table 7.1, Λ_n is given for equidistant knots in dependence of n. Obviously, Λ_n grows rapidly beyond all reasonable bounds. However, this is not true for any choice of knots. As a counterexample, in Table 7.1 are also listed the Lebesgue constants for the *Chebyshev knots* (see Section 7.1.4)

$$t_i = \cos\left(\frac{2i+1}{2n+2}\pi\right) \quad \text{for} \quad i = 0, \ldots, n$$

(where the maximum was taken over $[-1,1]$). They grow only very slowly.

Table 7.1: Lebesgue constant Λ_n for equidistant and for Chebyshev knots

n	Λ_n for equidistant knots	Λ_n for Chebyshev knots
5	3.106292	2.104398
10	29.890695	2.489430
15	512.052451	2.727778
20	10986.533993	2.900825

7.1.2 Hermite interpolation and divided differences

If one is only interested in the interpolating polynomial P at a single position t, then the recursive computation of $P(t)$ turns out to be the most effective method. It is based on the following simple observation, the *Aitken Lemma*.

Lemma 7.4 *The interpolating polynomial $P = P(f \mid t_0, \ldots, t_n)$ satisfies the recurrence relation*

$$P(f \mid t_0, \ldots, t_n) = \frac{(t_0 - t)P(f \mid t_1, \ldots, t_n) - (t_n - t)P(f \mid t_0, \ldots, t_{n-1})}{t_0 - t_n} \,.$$

$$(7.3)$$

Proof. Let $\varphi(t)$ be defined as the expression on the right hand side of (7.3). Then $\varphi \in \mathbf{P}_n$, and

$$\varphi(t_i) = \frac{(t_0 - t_i)f_i - (t_n - t_i)f_i}{t_0 - t_n} = f_i \quad \text{for} \quad i = 1, \ldots, n-1 .$$

Similarly, it is simple to conclude that $\varphi(t_0) = f_0$ and $\varphi(t_n) = f_n$, and the statement therefore follows. $\qquad\square$

The interpolation polynomials for only one single node are nothing else than the constants

$$P(f \,|\, t_i) = f_i \quad \text{for} \quad i = 0, \ldots, n .$$

If we simplify the notation for fixed t by

$$P_{ik} := P(f \,|\, t_{i-k}, \ldots, t_i)(t) \quad \text{for} \quad i \geq k ,$$

then the value $P_{nn} = P(f \,|\, t_0, \ldots, t_n)(t)$ can be computed according to the *Neville scheme*

$$
\begin{array}{ccccccc}
P_{00} & & & & & & \\
& \searrow & & & & & \\
P_{10} & \rightarrow & P_{11} & & & & \\
\vdots & & & \ddots & & & \\
P_{n-1,0} & \rightarrow & \cdots & \rightarrow & P_{n-1,n-1} & & \\
& & & & & \searrow & \\
P_{n0} & \rightarrow & \cdots & \rightarrow & P_{n,n-1} & \rightarrow & P_{nn}
\end{array}
$$

given by

$$
\begin{aligned}
P_{i0} &= f_i \quad \text{for} \quad i = 0, \ldots, n \\
P_{ik} &= P_{i,k-1} + \frac{t - t_i}{t_i - t_{i-k}}(P_{i,k-1} - P_{i-1,k-1}) \quad \text{for} \quad i \geq k . \quad (7.4)
\end{aligned}
$$

Example 7.5 Computation of $\sin 62^o$ from the nodes

$$(50^o, \sin 50^o), (55^o, \sin 55^o), \ldots, (70^o, \sin 70^o)$$

by the Aitken-Neville algorithm.

t_i	$\sin t_i$				
$50°$	0.7660444				
$55°$	0.8191520	0.8̲935027			
$60°$	0.8660254	0.8̲847748	0.8̲830292		
$65°$	0.9063078	0.8̲821384	0.8̲829293	0.8̲829493	
$70°$	0.9396926	0.8̲862768	0.8̲829661	0.8̲829465	0.8̲829476

The recursive structure of the interpolation polynomials according to Lemma 7.4 can also be utilized for the determination of the entire polynomial $P(f\,|\,t_0,\ldots,t_n)$. This is also true for the generalized interpolation problem, where besides function values $f(t_i)$, also derivatives at the knots are given, the *Hermite interpolation*. For this we introduce the following practical notation. We allow for the occurence of multiple knots in the sequence $\Delta := \{t_i\}_{i=0,\ldots,n}$ with

$$a = t_0 \leq t_1 \leq \cdots \leq t_n = b \,. \tag{7.5}$$

If at a knot t_i, the value $f(t_i)$ and the derivatives $f'(t_i),\ldots,f^{(k)}(t_i)$ are given up to an order k, then t_i shall occur $(k+1)$-times in the sequence Δ. The same knots are enumerated from the left to the right by

$$d_i := \max\{j \mid t_i = t_{i-j}\} \,,$$

e.g.

t_i	t_0	$=$	t_1	$<$	t_2	$=$	t_3	$=$	t_4	$<$	t_5	$<$	t_6	$=$	t_7
d_i	0		1		0		1		2		0		0		1

.

By defining now for $i = 0,\ldots,n$ the linear mappings

$$\mu_i : C^n[a,b] \to \mathbf{R} \,, \quad \mu_i(f) := f^{(d_i)}(t_i) \,,$$

then the problem of the *Hermite interpolation* can be phrased as follows: Find a polynomial $P \in \mathbf{P}_n$, such that

$$\mu_i(P) = \mu_i(f) \quad \text{for all} \quad i = 0,\ldots,n \,. \tag{7.6}$$

The solution $P = P(f\,|\,t_0,\ldots,t_n) \in \mathbf{P}_n$ of the interpolation problem (7.6) is called *Hermite interpolation* of f at the knots $t_0,\ldots,t_n$. Existence and uniqueness follows as in Theorem 7.1.

Theorem 7.6 *For each function $f \in C^n[a,b]$ and each monotone sequence*

$$a = t_0 \leq t_1 \leq \cdots \leq t_n = b$$

of (not necessarily distinct) knots, there exists a unique polynomial $P \in \mathbf{P}_n$ such that

$$\mu_i P = \mu_i f \quad for \ all \quad i = 0, \ldots, n \ .$$

Proof. The mapping

$$\mu : \mathbf{P}_n \to \mathbf{R}^{n+1} \ , \quad P \mapsto (\mu_0 P, \ldots, \mu_n P)$$

is obviously linear and also injective. Now $\mu(P) = 0$ implies that P possesses at least $n + 1$ roots (counted with multiplicity), and it is therefore the null-polynomial. Since $\dim \mathbf{P}_n = \dim \mathbf{R}^{n+1} = n + 1$, this implies again the existence. $\qquad\qquad\square$

If all knots are pairwise distinct, then we recover the Lagrange interpolation

$$P(f \,|\, t_0, \ldots, t_n) = \sum_{i=0}^{n} f(t_i) L_i \ .$$

If all knots coincide, $t_0 = t_1 = \cdots = t_n$, then the interpolation polynomial is the Taylor polynomial centered at $t = t_0$,

$$P(f \,|\, t_0, \ldots, t_n)(t) = \sum_{j=0}^{n} \frac{(t - t_0)^j}{j!} f^{(j)}(t_0) \ , \qquad (7.7)$$

also called the *Taylor interpolation*.

Remark 7.7 An important application is the *cubic Hermite interpolation*, where function values f_0, f_1 and derivatives f_0', f_1' are given at two knots t_0, t_1. According to Theorem 7.6, this determines uniquely a cubic polynomial $P \in \mathbf{P}_3$. If the *Hermite polynomials* $H_0^3, \ldots, H_3^3 \in \mathbf{P}_3$ are defined by

$$H_0^3(t_0) = 1, \quad \frac{d}{dt} H_0^3(t_0) = 0, \quad H_0^3(t_1) = 0, \quad \frac{d}{dt} H_0^3(t_1) = 0 \ ,$$

$$H_1^3(t_0) = 0, \quad \frac{d}{dt} H_1^3(t_0) = 1, \quad H_1^3(t_1) = 0, \quad \frac{d}{dt} H_1^3(t_1) = 0 \ ,$$

$$H_2^3(t_0) = 0, \quad \frac{d}{dt} H_2^3(t_0) = 0, \quad H_2^3(t_1) = 1, \quad \frac{d}{dt} H_2^3(t_1) = 0 \ ,$$

$$H_3^3(t_0) = 0, \quad \frac{d}{dt} H_3^3(t_0) = 0, \quad H_3^3(t_1) = 0, \quad \frac{d}{dt} H_3^3(t_1) = 1 \ ,$$

then the polynomials

$$\{H_0^3(t), H_1^3(t), H_2^3(t), H_3^3(t)\}$$

form a basis of $\mathbf{P}_3$, the so-called *cubic Hermite basis*, with respect to the knots t_0, t_1. The Hermite polynomial corresponding to the values $\{f_0, f_0', f_1, f_1'\}$ is thus formally given by

$$P(t) = f_0 H_0^3(t) + f_0' H_1^3(t) + f_1 H_2^3(t) + f_1' H_3^3(t) \ .$$

If an entire series $t_0, \ldots, t_n$ of knots is given, with function values f_i and derivatives f_i', then on each interval $[t_i, t_{i+1}]$, we can consider the cubic Hermite interpolation and join these polynomials at the knots. Because of the data, it is clear that this piecewise defined function is C^1. This kind of interpolation is called *locally cubic Hermite interpolation*. We have already seen an application in Section 4.4.2. When computing a solution curve by a tangential continuation method, then we obtain solution points (x_i, λ_i), together with the slopes x_i'. In order to get an impression of the entire solution curve from this discrete information, we have joined the points by their locally cubic Hermite interpolation.

Similar to the Aitken Lemma, the following recurrence relation is valid for two distinct knots $t_i \neq t_j$.

Lemma 7.8 *If $t_i \neq t_j$, then the Hermite interpolation polynomial $P = P(f \,|\, t_0, \ldots, t_n)$ satisfies*

$$P = \frac{(t_i - t)P(f \,|\, t_1, \ldots, \widehat{t_j}, \ldots, t_n) - (t_j - t)P(f \,|\, t_1, \ldots, \widehat{t_i}, \ldots, t_n)}{t_i - t_j} \ ,$$

where $\widehat{}$ indicates that the corresponding knot is omitted. ('has to lift its hat').

Proof. Verification of the interpolation property by inserting the definitions.
□

For the representation of the interpolation polynomial, we use the so-called *Newton basis* $\omega_0, \ldots, \omega_n$ of the space of polynomials $\mathbf{P}_n$:

$$\omega_i(t) := \prod_{j=0}^{i-1}(t - t_i) \ , \quad \omega_i \in \mathbf{P}_i \ .$$

The coefficients with respect to this basis are the so-called *divided differences*, which we now define.

Definition 7.9 The leading coefficient a_n of the interpolation polynomial

$$P(f \mid t_0, \ldots, t_n)(t) = a_n t^n + a_{n-1} t^{n-1} + \cdots + a_0$$

of f corresponding to the (not necessarily distinct) knots $t_0 \leq t_1 \leq \cdots \leq t_n$ is called n-th *divided difference* of f at $t_0, \ldots, t_n$, and it is denoted by

$$[t_0, \ldots, t_n]f := a_n \ .$$

Theorem 7.10 *For each function $f \in C^n$ and for given (not necessarily distinct) knots $t_0 \leq \cdots \leq t_n$, the interpolation polynomial $P(f \mid t_0, \ldots, t_n)$ of f at $t_0, \ldots, t_n$ is given by*

$$P := \sum_{i=0}^{n} [t_0, \ldots, t_i]f \cdot \omega_i \ .$$

If $f \in C^{n+1}$, then

$$f(t) = P(t) + [t_0, \ldots, t_n, t]f \cdot \omega_{n+1}(t) \ . \tag{7.8}$$

Proof. We show the first statement by induction over n. The statement is trivial for $n = 0$. Thus let $n > 0$, and let

$$P_{n-1} := P(f \mid t_0, \ldots, t_{n-1}) = \sum_{i=0}^{n-1} [t_0, \ldots, t_i]f \cdot \omega_i$$

be the interpolation polynomial of f at $t_0, \ldots, t_{n-1}$. Then the interpolation polynomial $P_n = P(f \mid t_0, \ldots, t_n)$ of f at $t_0, \ldots, t_n$ can be written in the form

$$\begin{aligned} P_n(t) &= [t_0, \ldots, t_n]f \cdot t^n + a_{n-1} t^{n-1} + \cdots + a_0 \\ &= [t_0, \ldots, t_n]f \cdot \omega_n(t) + Q_{n-1}(t) \ , \end{aligned}$$

with a polynomial $Q_{n-1} \in \mathbf{P}_{n-1}$. But

$$Q_{n-1} = P_n - [t_0, \ldots, t_n]f \cdot \omega_n$$

obviously satisfies the interpolation conditions for $t_0, \ldots, t_{n-1}$, so that

$$Q_{n-1} = P_{n-1} = \sum_{i=0}^{n-1} [t_0, \ldots, t_i]f \cdot \omega_i \ .$$

This proves the first statement. In particular, it follows that

$$P_n + [t_0, \ldots, t_n, t]f \cdot \omega_{n+1}$$

interpolates the function f at the knots $t_0, \ldots, t_n$ and t, which proves (7.8).

$\square$

From the properties of the Hermite interpolation one can immediately deduce the following statements about the divided differences of f.

Lemma 7.11 *The divided differences* $[t_0, \ldots, t_n]f$ *satisfy the following properties* $(f \in C^n)$:

i) $[t_0, \ldots, t_n]P = 0$ *for all* $P \in \mathbf{P}_{n-1}$.

ii) *For multiple knots* $t_0 = \cdots = t_n$,

$$[t_0, \ldots, t_n]f = f^{(n)}(t_0)/n! \ . \tag{7.9}$$

iii) *The following recurrence relation holds for* $t_i \neq t_j$:

$$[t_0, \ldots, t_n]f = \frac{[t_0, \ldots, \widehat{t_i}, \ldots, t_n]f - [t_0, \ldots, \widehat{t_j}, \ldots, t_n]f}{t_j - t_i} \ . \tag{7.10}$$

Proof. i) is true, because the n-th coefficient of a polynomial of degree less than or equal to $n-1$ vanishes. ii) follows from the Taylor interpolation (7.7) and iii) from Lemma 7.8 and the uniqueness of the leading coefficient.

$\square$

Using the properties ii) and iii), the divided differences can be computed recursively from the function values and derivatives $f^{(j)}(t_i)$ of f at the knots t_i. We also need the recurrence relation in the proof of the following theorem, which states a surprising interpretation of the divided differences: The n-th divided difference of a function $f \in C^n$ with respect to the knots $t_0, \ldots, t_n$ is the integral of the n-th derivative over the n-dimensional standard simplex

$$\Sigma^n := \left\{ s = (s_0, \ldots, s_n) \in \mathbf{R}^{n+1} \ \Big| \ \sum_{i=0}^{n} s_i = 1 \ \text{ and } \ s_i \geq 0 \right\} \ .$$

Theorem 7.12 *Hermite-Genocchi Formula. The n-th divided difference of a n-times continuously differentiable function* $f \in C^n$ *satisfies*

$$[t_0, \ldots, t_n]f = \int_{\Sigma^n} f^{(n)} \Big(\sum_{i=0}^{n} s_i t_i \Big) \, ds \ . \tag{7.11}$$

Proof. We prove the formula by induction over n. The statement is trivial for $n = 0$. The induction step from n to $n+1$ is as follows: If all knots coincide, then the statement follows from (7.9). We can therefore assume without loss of generality that $t_0 \neq t_{n+1}$. We then have

$$
\int\limits_{\substack{n+1 \\ \sum\limits_{i=0} s_i = 1}} f^{(n+1)}\Big(\sum_{i=0}^{n} s_i t_i\Big)\, ds
$$

$$
= \int\limits_{\substack{n+1 \\ \sum\limits_{i=1} s_i \leq 1}} f^{(n+1)}\Big(t_0 + \sum_{i=1}^{n} s_i(t_i - t_0)\Big)\, ds
$$

$$
= \int\limits_{\substack{n \\ \sum\limits_{i=1} s_i \leq 1}} \int\limits_{s_0 = 0}^{1 - \sum\limits_{i=1}^{n} s_i} f^{(n+1)}\Big(t_0 + \sum_{i=1}^{n} s_i(t_i - t_0) + s_{n+1}(t_{n+1} - t_0)\Big)\, ds
$$

$$
= \frac{1}{t_{n+1} - t_0} \int\limits_{\substack{n \\ \sum\limits_{i=1} s_i \leq 1}} \Bigg\{ f^{(n)}\Big(t_{n+1} + \sum_{i=1}^{n} s_i(t_i - t_{n+1})\Big) - f^{(n)}\Big(t_0 + \sum_{i=1}^{n} s_i(t_i - t_0)\Big) \Bigg\}\, ds
$$

$$
= \frac{1}{t_{n+1} - t_0} \big([t_1, \ldots, t_{n+1}]f - [t_0, \ldots, t_n]f\big)
$$

$$
= [t_0, \ldots, t_{n+1}]f
$$

$\square$

Corollary 7.13 *Let $g : \mathbf{R}^{n+1} \to \mathbf{R}$ be the mapping, which is given by the n-th divided difference of a function $f \in C^n$ with*

$$
g(t_0, \ldots, t_n) := [t_0, \ldots, t_n]f \ .
$$

Then g is continuous in its arguments t_i. Furthermore, for all knots $t_0 \leq \cdots \leq t_n$, there exists a $\tau \in [t_0, t_n]$ such that

$$
[t_0, \ldots, t_n]f = \frac{f^{(n)}(\tau)}{n!} \ . \tag{7.12}
$$

Proof. The integral representation (7.11) yields immediately the continuity; and (7.12) follows from the integral mean-value theorem, because the volume of the n-dimensional standard simplex is $\mathrm{vol}(\Sigma^n) = 1/n!$. □

For pairwise distinct knots $t_0 < \cdots < t_n$, the divided differences can be arranged similar to the Neville-scheme because of the recurrence relation (7.10).

$$
\begin{array}{lll}
f_0 &=& [t_0]f \\[2mm]
&& \searrow \\[2mm]
f_1 &=& [t_1]f \quad \rightarrow \quad [t_0, t_1]f \\[1mm]
\vdots && \qquad\qquad\qquad \ddots \\[2mm]
f_{n-1} &=& [t_{n-1}]f \;\rightarrow\; \cdots \;\rightarrow\; [t_0, \ldots, t_{n-1}]f \\[2mm]
&& \qquad\qquad\qquad\qquad\qquad \searrow \\[2mm]
f_n &=& [t_n]f \;\rightarrow\; \cdots \;\rightarrow\; [t_1, \ldots, t_n]f \;\rightarrow\; [t_0, \ldots, t_n]f
\end{array}
$$

Example 7.14 We compute the interpolation polynomial corresponding to the values

t_i	0	1	2	3
f_i	1	2	0	1

by employing Newton's divided differences,

$$
\begin{array}{llll}
f[t_0] = 1 \\
f[t_1] = 2 & f[t_0, t_1] = 1 \\
f[t_2] = 0 & f[t_1, t_2] = -2 & f[t_0, t_1, t_2] = -3/2 \\
f[t_3] = 1 & f[t_2, t_3] = 1 & f[t_1, t_2, t_3] = 3/2 & f[t_0, t_1, t_2, t_3] = 1 \;,
\end{array}
$$

i.e.

$$
(\alpha_0, \alpha_1, \alpha_2, \alpha_3) = (1, 1, -3/2, 1) \;.
$$

The interpolation polynomial is therefore

$$
\begin{aligned}
P(t) &= 1 + 1(t - 0) + (-3/2)(t - 0)(t - 1) + 1(t - 0)(t - 1)(t - 2) \\
&= t^3 - 4.5\,t^2 + 4.5\,t + 1 \;.
\end{aligned}
$$

A further important property of the divided differences is the following *Leibniz formula*.

Lemma 7.15 *Let $g, h \in C^n$, and let $t_0 \leq t_1 \leq \cdots \leq t_n$ be an arbitrary sequence of knots. Then*

$$[t_0, \ldots, t_n]gh = \sum_{i=0}^{n} [t_0, \ldots, t_i]g \cdot [t_i, \ldots, t_n]h \ .$$

Proof. First suppose that the knots $t_0, \ldots, t_n$ are pairwise distinct. Set

$$\omega_i(t) := \prod_{k=0}^{i-1} (t - t_k) \quad \text{and} \quad \bar{\omega}_j(t) := \prod_{l=j+1}^{n} (t - t_l) \ .$$

Then, according to Theorem 7.10, the interpolation polynomials $P, Q \in \mathbf{P}_n$ of g respectively h are given by

$$P = \sum_{i=0}^{n} [t_0, \ldots, t_i]g \cdot \omega_i \quad \text{and} \quad Q = \sum_{j=0}^{n} [t_j, \ldots, t_n]h \cdot \bar{\omega}_j \ .$$

The product

$$PQ = \sum_{i,j=0}^{n} [t_0, \ldots, t_i]g \ [t_j, \ldots, t_n]h \cdot \omega_i \bar{\omega}_j$$

thus interpolates the function $f := gh$ in $t_0, \ldots, t_n$. Since $\omega_i(t_k)\bar{\omega}_j(t_k) = 0$ for all k and $i > j$, it follows that

$$F := \sum_{\substack{i,j=0 \\ i \leq j}}^{n} [t_0, \ldots, t_i]g \ [t_j, \ldots, t_n]h \cdot \omega_i \bar{\omega}_j \in \mathbf{P}_n$$

is the interpolation polynomial of gh in $t_0, \ldots, t_n$. As claimed, the leading coefficient is

$$\sum_{i=0}^{n} [t_0, \ldots, t_i]g \ [t_i, \ldots, t_n]h \ .$$

For arbitrary, not necessarily distinct knots t_i, the statement now follows from the continuity of the divided differences at the knots t_i. $\qquad\square$

7.1.3   Approximation error

We shall now turn our attention towards the second requirement, namely the *approximation property*, and analyze to what extent the polynomials $P(f \,|\, t_0, \ldots, t_n)$ approximate the original function f. By employing previously derived properties of the divided differences, it is simple to find a representation of the approximation error.

Theorem 7.16 *Suppose that $f \in C^{n+1}$. Then for the approximation error of the Hermite interpolation $P(f \mid t_0, \ldots, t_n)$ and $t_i, t \in [a, b]$ we have*

$$f(t) - P(f \mid t_0, \ldots, t_n)(t) = \frac{f^{(n+1)}(\tau)}{(n+1)!} \omega_{n+1}(t) \qquad (7.13)$$

for some $\tau = \tau(t) \in]a, b[$.

Proof. According to theorem 7.10 and theorem 7.13, we have for $P := P(f \mid t_0, \ldots, t_n)$ that

$$f(t) - P(t) = [t_0, \ldots, t_n, t] f \cdot \omega_{n+1}(t) = \frac{f^{(n+1)}(\tau)}{(n+1)!} \omega_{n+1}(t)$$

for some $\tau \in]a, b[$. $\qquad\qquad\qquad\qquad\qquad\qquad\qquad\qquad\square$

Example 7.17 In the case of the Taylor interpolation (7.7), i.e. $t_0 = \cdots = t_n$, the error formula (7.13) is just the Lagrange remainder of the Taylor expansion

$$f(t) - P(f \mid t_0, \ldots, t_n)(t) = \frac{f^{(n+1)}(\tau)}{(n+1)!} (t - t_0)^{n+1}.$$

If we consider the class of functions

$$\mathcal{F} := \{ f \in C^{n+1}[a, b] \mid \sup_{\tau \in [a,b]} |f^{n+1}(\tau)| \leq M(n+1)! \}$$

for a constant $M > 0$, then the approximation error obviously depends crucially on the choice of the knots $t_0, \ldots, t_n$ via the expression

$$\omega_{n+1}(t) = (t - t_0) \cdots (t - t_n). \qquad (7.14)$$

In the next section (see example 7.20) we will show that, in the case of pairwise distinct knots, the expression (7.14) can be minimized on an interval $[a, b]$,

$$\max_{t \in [a,b]} |\omega_{n+1}(t)| = \min,$$

by choosing again the Chebyshev knots for the knots t_i. We now turn our attention to the question of whether the polynomial interpolation satisfies the approximation property. For merely continuous functions $f \in C[a, b]$, and the supremum-norm $\|f\| = \sup_{t \in [a,b]} f(t)$, the approximation error can in principle grow beyond all bounds. More precisely, according to FABER, for each sequence $\{T_k\}$ of sets of nodes $T_k = \{t_{k,0}, \ldots, t_{k,n_k}\} \subset [a, b]$, there exists a continuous function $f \in C[a, b]$, such that the sequence $\{P_k\}$ of the interpolation polynomials, which belong to the T_k, does not converge uniformly to f.

7.1.4   Min-max property of Chebyshev polynomials

In the previous chapter we have repeatedly mentioned the Chebyshev knots, for which the polynomial interpolation has particularly nice properties (such as bounded condition and optimal approximation of f over the class $\mathcal{F}$). The Chebyshev knots are the roots of the Chebyshev polynomials, which we already encountered in Chapter 6.1.1. In the investigation of the approximation error, as well as in the condition analysis of the polynomial interpolation, we were lead to the following approximation problem: Find the polynomial $P_n \in \mathbf{P}_n$ of degree $\deg P_n = n$ with leading coefficient 1, and the smallest supremum-norm over the interval $[a, b]$, i.e.

$$\max_{t \in [a,b]} |P_n(t)| = \min . \tag{7.15}$$

We shall now see that the *Chebyshev polynomials* T_n, which we already encountered as orthogonal polynomials with respect to the weight function $\omega(x) = (1 - x^2)^{-\frac{1}{2}}$ over $[-1, 1]$, solve this *min-max problem* (up to a scalar factor and an affine transformation). In order to show this, we first reduce the problem to the interval $[-1, 1]$, which is suitable for the Chebyshev polynomials, with the help of the affine mapping

$$x : [a, b] \stackrel{\cong}{\longrightarrow} [-1, 1]$$
$$t \longmapsto x = x(t) = 2\,\frac{t - a}{b - a} - 1 = \frac{2\,t - a - b}{b - a}$$

whose inverse mapping is $t : [-1, 1] \stackrel{\cong}{\longrightarrow} [a, b]$,

$$t = t(x) = \frac{1 - x}{2}a + \frac{1 + x}{2}b .$$

If $P_n \in \mathbf{P}_n$ with $\deg P = n$ and leading coefficient 1 is the solution of the min-max problem

$$\max_{x \in [-1,1]} |P_n(x)| = \min , \tag{7.16}$$

then $\hat{P}_n(t) := P_n(t(x))$ is the solution of the original problem (7.15) with leading coefficient $2^n/(b - a)^n$.

In Example 6.3, we introduced the Chebyshev polynomials via

$$T_n(x) = \cos(n \arccos x) \quad \text{for} \quad x \in [-1, 1]$$

and more generally for $x \in \mathbf{R}$ via the three-term recurrence relation

$$T_k(x) = 2xT_{k-1}(x) - T_{k-2}(x) , \quad T_0(x) = 1 , \quad T_1(x) = x .$$

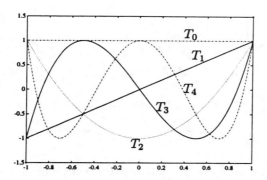

Figure 7.3: Chebyshev-polynomials $T_0, \ldots, T_4$

The following properties of the Chebyshev polynomials are either obvious or easily verified. In particular we can directly give the roots $x_1, \ldots, x_n$ of $T_n(x)$, which are real and simple according to theorem 6.5 (see property 7 below).

Remark 7.18

1. The Chebyshev polynomials have integer coefficients.

2. The leading coefficient of T_n is $a_n = 2^{n-1}$.

3. T_n is an even function if n is even, and an odd one if n is odd.

4. $T_n(1) = 1$, $T_n(-1) = (-1)^n$.

5. $|T_n(x)| \leq 1$ for $x \in [-1, 1]$.

6. $|T_n(x)|$ *takes on the value* 1 *at the* Chebyshev abscissae $\bar{x}_k = \cos(k\pi/n)$, *i.e.*

$$|T_n(x)| = 1 \iff x = \bar{x}_k = \cos\frac{k\pi}{n} \quad \text{for a } k = 0, \ldots, n.$$

7. *The roots of* $T_n(x)$ *are*

$$x_k := \cos\left(\frac{2k-1}{2n}\pi\right) \quad \text{for } k = 1, \ldots, n .$$

8. *We have*

$$
T_k(x) = \begin{cases}
\cos(k \arccos x) & , \quad if \ -1 \leq x \leq 1 \\
\cosh(k \operatorname{arccosh} x) & , \quad if \ x \geq 1 \\
(-1)^k \cosh(k \operatorname{arccosh}(-x)) & , \quad if \ x \leq -1
\end{cases}
$$

9. *The Chebyshev polynomials have the global representation*

$$
T_k(x) = \frac{1}{2}((x + \sqrt{x^2 - 1})^k + (x - \sqrt{x^2 - 1})^k) \ \ for \ \ x \in \mathbf{R} \ . \quad (7.17)
$$

Properties 8 and 9 are most easily checked by verifying that they satisfy the three-term recurrence (including the starting values). The min-max property of the Chebyshev polynomials follows from the intermediate-value theorem:

Theorem 7.19 *Every polynomial $P_n \in \mathbf{P}_n$ with leading coefficient $a_n \neq 0$ attains a value of absolute value $\geq |a_n|/2^{n-1}$ in the interval $[-1,1]$. In particular the Chebyshev polynomials $T_n(x)$ are minimal with respect to the maximum-norm $\|f\|_\infty = \max_{x \in [-1,1]} |f(x)|$ among the polynomials of degree n with leading coefficient 2^{n-1}.*

Proof.   Let $P_n \in \mathbf{P}_n$ be a polynomial with leading coefficient $a_n = 2^{n-1}$ and $|P_n(x)| < 1$ for $x \in [-1,1]$. Then $T_n - P_n$ is a polynomial of degree less then or equal to $n - 1$. At the Chebyshev abscissas $\bar{x}_k := \cos \frac{k\pi}{n}$ we have

$$
\begin{aligned}
T_n(\bar{x}_{2k}) = 1, & \quad P_n(\bar{x}_{2k}) < 1 & \implies & \quad P_n(\bar{x}_{2k}) - T_n(\bar{x}_{2k}) < 0 \\
T_n(\bar{x}_{2k+1}) = -1, & \quad P_n(\bar{x}_{2k+1}) > -1 & \implies & \quad P_n(\bar{x}_{2k+1}) - T_n(\bar{x}_{2k+1}) > 0 \ ,
\end{aligned}
$$

i.e. at the $n + 1$ Chebyshev abscissae, the difference $T_n - P_n$ is alternating between positive and negative and has therefore at least n roots in $[-1,1]$, which contradicts $0 \neq T_n - P_n \in \mathbf{P}_{n-1}$. Thus for each polynomial $P_n \in \mathbf{P}_n$ with leading coefficient $a_n = 2^{n-1}$, there must exist an $x \in [-1,1]$ such that $|P_n(x)| \geq 1$. For an arbitrary polynomial $P_n \in \mathbf{P}_n$ with leading coefficient $a_n \neq 0$, the statement follows from the fact that $\tilde{P}_n := \frac{2^{n-1}}{a_n} P_n$ is a polynomial with leading coefficient $\tilde{a}_n = 2^{n-1}$. $\square$

Example 7.20 When minimizing the approximation error of the polynomial interpolation, we try to find the knots $t_0, \ldots, t_n \in [a,b]$, which solve the min-max problem

$$
\max_{t \in [a,b]} |\omega(t)| = \max_{t \in [a,b]} |(t - t_0) \cdots (t - t_n)| = \min \ .
$$

To put it differently, the goal is to determine the normalized polynomial $\omega(t) \in \mathbf{P}_{n+1}$ with the real roots $t_0, \ldots, t_n$, for which $\max_{t \in [a,b]} |\omega(t)| = \min$. According to theorem 7.19, for the interval $[a, b] = [-1, 1]$, this is just the $(n+1)$-th Chebyshev polynomial $\omega(t) = T_{n+1}(t)$, whose roots

$$t_i = \cos\left(\frac{2i+1}{2n+2}\pi\right) \quad \text{for} \quad i = 0, \ldots, n$$

are just the *Chebyshev knots*.

We shall now derive a second property of the Chebyshev polynomials, which we will be useful in Chapter 8.

Theorem 7.21 *Let $[a, b]$ be an arbitrary interval, and let $t_0 \notin [a, b]$. Then the modified Chebyshev polynomial*

$$\hat{T}_n(t) := \frac{T_n(x(t))}{T_n(x(t_0))} \quad \text{with} \quad x(t) := 2\frac{t-a}{b-a} - 1$$

is minimal with respect to the maximum-norm $\|f\|_\infty = \max_{t \in [a,b]} |f(t)|$ among the polynomials $P_n \in \mathbf{P}_n$ with $P_n(t_0) = 1$.

Proof. Since all roots of $T_n(x(t))$ lie in $[a, b]$, we have $c := T_n(x(t_0)) \neq 0$ and $\hat{T}_n$ is well defined. Furthermore $\hat{T}_n(t_0) = 1$, and $|\hat{T}_n(t)| \leq |c|^{-1}$ for all $t \in [a, b]$. Suppose now that there is a polynomial $P_n \in \mathbf{P}_n$ such that $P_n(t_0) = 1$, and $|P_n(t)| < |c|^{-1}$ for all $t \in [a, b]$, then t_0 is a root of the difference $\hat{T}_n - P_n$, i.e.

$$\hat{T}_n(t) - P_n(t) = Q_{n-1}(t)(t - t_0) \quad \text{for a polynomial } Q_{n-1} \in \mathbf{P}_{n-1}.$$

As in the proof of theorem 7.19, Q_{n-1} changes sign at the Chebyshev abscissae $t_k = t(\bar{x}_k)$ for $k = 0, \ldots, n$ and has therefore at least n distinct roots in $[a, b]$. This contradicts $0 \neq Q_{n-1} \in \mathbf{P}_{n-1}$. $\qquad\square$

7.2 Trigonometric Interpolation

In this section we shall interpolate periodic functions by trigonometric polynomials, i.e. linear combinations of trigonometric functions. Next to the polynomials, this class is one of the most important as far as interpolation is concerned, not least because there is an extremely effective algorithm for the solution of the interpolation problem, the *fast Fourier transform*.

In example 6.20, we have already encountered the algorithm of Goertzel and Reinsch for the evaluation of a trigonometric polynomial

$$f_n(t) = \frac{a_0}{2} + \sum_{j=1}^{n}(a_j \cos jt + b_j \sin jt) = \sum_{j=-n}^{n} c_j e^{ijt} \ .$$

First, we define the N-dimensional spaces of trigonometric polynomials, real as well as complex. In the real case it is necessary to distinguish between the cases of even and odd N. The following considerations will therefore be carried out mostly in the complex case only.

Definition 7.22 By $\mathbf{T}_C^N$ we denote the N-dimensional space of the complex trigonometric polynomials

$$\phi_N(t) = \sum_{j=0}^{N-1} c_j e^{ijt} \quad \text{with} \quad c_j \in \mathbf{C}$$

of degree $N-1$. The N-dimensional spaces $\mathbf{T}_R^N$ contain all real trigonometric polynomials $\phi_N(t)$ of the form

$$\phi_{2n+1}(t) = \frac{a_0}{2} + \sum_{j=1}^{n}(a_j \cos jt + b_j \sin jt) \tag{7.18}$$

for odd $N = 2n + 1$ respectively

$$\phi_{2n}(t) = \frac{a_0}{2} + \sum_{j=1}^{n-1}(a_j \cos jt + b_j \sin jt) + \frac{a_n}{2} \cos nt \tag{7.19}$$

for even $N = 2n$, where $a_j, b_j \in \mathbf{R}$.

The thus defined trigonometric polynomials are apparently 2π-periodic, i.e. $\varphi(t + 2\pi) = \varphi(t)$. Hence, considering interpolation, we choose N distinct nodes

$$0 \le t_0 < t_1 < \ldots < t_{N-1} < 2\pi$$

in the half-open interval $[0, 2\pi[$. Given nodal values $f_0, \ldots, f_{N-1}$ we look for trigonometric polynomials $\phi_N \in T_C^N$ resp. T_R^N satisfying

$$\phi_N(t_j) = f_j \quad \text{for} \quad j = 0, \ldots, N - 1 \ . \tag{7.20}$$

Here, the nodes f_j are of course complex in one case and real in the other. The complex trigonometric interpolation may be easily derived from the standard (complex) polynomial interpolation since we have the bijection

$$T_C^N \longrightarrow \mathbf{P}_{N-1}$$

$$\phi(t) = \sum_{h=0}^{N-1} c_j e^{ijt} \longmapsto P(\omega) = \sum_{j=0}^{N-1} c_j \omega^j .$$

Hence, there is a unique polynomial $\phi_N \in T_C^N$ satisfying the interpolation condition (7.20). For equidistant knots

$$t_j = 2\pi j/N , \ j = 0, \ldots, N-1 ,$$

to which we shall restrict ourselves, we introduce the N-th unit roots $\omega_j := e^{it_j} = e^{2\pi ij/N}$. Thus for the coefficients c_j of the complex trigonometric interpolation $\phi_N \in \mathbf{T}_C^N$, we obtain the Vandermonde system:

$$\underbrace{\begin{bmatrix} 1 & \omega_0 & \omega_0^2 & \cdots & \omega_0^{N-1} \\ \vdots & \vdots & \vdots & & \vdots \\ 1 & \omega_{N-1} & \omega_{N-1}^2 & \cdots & \omega_{N-1}^{N-1} \end{bmatrix}}_{=: V_{N-1}} \begin{pmatrix} c_0 \\ \vdots \\ c_{N-1} \end{pmatrix} = \begin{pmatrix} f_0 \\ \vdots \\ f_{N-1} \end{pmatrix} .$$

Using the complex solution $\phi_N \in T_C^N$ of a *real* interpolation problem, $f_j \in \mathbf{R}$, we can compute the real solution $\phi_N \in T_R^N$ by the following lemma.

Lemma 7.23 *Let*

$$\phi_N(t) = \sum_{j=0}^{N-1} c_j e^{ijt} \in T_C^N$$

be the solution of the interpolation problem (7.20) with real nodes $f_j \in \mathbf{R}$ and equidistant knots $t_j = 2\pi j/N$. Then the real trigonometric polynomial $\psi_N \in T_R^N$, given by the coefficients

$$a_j = 2\Re c_j = c_j + c_{N-j} \quad and \quad b_j = -2\Im c_j = i(c_j - c_{N-j})$$

from the representations (7.18) resp. (7.19), also satisfies the interpolation conditions, i.e. $\varphi_N(t_j) = f_j$.

Proof. We evaluate the polynomial at the N equidistant knots t_k. Since $e^{2\pi i(N-j)/N} = e^{-2\pi ij/N}$, we have

$$\phi_N(t_k) = \sum_{j=0}^{N-1} c_j e^{ijt_k} = \overline{\phi_N(t_k)} = \sum_{j=0}^{N-1} \bar{c}_j e^{-ijt_k} = \sum_{j=0}^{N-1} \bar{c}_{N-j} e^{ijt_k} .$$

The unique interpolation property therefore implies that $c_j = \bar{c}_{N-j}$. In particular, c_0 is real and for even $N = 2n$ also c_{2n}. For odd $N = 2n+1$, we obtain

$$
\begin{aligned}
\phi_N(t_k) &= c_0 + \sum_{j=1}^{2n} c_j e^{ijt_k} = c_0 + \sum_{j=1}^{n}\left(c_j e^{ijt_k} + \bar{c}_j e^{-ijt_k}\right)\\
&= c_0 + \sum_{j=1}^{n} 2\Re(c_j e^{ijt_k}) = c_0 + \sum_{j=1}^{n}(2\Re c_j \cos jt - 2\Im c_j \sin jt)\,.
\end{aligned}
$$

Hence, the real trigonometric polynomial with the coefficients

$$
a_j = 2\Re c_j = c_j + \bar{c}_j = c_j + c_{N-j}
$$

and

$$
b_j = -2\Im c_j = i(c_j - \bar{c}_j) = i(c_j - c_{N-j})
$$

solves the interpolation problem. For even $N = 2n$, the statement follows similarly. □

Lemma 7.23 shows the existence of a real trigonometric interpolating polynomial. However, since the interpolation problem is linear and the numbers of nodes and coefficients coincide, this also implies uniqueness.

In the case at hand, the Vandermonde matrix V_{N-1} can easily be inverted analytically. In order to demonstrate this, we shall first show the orthonormality of the basis functions $\psi_j(t) := e^{ijt}$ with respect to the following scalar product, which is given by the equidistant knots t_j,

$$
\langle f,g\rangle := \frac{1}{N}\sum_{j=0}^{N-1} f(t_j)\overline{g(t_j)}\,. \tag{7.21}
$$

Lemma 7.24 *For the N-th unit roots $\omega_j := e^{2\pi ij/N}$ we have*

$$
\sum_{j=0}^{N-1}\omega_j^k\omega_j^{-l} = N\delta_{kl}\,.
$$

In particular, the functions $\psi_j(t) = e^{ijt}$ are orthonormal with respect to the scalar product (7.21), i.e. $\langle\psi_k,\psi_l\rangle = \delta_{kl}$.

Proof. The statement is obviously equivalent to

$$
\sum_{j=0}^{N-1}\omega_k^j = N\delta_{0k}\,.
$$

(Observe that $\omega_j^k = \omega_k^j$.) Now the N-th unit roots ω_k are solutions of the equation

$$0 = \omega^N - 1 = (\omega - 1)(\omega^{N-1} + \omega^{N-2} + \cdots + 1) = (\omega - 1) \sum_{j=0}^{N-1} \omega^j .$$

If $k \neq 0$, then $\omega_k \neq 1$ and therefore $\sum_{j=0}^{N-1} \omega_k^j = 0$. In the other case we obviously have $\sum_{j=0}^{N-1} \omega_0^j = \sum_{j=0}^{N-1} 1 = N$. $\square$

With the help of this orthogonality relation, we can easily give the solution of the interpolation problem.

Theorem 7.25 *The coefficients c_j of the trigonometric interpolation corresponding to the N nodes (t_k, f_k) with equidistant knots $t_k = 2\pi k/N$, i.e.*

$$\phi_N(t_k) = \sum_{j=0}^{N-1} c_j e^{ijt_k} = \sum_{j=0}^{N-1} c_j \omega_k^j = f_k \ \ for \ k = 0, \ldots, N-1 \,,$$

are given by

$$c_j = \frac{1}{N} \sum_{k=0}^{N-1} f_k \omega_k^{-j} \ \ for \ j = 0, \ldots, N-1.$$

Proof. We insert the given solution for the coefficients c_j and obtain

$$
\begin{aligned}
\sum_{j=0}^{N-1} c_j \omega_l^j &= \sum_{j=0}^{N-1} \left(\frac{1}{N} \sum_{k=0}^{N-1} f_k \omega_k^{-j} \right) \omega_l^j \\
&= \frac{1}{N} \sum_{k=0}^{N-1} f_k \sum_{j=0}^{N-1} \omega_k^{-j} \omega_l^j = \frac{1}{N} \sum_{k=0}^{N-1} f_k \delta_{kl} N = f_l \,.
\end{aligned}
$$

$\square$

Remark 7.26 For odd $N = 2n + 1$, and letting $c_{-j} := c_{N-j}$ for $j > 0$, we can rewrite the trigonometric interpolation polynomial in symmetric form:

$$\varphi_{N-1}(t_k) = \sum_{j=0}^{N-1} c_j e^{ijt_k} = \sum_{j=-n}^{n} c_j e^{ijt_k} \,.$$

In this form, it strongly resembles the truncated Fourier series

$$f_n(t) = \sum_{j=-n}^{n} \hat{f}(j) e^{ijt}$$

of a 2π-periodic function $f \in L^2(\mathbf{R})$ with coefficients

$$\hat{f}(j) = (f, e^{ijt}) = \frac{1}{2\pi} \int_0^{2\pi} f(t)e^{-ijt}\,dt \ . \tag{7.22}$$

In fact, the coefficients c_j can be considered as the approximation of the integral in (7.22) by the trapezoidal sum (compare Section 9.2) with respect to the nodes $t_k = 2\pi k/N$. If we insert this approximation

$$\int_0^{2\pi} g(t)\,dt \ \approx \ \frac{2\pi}{N} \sum_{k=0}^{N-1} g(t_k) \tag{7.23}$$

into (7.22), then this yields

$$\hat{f}(j) \ \approx \ \frac{1}{N} \sum_{k=0}^{N-1} f_k e^{-ijt_k} = \frac{1}{N} \sum_{k=0}^{N-1} e^{-2\pi ijk/N} f_k = c_j \ . \tag{7.24}$$

Observe that the formula (7.23) is actually exact for trigonometric polynomials $g \in \mathbf{T}_C^N$, and equality in (7.24) holds therefore also for $f \in \mathbf{T}_C^N$. For this reason the isomorphism

$$\mathcal{F}_N : \mathbf{C}^N \to \mathbf{C}^N, \quad (f_j) \mapsto (c_j)$$

with

$$c_j = \frac{1}{N} \sum_{k=0}^{N-1} f_k e^{-2\pi ijk/N} \quad \text{for} \quad j = 0, \dots, N-1 \tag{7.25}$$

is called a *discrete Fourier transform*. The inverse mapping $\mathcal{F}_N^{-1}$ is

$$f_j = \sum_{k=0}^{N-1} c_k e^{2\pi ijk/N} \quad \text{for} \quad j = 0, \dots, N-1 \ . \tag{7.26}$$

The computation of the coefficients c_j from the values f_j (or the other way around) is in principle a matrix-vector multiplication, for which we expect a cost of $O(N^2)$ operations. However there is an algorithm which requires only $O(N \log_2 N)$ operations, the *fast Fourier transform*, FFT for short. It is based on a separate analysis of the expressions for the coefficients c_j for odd respectively even indices j, called the *odd even reduction*. This way it is possible to transform the original problem into two similar partial problems of half dimension.

Lemma 7.27 *Let $N = 2M$ be even and $\omega = e^{\pm 2\pi i/N}$. Then the trigonometric sums*

$$\alpha_j = \sum_{k=0}^{N-1} f_k \omega^{kj} \quad \text{for } j = 0, \dots, N-1$$

can be computed as follows, where $\xi := \omega^2$ and $l = 0, \ldots, M - 1$:

$$\alpha_{2l} = \sum_{k=0}^{M-1} g_k \xi^{kl} \quad \text{with} \quad g_k = f_k + f_{k+M}$$

$$\alpha_{2l+1} = \sum_{k=0}^{M-1} h_k \xi^{kl} \quad \text{with} \quad h_k = (f_k - f_{k+M})\omega^k,$$

i.e. the computation of the α_j can be reduced to two similar problems of half dimension $M = N/2$.

Proof.   In the even case $j = 2l$, and because of $\omega^{Nl} = 1$, it follows that

$$\alpha_{2l} = \sum_{k=0}^{N-1} f_k \omega^{2kl}$$

$$= \sum_{k=0}^{N/2-1} \left(f_k \omega^{2kl} + f_{k+N/2}\, \omega^{2(k+N/2)l} \right)$$

$$= \sum_{k=0}^{M-1} (f_k + f_{k+M})(\omega^2)^{kl}.$$

Similarly, because of $\omega^{N/2} = -1$, we obtain for odd indices $j = 2l + 1$,

$$\alpha_{2l+1} = \sum_{k=0}^{N-1} f_k \omega^{k(2l+1)}$$

$$= \sum_{k=0}^{N/2-1} \left(f_k \omega^{k(2l+1)} + f_{k+N/2}\omega^{(k+N/2)(2l+1)} \right)$$

$$= \sum_{k=0}^{M-1} (f_k - f_{k+M})\omega^k (\omega^2)^{kl}.$$

$\square$

The lemma can be applied to the discrete Fourier analysis $(f_k) \mapsto (c_j)$, as well as to the synthesis $(c_j) \mapsto (f_k)$. If the number N of the given points is a power of two $N = 2^p$, $p \in \mathbf{N}$, then we can iterate the process. This algorithm is frequently called *algorithm of* COOLEY-TUKEY [77]. The computation can essentially be carried out on a single vector, if the current number-pairs are overwritten. In the algorithm 7.28, we simply overwrite the input values $f_0, \ldots, f_{N-1}$. However, here the order is interchanged in each reduction step because of the separation of even and odd indices. We have illustrated this permutation of indices in table 7.2. We obtain the right indices by reversing the order of the bits in the dual-representation of the indices.

Table 7.2: interchange of the indices of the fast Fourier transform for $N = 8$, i.e. $p = 3$

k	dual	1.reduction	2.reduction	dual
0	000	0	0	000
1	001	2	4	100
2	010	4	2	010
3	011	6	6	110
4	100	1	1	001
5	101	3	5	101
6	110	5	3	011
7	111	7	7	111

We therefore define a permutation σ,

$$\sigma : \{0, \ldots, N-1\} \longrightarrow \{0, \ldots, N-1\}$$
$$\sum_{j=0}^{p-1} a_j 2^j \longmapsto \sum_{j=0}^{p-1} a_{p-1-j} 2^j, \quad a_j \in \{0,1\},$$

which represents this operation, and which can be realized on a computer at little cost by a corresponding bit-manipulation.

Algorithm 7.28 *Fast Fourier transform.* For $N = 2^p$ and $\omega = e^{\pm 2\pi i/N}$, the algorithm computes from the input values $f_0, \ldots, f_{N-1}$ the transformed values $\alpha_0, \ldots, \alpha_{N-1}$ with $\alpha_j = \sum_{k=0}^{N-1} f_k \omega^{kj}$.

$N_{\text{red}} := N;$
$z := \omega;$
while $N_{\text{red}} > 1$ **do**
    $M_{\text{red}} := N_{\text{red}}/2;$
    for $j := 0$ **to** $N/N_{\text{red}} - 1$ **do**
        $l := jN_{\text{red}};$
        for $k := 0$ **to** $M_{\text{red}} - 1$ **do**

$$a := f_{l+k} + f_{l+k+M_{\text{red}}};$$
$$f_{l+k+M_{\text{red}}} := (f_{l+k} - f_{l+k+M_{\text{red}}})z^k;$$
$$f_{l+k} := a;$$

      end for
    end for
    $N_{\text{red}} := M_{\text{red}};$
    $z := z^2;$
end while
for $k := 0$ **to** $N - 1$ **do**
    $\alpha_{\sigma(k)} := f_k$
end for

In each reduction step, we need $2 \cdot 2^p = 2N$ multiplications, where the evaluation of the exponential function counts for one multiplication (recursive computation of $\cos jx$, $\sin jx$). After $p = \log_2 N$ steps, all $\alpha_0, \ldots, \alpha_{N-1}$ are computed at the cost of $2N \log_2 N$ multiplications.

7.3 Bézier Techniques

The topics which have so far been presented in this chapter belong to the classical part of Numerical Analysis, as the names Lagrange and Newton indicate. With the increasing importance of computer aided construction, new ground has recently been broken (i.e. in the last twenty years) in the interpolation and approximation theory, which we shall indicate in this section. It is interesting that geometric aspects gain a decisive importance here. A curve or surface has to be represented on a computer in a way that it can be drawn and manipulated quickly. In order to achieve this, parametrizations of the geometric objects are used, whose relevant parameters have geometric meaning.

In this introduction we can only illustrate these considerations in the simplest situations. In particular, we shall restrict ourselves to polynomial curves, i.e. one-dimensional geometric objects. The book by DE BOOR [14] and the newer textbook by FARIN [28] are recommended to those who want to familiarize themselves in more detail with this area. The lecture notes [12] also give a very good overview.

We start with a generalization of real-valued polynomials.

Definition 7.29 A *polynomial* (or a *polynomial curve*) *of degree* n in $\mathbf{R}^d$ is a function P of the form

$$P : \mathbf{R} \to \mathbf{R}^d, \quad P(t) = \sum_{i=0}^{n} a_i t^i \quad \text{with} \quad a_0, \dots, a_n \in \mathbf{R}^d, \quad a_n \neq 0 .$$

The space of polynomials of degree less than or equal to n in $\mathbf{R}^d$ is denoted by $\mathbf{P}_n^d$.

The most interesting cases for us are the curves in space $(d = 3)$ or in the plane $(d = 2)$. If $\{P_0, \dots, P_n\}$ is a basis of $\mathbf{P}_n$ and $\{e_1, \dots, e_d\}$ the standard basis of $\mathbf{R}^d$, then the polynomials

$$\{e_i P_j \mid i = 1, \dots, d \text{ and } j = 0, \dots, n\}$$

form a basis of $\mathbf{P}_n^d$. The graph Γ_P of a polynomial $P \in \mathbf{P}_n^d$

$$\Gamma_P : \mathbf{R} \longrightarrow \mathbf{R}^{d+1}, \quad t \longmapsto (t, P(t)),$$

can now again be considered as a polynomial $\Gamma_P \in \mathbf{P}_n^{d+1}$. If P is given in coefficient representation

$$P(t) = a_0 + a_1 t + \cdots + a_n t^n ,$$

then

$$\Gamma_P(t) = \binom{0}{a_0} + \binom{1}{a_1} t + \binom{0}{a_2} t^2 + \cdots + \binom{0}{a_n} t^n .$$

7.3.1   Bernstein polynomials and Bézier representation

So far, we have considered three different bases of the space P_n of polynomials of degree less than or equal to n.

a) Monomial basis $\{1, t, t^2, \dots, t^n\}$

b) Lagrange basis $\{L_0(t), \dots, L_n(t)\}$

c) Newton basis $\{\omega_0(t), \dots, \omega_n(t)\}$

The last two bases are already oriented towards interpolation and depend on the knots $t_0, \dots, t_n$. The basis polynomials, which we shall now present, apply to two parameters $a, b \in \mathbf{R}$. They are therefore very suitable for the *local* representation of a polynomial. In the following, the closed interval *between* the two points a and b is denoted by $[a, b]$ also when $a > b$, i.e. (compare Definition 7.37)

$$[a, b] := \{x = \lambda a + (1 - \lambda)b \mid \lambda \in [0, 1]\} .$$

The first step consists of an affine transformation onto the unit interval $[0, 1]$,

$$
\begin{aligned}
[a, b] &\longrightarrow [0, 1] \\
t &\longmapsto \lambda = \lambda(t) := \frac{t - a}{b - a} ,
\end{aligned}
\qquad (7.27)
$$

with the help of which we can usually restrict our consideration to $[0, 1]$. By virtue of the binomial theorem, we can represent the unit function as

$$1 = ((1 - \lambda) + \lambda)^n = \sum_{i=0}^{n} \binom{n}{i} (1 - \lambda)^{n-i} \lambda^i .$$

The terms of this partition of unity are just the *Bernstein polynomials* with respect to the interval $[0, 1]$. By composing these with the above affine transformation (7.27), we then obtain the Bernstein polynomials with respect to the interval $[a, b]$.

Definition 7.30 The *i-th Bernstein polynomial* of degree n with respect to the interval $[0, 1]$ is the polynomial $B_i^n \in \mathbf{P}_n$ with

$$B_i^n(\lambda) := \binom{n}{i} (1 - \lambda)^{n-i} \lambda^i ,$$

where $i = 0, \ldots, n$. Similarly, $B_i^n(\,\cdot\,; a, b) \in \mathbf{P}_n$ with

$$B_i^n(t; a, b) := B_i^n(\lambda(t)) = B_i^n\left(\frac{t - a}{b - a}\right) = \frac{1}{(b - a)^n} \binom{n}{i} (t - a)^i (b - t)^{n-i}$$

is the *i-th Bernstein polynomial* of degree n with respect to the interval $[a, b]$.

Instead of $B_i^n(t; a, b)$, we shall in the following often simply write $B_i^n(t)$, if confusion with the Bernstein polynomials $B_i^n(\lambda)$ with respect to $[0, 1]$ is impossible. In the following theorem we list the most important properties of the Bernstein polynomials.

Theorem 7.31 *The Bernstein polynomials $B_i^n(\lambda)$ satisfy the following properties:*

1. $\lambda = 0$ *is a multiplicity i root of B_i^n.*

2. $\lambda = 1$ *is a multiplicity $(n - i)$ root of B_i^n.*

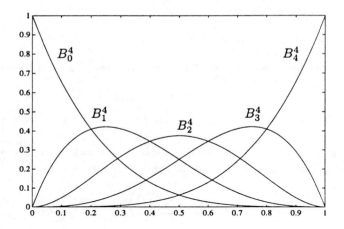

Figure 7.4: Bernstein polynomials for $n = 4$.

3. $B_i^n(\lambda) = B_{n-i}^n(1 - \lambda)$ for $i = 0, \ldots, n$ (symmetry).

4. $(1 - \lambda)B_0^n = B_0^{n+1}$ and $\lambda B_n^n = B_{n+1}^{n+1}$.

5. The Bernstein polynomials B_i^n are non-negative on $[0, 1]$ and form a partition of unity, i.e.

$$B_i^n(\lambda) \geq 0 \quad \text{for} \quad \lambda \in [0, 1] \quad \text{and} \quad \sum_{i=0}^{n} B_i^n(\lambda) = 1 \quad \text{for} \quad \lambda \in \mathbf{R} \, .$$

6. B_i^n has exactly one maximum value in the interval $[0, 1]$, namely at $\lambda = i/n$.

7. The Bernstein polynomials satisfy the recurrence relation

$$B_i^n(\lambda) = \lambda B_{i-1}^{n-1}(\lambda) + (1 - \lambda)B_i^{n-1}(\lambda) \tag{7.28}$$

for $i = 1, \ldots, n$ and $\lambda \in \mathbf{R}$.

8. The Bernstein polynomials form a basis $\mathcal{B} := \{B_0^n, \ldots, B_n^n\}$ of $\mathbf{P}_n$.

Proof.   The first five statements are either obvious or can be easily verified. Statement 6. follows from the fact that

$$\frac{d}{d\lambda}B_i^n(\lambda) = \binom{n}{i}(1 - \lambda)^{n-i-1}\lambda^{i-1}(i - n\lambda)$$

for $i = 1, \ldots, n$. The recurrence relation (7.28) follows from the definition and the formula

$$\binom{n}{i} = \binom{n-1}{i-1} + \binom{n-1}{i}$$

for the binomial coefficients. For the last statement, we show that the $n+1$ polynomials B_i^n are linearly independent. If

$$0 = \sum_{i=0}^{n} b_i B_i^n(\lambda) \ ,$$

then according to 1. and 2.,

$$0 = \sum_{i=0}^{n} b_i B_i^n(1) = b_n B_n^n(1) = b_n$$

and therefore inductively $b_0 = \cdots = b_n = 0$. $\qquad\qquad\square$

Similar statements are of course true for the Bernstein polynomials with respect to the interval $[a, b]$. Here the maximum value of $B_i^n(t; a, b)$ in $[a, b]$ is attained at

$$t = a + \frac{i}{n}(b - a) \ .$$

Remark 7.32 The property that the Bernstein polynomials form a partition of unity is equivalent to the fact that the Bézier points are *affine invariant*. If $\phi : \mathbf{R}^d \to \mathbf{R}^d$ is an affine mapping,

$$\phi : \mathbf{R}^d \longrightarrow \mathbf{R}^d \ \text{ with } \ A \in \text{Mat}_d(\mathbf{R}) \ \text{ and } \ v \in \mathbf{R}^d$$
$$u \longmapsto Au + v,$$

then the images $\phi(b_i)$ of the Bézier points b_i of a polynomial $P \in \mathbf{P}_n^d$ are the Bézier points of $\phi \circ P$.

We now know that we can write any polynomial $P \in \mathbf{P}_n^d$ as a linear combination with respect to the Bernstein basis

$$P(t) = \sum_{i=0}^{n} b_i B_i^n(t; a, b), \quad b_i \in \mathbf{R}^d \ . \tag{7.29}$$

Remark 7.33 The symmetry $B_i^n(\lambda) = B_{n-i}^n(1 - \lambda)$ of the Bernstein polynomials yields in particular

$$\sum_{i=0}^{n} b_i B_i^n(t; a, b) = \sum_{i=0}^{n} b_{n-i} B_i^n(t; b, a) \ ,$$

i.e. the Bézier coefficients with respect to b, a are just the ones of a, b in reverse order.

The coefficients $b_0, \ldots, b_n$ are called *control* or *Bézier points* of P, the corresponding polygonal path a *Bézier polygon*. Because of

$$\lambda = \sum_{i=0}^{n} \frac{i}{n} B_i^n(\lambda) \quad \Longrightarrow \quad t = \sum_{i=0}^{n} \left(a + \frac{i}{n}(b-a)\right) B_i^n(t; a, b)$$

the Bézier points of the polynomial $P(t) = t$ are, just the maxima $b_i = a + \frac{i}{n}(b-a)$ of the Bernstein polynomials. The Bézier representation of the graph Γ_P of a polynomial P as in (7.29) is therefore just

$$\Gamma_P(t) = \begin{pmatrix} t \\ P(t) \end{pmatrix} = \sum_{i=0}^{n} \begin{pmatrix} a + \frac{i}{n}(b-a) \\ b_i \end{pmatrix} B_i^n(t; a, b) . \tag{7.30}$$

In Figure 7.5 we have plotted the graph of a cubic polynomial together with

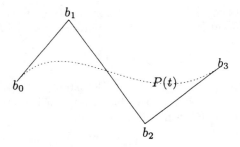

Figure 7.5: Cubic polynomial with its Bézier points

its Bézier polygon. It is striking that the shape of the curve is closely related to the shape of the Bézier polygon. In the following we shall more closely investigate this geometric meaning of the Bézier points. First, it is clear from Theorem 7.31 that the beginning and ending points of the polynomial curve and the Bézier polygon coincide. Furthermore, it appears that the tangents at the boundary points also coincide with the straight lines at the end of the Bézier polygon. In order to verify this property, we compute the derivatives of a polynomial in Bézier representation. We shall restrict ourselves to the derivatives of the Bézier representation with respect to the unit interval $[0, 1]$. Together with the derivative of the affine transformation $\lambda(t)$ from $[a, b]$ onto $[0, 1]$,

$$\frac{d}{dt}\lambda(t) = \frac{1}{b-a},$$

one immediately obtains the derivatives in the general case also.

Lemma 7.34 *The derivative of the Bernstein polynomials $B_i^n(\lambda)$ with respect to $[0,1]$ satisfies*

$$\frac{d}{d\lambda} B_i^n(\lambda) = \begin{cases} -nB_0^{n-1} & \text{for} \quad i = 0 \\ n(B_{i-1}^{n-1}(\lambda) - B_i^{n-1}(\lambda)) & \text{for} \quad i = 1, \ldots, n-1 \\ nB_{n-1}^{n-1}(\lambda) & \text{for} \quad i = n. \end{cases}$$

Proof. The statement follows from

$$\frac{d}{d\lambda} B_i^n(\lambda) = \binom{n}{i} \left[i(1-\lambda)^{n-i}\lambda^{i-1} - (n-i)(1-\lambda)^{n-i-1}\lambda^i \right]$$

by virtue of the identities of theorem 7.31. □

Theorem 7.35 *Let $P(\lambda) = \sum_{i=0}^n b_i B_i^n(\lambda)$ be a polynomial in Bézier representation with respect to $[0,1]$. Then the k-th derivative of P satisfies*

$$P^{(k)}(\lambda) = \frac{n!}{(n-k)!} \sum_{i=0}^{n-k} \Delta^k b_i B_i^{n-k}(\lambda),$$

where the forward difference operator Δ operates on the lower index, i.e.

$$\Delta^1 b_i := b_{i+1} - b_i \quad and \quad \Delta^k b_i := \Delta^{k-1} b_{i+1} - \Delta^{k-1} b_i \quad for \quad k > 1.$$

Proof. Induction over k, see Exercise 7.6. □

Corollary 7.36 *For the boundary points $\lambda = 0, 1$ one obtains the values*

$$P^{(k)}(0) = \frac{n!}{(n-k)!} \Delta^k b_0 \quad and \quad P^{(k)}(1) = \frac{n!}{(n-k)!} \Delta^k b_{n-k},$$

thus in particular up to the second derivative,

a) $P(0) = b_0$ *and* $P(1) = b_n$,

b) $P'(0) = n(b_1 - b_0)$ *and* $P'(1) = n(b_n - b_{n-1})$,

c) $P''(0) = n(n-1)(b_2 - 2b_1 + b_0)$ *and* $P''(1) = n(n-1)(b_n - 2b_{n-1} + b_{n-2})$.

Proof.   Note that $B_i^{n-k}(0) = \delta_{0,i}$ and $B_i^{n-k}(1) = \delta_{n-k,i}$.   □

Corollary 7.36 therefore confirms the geometric observations which we described above. It is important that at a boundary point, the curve is determined up to the k-th derivative by the k closest Bézier points. This property will be crucial later on for the purpose of joining several pieces together. The Bézier points are of further geometric importance. In order to describe this, we need the notion of a *convex hull* of a set $A \subset \mathbf{R}^d$, which we shall briefly review.

Definition 7.37 A set $A \subset \mathbf{R}^d$ is called a *convex*, if together with any two points $x, y \in A$, the straight line which joins them is also contained in A, i.e.

$$[x, y] := \left\{ \lambda x + (1 - \lambda)y \mid \lambda \in [0, 1] \right\} \subset A \quad \text{for all} \quad x, y \in A .$$

The *convex hull* co(A) of a set $A \subset \mathbf{R}^d$ is the smallest convex subset of $\mathbf{R}^d$ which contains A. A linear combination of the form

$$x = \sum_{i=1}^{k} \lambda_i x_i, \quad \text{with} \quad x_i \in \mathbf{R}^d, \quad \lambda_i \geq 0 \quad \text{and} \quad \sum_{i=1}^{k} \lambda_i = 1$$

is called *convex combination* of $x_1, \ldots, x_k$.

Remark 7.38 The convex hull co(A) of $A \subset \mathbf{R}^d$ is the set of all convex combinations of points of A, i. e.

$$\text{co}(A) = \bigcap \{ B \subset \mathbf{R}^d \mid B \text{ convex with } A \subset B \}$$

$$= \left\{ x = \sum_{i=1}^{m} \lambda_i x_i \;\middle|\; m \in \mathbf{N}, \; x_i \in A, \; \lambda_i \geq 0, \; \sum_{i=1}^{m} \lambda_i = 1 \right\} .$$

The following theorem states that a polynomial curve is always contained in the convex hull of its Bézier points.

Theorem 7.39 *The image $P([a, b])$ of a polynomial $P \in \mathbf{P}_n^d$ in Bernstein representation $P(t) = \sum_{i=0}^{n} b_i B_i^n(t; a, b)$ with respect to $[a, b]$ is contained in the convex hull of the Bézier points b_i, i.e.*

$$P(t) \in \text{co}(b_0, \ldots, b_n) \quad \text{for} \quad t \in [a, b] .$$

In particular, the graph of the polynomial for $t \in [a, b]$ is contained in the convex hull of the points $\mathbf{b}_i$.

Proof.   On $[a, b]$, the Bernstein polynomials form a non-negative partition of unity, i.e. $B_i^n(t; a, b) \geq 0$ for $t \in [a, b]$ and $\sum_{i=0}^{n} B_i^n(t) = 1$. Therefore

$$P(t) = \sum_{i=0}^{n} b_i B_i^n(t; a, b)$$

is a convex combination of the Bézier points $b_0, \ldots, b_n$. The second statement follows from the Bézier representation (7.30) of the graph Γ_P of P. $\qquad\square$

As one can already see in Figure 7.5, for a cubic polynomial $P \in \mathbf{P}_3$, this means that the graph of P for $t \in [a, b]$ is completely contained in the convex hull of the four Bézier points $\mathbf{b}_1, \mathbf{b}_2, \mathbf{b}_3$ and $\mathbf{b}_4$. The name control point is explained by the fact that, because of their geometric significance, the points $\mathbf{b}_i$ can be used to control a polynomial curve. Because of Theorem 7.31, at the position $\lambda = i/n$ over which it is plotted, the control point b_i has the greatest 'weight' $B_i^n(\lambda)$. This is another reason that the curve between a and b is closely related to the Bézier polygon, as the figure indicates.

7.3.2 De Casteljau's algorithm

Besides the geometric interpretation of the Bézier points, the importance of the Bézier representation rests mainly on the fact that there is an algorithm which builds on continued convex combinations, and which, besides the function value $P(t)$ at an arbitrary position t, also yields information about the derivatives. Furthermore, the same algorithm can be used to subdivide the Bézier curve into two segments. By repeating this partitioning into segments, then the sequence of the Bézier polygons converges extremely fast to the curve (exponentially when dividing the interval into halves), so that this method is very well suited to effectively plot a curve, e.g. in computer graphics. This construction principle is also tailor-made to control a milling cutter, which can only 'remove' material. We start with the definition of the *partial polynomials* of P.

Definition 7.40 Let $P(\lambda) = \sum_{i=0}^{n} b_i B_i^n(\lambda)$ be a polynomial in Bézier representation with respect to $[0, 1]$. Then we define the *partial polynomials* $b_i^k \in \mathbf{P}_k^d$ of P for $i = 0, \ldots, n - k$ by

$$b_i^k(\lambda) := \sum_{j=0}^{k} b_{i+j} B_j^k(\lambda) = \sum_{j=i}^{i+k} b_j B_{j-i}^k(\lambda) \ .$$

For a polynomial $P(t) = \sum_{i=0}^{n} b_i B_i^n(t; a, b)$ in Bézier representation with respect to $[a, b]$, the partial polynomials b_i^k are similarly defined by

$$b_i^k(t; a, b) := b_i^k(\lambda(t)) = \sum_{j=0}^{k} b_{i+j} B_j^k(t; a, b) \ .$$

Thus the partial polynomial $b_i^k \in \mathbf{P}_k^d$ is just the polynomial which is defined by the Bézier points $b_i, \ldots, b_{i+k}$ (see figure 7.6). If no confusion arises,

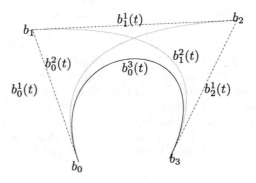

Figure 7.6: Cubic polynomial with its partial polynomials

then we simply write $b_i^k(t)$ for $b_i^k(t; a, b)$. In particular, $b_0^n(t) = P(t)$ is the starting polynomial, and the $b_i^0(t) = b_i$ are its Bézier points for all $t \in \mathbf{R}$. Furthermore, for all boundary points,

$$b_i^k(a) = b_i \quad \text{and} \quad b_i^k(b) = b_{i+k} \ .$$

Similar to the Aitken lemma, the following recurrence relation is true, which is the base for the algorithm of de Casteljau.

Lemma 7.41 *The partial polynomials* $b_i^k(t)$ *of* $P(t) = \sum_{i=0}^n b_i B_i^n(t; a, b)$ *satisfy the recurrence relation*

$$b_i^k = (1 - \lambda)b_i^{k-1} + \lambda b_{i+1}^{k-1} \quad \text{with} \quad \lambda = \lambda(t) = \frac{t - a}{b - a} \qquad (7.31)$$

for $k = 0, \dots, n$ *and* $i = 0, \dots, n - k$.

Proof. We insert the recurrence relation (7.28) into the definition of the partial polynomials b_i^k and obtain

$$
\begin{aligned}
b_i^k &= \sum_{j=0}^k b_{i+j} B_j^k \\
&= b_i B_0^k + b_{i+k} B_k^k + \sum_{j=1}^{k-1} b_{i+j} B_j^k \\
&= b_i (1 - \lambda) B_0^{(k-1)} + b_{i+k} \lambda B_{k-1}^{k-1} + \sum_{j=1}^{k-1} b_{i+j} \left((1 - \lambda) B_j^{k-1} + \lambda B_{j-1}^{k-1} \right)
\end{aligned}
$$

$$= \sum_{j=0}^{k-1} b_{i+j}(1-\lambda)B_j^{k-1} + \sum_{j=1}^{k} b_{i+j}\lambda B_{j-1}^{k-1}$$

$$= (1-\lambda)b_i^{k-1} + \lambda b_{i+1}^{k-1} .$$

$\square$

Because of $b_i^0(t) = b_i$, by continued convex combination (which, for $t \notin [a,b]$ is only an affine combination) we can compute the function value $P(t) = b_0^n(t)$ from the Bézier points. The auxiliary points b_i^k can, similar to the scheme of Neville, be arranged in the *de Casteljau scheme*.

$$
\begin{array}{ccccccccc}
b_n & = & b_n^0 & & & & & & \\
& & & \searrow & & & & & \\
b_{n-1} & = & b_{n-1}^0 & \rightarrow & b_{n-1}^1 & & & & \\
& \vdots & & & & \ddots & & & \\
b_1 & = & b_1^0 & \rightarrow & \cdots & \rightarrow & b_1^{n-1} & & \\
& & & \searrow & & & & \searrow & \\
b_0 & = & b_0^0 & \rightarrow & \cdots & \rightarrow & b_0^{n-1} & \rightarrow & b_0^n
\end{array}
\qquad (7.32)
$$

In fact, the derivatives of P are hidden behind the auxiliary points b_i^k of de Casteljau's scheme, as the following theorem shows. Here we again only consider the Bézier representation with respect to the unit interval $[0,1]$.

Theorem 7.42 *Let $P(\lambda) = \sum_{i=0}^{n} b_i B_i^n(\lambda)$ be a polynomial in Bézier representation with respect to $[0,1]$. Then the derivatives $P^{(k)}(\lambda)$ for $k = 0,\ldots,n$ can be computed from the partial polynomials $b_i^k(\lambda)$ via the relation*

$$P^{(k)}(\lambda) = \frac{n!}{(n-k)!}\Delta^k b_0^{n-k}(\lambda),$$

where $\Delta b_i^k = b_{i+1}^k - b_i^k$.

Proof. The statement follows from Theorem 7.35 and from the fact that the forwards difference operator commutes with the sum:

$$P^{(k)}(\lambda) = \frac{n!}{(n-k)!}\sum_{i=0}^{n-k}\Delta^k b_i B_i^{n-k}(\lambda) = \frac{n!}{(n-k)!}\Delta^k \sum_{i=0}^{n-k} b_i B_i^{n-k}(\lambda)$$

$$= \frac{n!}{(n-k)!}\Delta^k b_0^{n-k}(\lambda) .$$

□

Thus the k-th derivative $P^{(k)}(\lambda)$ at the position λ is computed from the $(n-k)$-th column of the de Casteljau scheme. In particular

$$
\begin{aligned}
P(t) &= b_0^n \,, \\
P'(t) &= n(b_1^{n-1} - b_0^{n-1}) \,, \\
P''(t) &= n(n-1)(b_2^{n-2} - 2b_1^{n-2} + b_0^{n-2}) \,.
\end{aligned}
\tag{7.33}
$$

So far we have only considered the Bézier representation of a single polynomial with respect to a fixed reference interval. Here the question remains open on how the Bézier points are transformed when we change the reference interval (see Figure 7.7). It would also be interesting to know how to

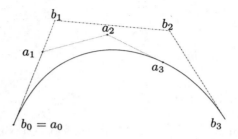

Figure 7.7: cubic polynomial with two Bézier representations

join several pieces of polynomial curves continuously or smoothly (see Figure 7.8). Finally, we would be interested in the possibility of subdividing curves, in the sense that we subdivide the reference interval and compute the Bézier points for the subintervals (see Figure 7.9). According to Theorem 7.39, the curve is contained in the convex hull of the Bézier points. Hence, it is clear that the Bézier polygons approach more and more closely the curve, when the subdivision is refined. These three questions are closely related, which can readily be seen from the figures. We shall see that they can be easily resolved in the context of the Bézier technique. The connecting elements are the partial polynomials. We have already seen in Corollary 7.36 that, at a boundary point, a Bézier curve P is determined to the k-th derivative by the k closest Bézier points. The opposite is also true: The values of P up to the k-th derivative at the position $\lambda = 0$ already determine the Bézier points $b_0, \ldots, b_k$. More precisely, this is even true for the partial polynomials $b_0^0(\lambda), \ldots, b_0^k(\lambda)$, as we shall prove in the following lemma.

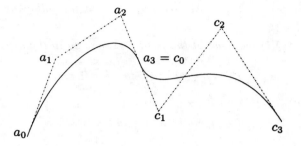

Figure 7.8: Two cubic Bézier curves with C^1-smoothness at $a_3 = c_0$

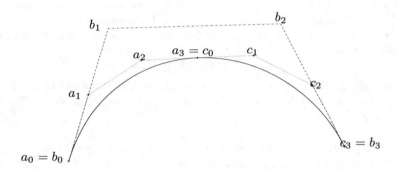

Figure 7.9: Subdivision of a cubic Bézier curve

Lemma 7.43 *The partial polynomial $b_0^k(\lambda)$ of a Bézier curve $P(\lambda) = b_0^n(\lambda)$*
$= \sum_{i=0}^n b_i B_i^n(\lambda)$ is completely determined by the values of P up to and in-
cluding the k-th derivative at the position $\lambda = 0$.

Proof. According to Theorem 7.42, the derivatives at the position $\lambda = 0$
satisfy

$$\frac{d^l}{d\lambda^l} b_0^k(0) = \frac{k!}{(k-l)!} \Delta^l b_0 = \frac{(n-l)!\,k!}{(k-l)!\,n!} \frac{d^l}{d\lambda^l} b_0^n(0)$$

for $l = 0, \ldots, k$. The statement follows, because a polynomial is completely
determined by all derivatives at one position. □

Together with Corollary 7.36, we obtain the following theorem:

Theorem 7.44 *Let* $P(t) = a_0^n(t; a, b)$ *and* $Q(t) = b_0^n(t; a, c)$ *be two Bézier curves with respect to* a, b *respectively* a, c. *Then the following statements are equivalent:*

i) $P(t)$ *and* $Q(t)$ *coincide at the position* $t = a$ *up to the* k-*th derivative, i.e.*

$$P^{(l)}(a) = Q^{(l)}(a) \quad for \quad l = 0, \ldots, k .$$

ii) $a_0^k(t; a, b) = b_0^k(t; a, c)$ *for all* $t \in \mathbf{R}$.

iii) $a_0^l(t; a, b) = b_0^l(t; a, c)$ *for all* $t \in \mathbf{R}$ *and* $l = 0, \ldots, k$.

iv) $a_l = b_0^l(b; a, c)$ *for* $l = 0, \ldots, k$.

Proof.   We show $i) \Leftrightarrow ii) \Rightarrow iii) \Rightarrow iv) \Rightarrow ii)$. According to Corollary 7.36 and Lemma 7.43, the two curves $P(t)$ and $Q(t)$ coincide at the position $t = a$ up to the k-th derivative, iff they have the same partial polynomials $a_0^k(t; a, b) = b_0^k(t; a, c)$. The two first statements are therefore equivalent. If a_0^k and b_0^k coincide, then so do their partial polynomials a_0^l and b_0^l for $l = 0, \ldots, k$, i.e. ii) implies iii). By inserting $t = b$ into iii), it follows in particular that

$$a_l = a_0^l(1) = a_0^l(b; a, b) = b_0^l(b; a, c) ,$$

and therefore iv). Since a polynomial is uniquely determined by its Bézier coefficients, iv) therefore implies ii) and thus the equivalence of the four statements.   □

With this result in hand, we can easily answer our three questions. As a first corollary we compute the Bézier points which are created when *subdividing* the reference interval. At the same time, this answers the question regarding the change of the reference interval.

Corollary 7.45 *Let*

$$a_0^n(t; a, b) = b_0^n(t; a, c) = c_0^n(t; b, c)$$

be the Bézier representations of a polynomial curve $P(t)$ *with respect to the intervals* $[a, b]$, $[a, c]$ *and* $[b, c]$, *i.e.*

$$P(t) = \sum_{i=0}^{n} a_i B_i^n(t; a, b) = \sum_{i=0}^{n} b_i B_i^n(t; a, c) = \sum_{i=0}^{n} c_i B_i^n(t; b, c)$$

(see Figure 7.9). Then the Bézier coefficients a_i *and* c_i *of the partial curves can be computed from the Bézier coefficients* b_i *with respect to the entire interval via*

$$a_k = b_0^k(b; a, c) \quad and \quad c_k = b_k^{n-k}(b; a, c)$$

for $k = 0, \ldots, n$.

Proof. Because a polynomial of degree n is completely determined by its derivatives at one point, the statement follows from Theorem 7.44 for $k = n$ and the symmetry of the Bézier representation, see Remark 7.33. □

Since the curve pieces always lie in the convex hull of their Bézier points, the corresponding Bézier polynomials converge to the curve when continuously subdivided. By employing this method, the evaluation of a polynomial is very stable, since only convex combinations are computed in the algorithm of de Casteljau. In Figure 7.10, we have always divided the reference interval of a Bézier curve of degree 4 in half, and we have plotted the Bézier polygon of the first three subdivisions. After only a few subdivisions, it is almost impossible to distinguish the curve from the polygonal path.

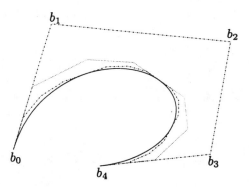

Figure 7.10: Threefold subdivision of a Bézier curve of degree $n = 4$

If we do utilize the fact that only the derivatives at one position must coincide, then we can solve the problem of continuously joining two polygonal curves:

Corollary 7.46 *A joined Bézier curve*

$$R(t) = \begin{cases} a_0^n(t; a, b) & \text{if} \quad a \le t < b \\ c_0^n(t; b, c) & \text{if} \quad b \le t \le c \end{cases}$$

is C^k-smooth, iff

$$c_l = a_{n-l}^l(c; a, b) \quad \text{for} \quad l = 0, \dots, k$$

or, equivalently,

$$a_{n-l} = c_0^l(a; b, c) \quad \text{for} \quad l = 0, \dots, k .$$

Therefore, through the C^k-smoothness, the first $k+1$ Bézier points of the second partial curve are determined by the last $k+1$ Bézier points of the first and vice versa. A polynomial $a_0^n(t; a, b)$ over $[a, b]$ can therefore be continued C^k-smoothly by a polynomial $c_0^n(t; b, c)$ over $[b, c]$, by determining the Bézier points $c_0, \ldots, c_k$ according to Corollary 7.45 by employing the algorithm of de Casteljau, whereas the remaining $c_{k+1}, \ldots, c_n$ can be chosen freely.

In particular, the joined curve $R(t)$ is *continuous*, iff

$$a_n = c_0 .$$

It is *continuously differentiable*, iff in addition

$$
\begin{aligned}
c_1 &= a_{n-1}^1(c; a, b) \\
&= a_{n-1}^1(\lambda) = (1 - \lambda)a_{n-1} + \lambda a_n \quad \text{with} \quad \lambda = \frac{c - a}{b - a}
\end{aligned}
$$

or, equivalently,

$$
\begin{aligned}
a_{n-1} &= c_0^1(a; b, c) \\
&= c_0^1(\mu) = (1 - \mu)c_0 + \mu c_1 \quad \text{with} \quad \mu = \frac{a - b}{c - b} .
\end{aligned}
$$

This implies that

$$a_n = c_0 = \frac{c - b}{c - a}a_{n-1} + \frac{b - a}{c - a}c_1 , \tag{7.34}$$

i.e. the point $a_n = c_0$ has to divide the segment $[a_{n-1}, c_1]$ in the proportion $c - b$ to $b - a$. If the pieces of curves fit C^2-*smoothly*, then a_{n-2}, a_{n-1} and a_n describe the same parabola as c_0, c_1 and c_2, namely with respect to $[a, b]$ respectively $[b, c]$. According to Corollary 7.46, the Bézier points of this parabola with respect to the entire interval $[a, c]$ are a_{n-2}, d and c_2, where d is the auxiliary point

$$d := a_{n-2}^1(c; a, b) = a_{n-2}^1(\lambda) = c_1^1(a; b, c) = c_1^1(\mu)$$

(see Figure 7.11). Furthermore, according to Corollary 7.46, it follows from the C^2-smoothness that

$$c_2 = a_{n-2}^2(\lambda) = (1 - \lambda) \underbrace{a_{n-2}^1(\lambda)}_{= d} + \lambda \underbrace{a_{n-1}^1(\lambda)}_{= c_1}$$

and

$$a_{n-2} = c_0^2(\mu) = (1 - \mu) \underbrace{c_0^1(\mu)}_{= a_{n-1}} + \mu \underbrace{c_1^1(\mu)}_{= d} .$$

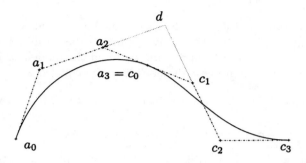

Figure 7.11: Two cubic Bézier curves with C^2-transition at $a_3 = c_0$

The joined curve is therefore C^2-smooth iff there is a point d such that

$$c_2 = (1 - \lambda)d + \lambda c_1 \quad \text{and} \quad a_{n-2} = (1 - \mu)a_{n-1} + \mu d .$$

The auxiliary point d, the *de Boor point*, will play an important role in the next section in the construction of cubic splines.

7.4   Splines

As we have seen, the classical polynomial interpolation is incapable of solving the approximation problem with a large number of equidistant knots. Polynomials of high degree tend to oscillate a lot, as the sketches of the Lagrange polynomials indicate (see Figure 7.2). They may thus not only spoil the condition number (small changes of the nodes f_i induce large changes of the interpolation polynomial $P(t)$ at intermediate values $t \neq t_i$), but also lead to large oscillations of the interpolating curve between the nodes. As one can imagine, such oscillations are highly undesirable. One need only think of the induced vibrations of an airfoil formed according to such an interpolation curve. If we require that an interpolating curve passes 'as smooth as possible' through given nodes (t_i, f_i), then it is obvious to locally use polynomials of *lower degree* and to join these at the nodes. As a first possibility, we have encountered the cubic Hermite interpolation in Example 7.7, which was however dependent on the special prescription of function values and derivatives at the nodes. A second possibility are the *spline functions*, with which we shall be concerned in this chapter.

7.4.1   Spline spaces and B-splines

We start with the definition of k-th order splines over a grid $\Delta = \{t_0, \ldots, t_{l+1}\}$ of node points. These functions have proven to be an extremely versatile tool, from interpolation and approximation and the modelling in CAGD to collocation and Galerkin methods for differential equations.

Definition 7.47 Let $\Delta = \{t_0, \ldots, t_{l+1}\}$ be a *grid* of $l + 2$ pairwise distinct node points

$$a = t_0 < t_1 < \cdots < t_{l+1} = b .$$

A *spline* of degree $k - 1$ (order k) with respect to Δ is a function $s \in C^{k-2}[a, b]$, which on each interval $[t_i, t_{i+1}]$ for $i = 0, \ldots, l$ coincides with a polynomial $s_i \in \mathbf{P}_{k-1}$ of degree $\leq k - 1$. The space of splines of degree $k - 1$ with respect to Δ is denoted by $\mathbf{S}_{k,\Delta}$.

The most important spline functions are the *linear splines* of order $k = 2$ (see Figure 7.12) and the *cubic splines* of order $k = 4$ (see Figure 7.13). The linear splines are the continuous, piecewise linear functions with respect to the intervals $[t_i, t_{i+1}]$.   The cubic splines are best suited for the graphic

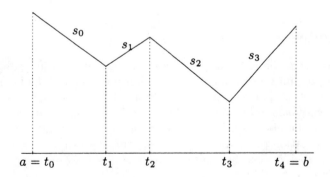

Figure 7.12: Linear splines, order $k = 2$

representation of curves, since the eye can still recognize discontinuities of curvature, i.e. of the second derivative. Thus the C^2-smooth cubic splines are recognized as 'smooth'.

It is obvious that $\mathbf{S}_{k,\Delta}$ is a real vector space, which, in particular, contains all polynomials of degree $\leq k - 1$, i.e. $\mathbf{P}_{k-1} \subset \mathbf{S}_{k,\Delta}$. Furthermore, the

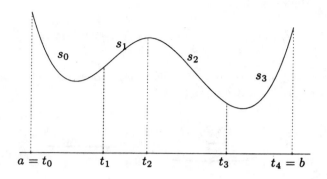

Figure 7.13: Cubic splines, order $k = 4$

truncated powers of degree k,

$$(t - t_i)_+^k := \begin{cases} (t - t_i)^k & \text{if} \quad t \geq t_i \\ 0 & \text{if} \quad t < t_i \end{cases}$$

are contained in $\mathbf{S}_{k,\Delta}$. Together with the monomials $1, t, \ldots, t^{k-1}$, they form a basis of $\mathbf{S}_{k,\Delta}$, as we shall show in the following theorem:

Theorem 7.48 *The monomials and truncated powers form a basis*

$$\mathcal{B} := \{1, t, \ldots, t^{k-1}, (t - t_1)_+^{k-1}, \ldots, (t - t_l)_+^{k-1}\} \quad (7.35)$$

of the spline space $\mathbf{S}_{k,\Delta}$. *In particular, the dimension of* $\mathbf{S}_{k,\Delta}$ *is*

$$\dim \mathbf{S}_{k,\Delta} = k + l .$$

Proof. We first show that one has at most $k + l$ degrees of freedom for the construction of a spline $s \in \mathbf{S}_{k,\Delta}$. On the interval $[t_0, t_1]$, we can choose any polynomial of degree $\leq k - 1$; these are k free parameters. Because of the smoothness requirement $s \in C^{k-2}$, the polynomials on the following intervals $[t_1, t_2], \ldots, [t_l, t_{l+1}]$ are determined by their predecessor up to one parameter. Thus, we have another l parameters. Therefore $\dim \mathbf{S}_{k,\Delta} \leq k + l$. The remaining claim is that the $k + l$ functions in $\mathcal{B}$ are linearly independent. To prove this, let

$$s(t) := \sum_{i=0}^{k-1} a_i t^i + \sum_{i=1}^{l} c_i (t - t_i)_+^{k-1} = 0 \quad \text{for all} \quad t \in [a, b] .$$

By applying the linear functionals

$$G_i(f) := \frac{1}{(k-1)!} \left(f^{(k-1)}(t_i^+) - f^{(k-1)}(t_i^-) \right)$$

to s (where $f(t^+)$ and $f(t^-)$ denote the right respectively left-sided limits), then for all $i = 1, \ldots, l$, it follows that

$$0 = G_i(s) = \underbrace{G_i\left(\sum_{j=0}^{k-1} a_j t^j \right)}_{=0} + \sum_{j=1}^{l} c_j \underbrace{G_i(t - t_j)_+^{k-1}}_{=\delta_{ij}} = c_i \;.$$

Thus $s(t) = \sum_{i=0}^{k-1} a_i t^i = 0$ for all $t \in [a, b]$, and therefore also $a_0 = \cdots = a_{k-1} = 0$. $\square$

However, the basis $\mathcal{B}$ of $\mathbf{S}_{k,\Delta}$ given in (7.35) has several disadvantages. For one, the basis elements are not local; e.g., the support of the monomials t^i is the whole of $\mathbf{R}$. Second, the truncated powers are "almost" linearly dependent for close knots t_i, t_{i+1}. This results in the fact that the evaluation of a spline in the representation

$$s(t) = \sum_{i=0}^{k-1} a_i t^i + \sum_{i=1}^{l} c_i (t - t_i)_+^{k-1}$$

is poorly conditioned with respect to perturbations in the coefficients c_i. Third, the coefficients a_i and c_i have no geometric meaning like e.g. the Bézier points b_i have. In the following, we shall therefore construct a basis for the spline space $\mathbf{S}_{k,\Delta}$, which has properties as good as the Bernstein basis has for $\mathbf{P}_k$. In order to achieve this, we define recursively the following generalization of the characteristic function $\chi_{[\tau_i, \tau_{i+1}[}$ of an interval and also the 'hat function' (see Figure 7.14).

Definition 7.49 Let $\tau_1 \leq \cdots \leq \tau_n$ be an arbitrary sequence of knots. Then the *B-splines* $N_{ik}(t)$ of order k for $k = 1, \ldots, n$ and $i = 1, \ldots, n - k$ are recursively defined by

$$N_{i1}(t) := \chi_{[\tau_i, \tau_{i+1}[}(t) = \begin{cases} 1 & \text{if} \quad \tau_i \leq t < \tau_{i+1} \\ 0 & \text{else} \end{cases}, \tag{7.36}$$

$$N_{ik}(t) := \frac{t - \tau_i}{\tau_{i+k-1} - \tau_i} N_{i,k-1}(t) + \frac{\tau_{i+k} - t}{\tau_{i+k} - \tau_{i+1}} N_{i+1,k-1}(t) \;. \tag{7.37}$$

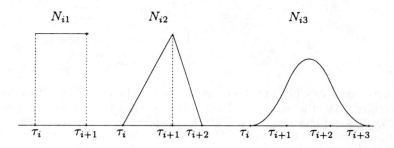

Figure 7.14: B-splines of order $k = 1, 2, 3$

Note that the characteristic function in (7.36) vanishes if the knots coincide, i.e.

$$N_{i1} = \chi_{[\tau_i, \tau_{i+1}[} = 0 \quad \text{if} \quad \tau_i = \tau_{i+1} .$$

The corresponding terms are omitted according to our convention $0/0 = 0$ in the recurrence relation (7.37). Thus, even if the knots coincide, the B-splines N_{ik} are well-defined by (7.36) and (7.37); furthermore, $N_{ik} = 0$ if $\tau_i = \tau_{i+k}$. The following properties are obvious because of the recursive definition.

Remark 7.50 The B-splines satisfy:

a) $\text{supp} N_{ik} \subset [\tau_i, \ldots, \tau_{i+k}]$ (local support),

b) $N_{ik}(t) \geq 0$ for all $t \in \mathbf{R}$ (non-negative),

c) N_{ik} is a piecewise polynomial of degree $\leq k - 1$ with respect to the intervals $[\tau_j, \tau_{j+1}]$.

In order to derive further properties, it is convenient to represent the B-splines in closed form. In fact, they can be written as an application of a k-th divided difference $[\tau_i, \ldots, \tau_{i+k}]$ to the truncated powers $f(s) = (s - t)_+^{k-1}$.

Lemma 7.51 *If* $\tau_i < \tau_{i+k}$, *then the B-spline* N_{ik} *satisfies*

$$N_{ik}(t) = (\tau_{i+k} - \tau_i)[\tau_i, \ldots, \tau_{i+k}](\cdot - t)_+^{k-1} .$$

Proof. For $k = 1$, we obtain for the right hand side

$$(\tau_{i+1} - \tau_i)[\tau_i, \tau_{i+1}](\cdot - t)_+^{k-1} \quad = \quad (\tau_{i+1} - \tau_i)\frac{(\tau_i - t)_+^0 - (\tau_{i+1} - t)_+^0}{\tau_i - \tau_{i+1}}$$

$$= \quad \begin{cases} 1 & , \quad \text{if} \quad \tau_i \leq t < \tau_{i+1} \\ 0 & , \quad \text{else} \end{cases} .$$

Furthermore, by employing the Leibniz formula (Lemma 7.15), it can easily be verified that the right hand side also satisfies the recurrence relation (7.37). The statement now follows inductively. $\qquad\square$

Corollary 7.52 *If τ_j is a m-fold knot,i.e.*

$$\tau_{j-1} < \tau_j = \cdots = \tau_{j+m-1} < \tau_{j+m} ,$$

then, at the position τ_j, N_{ik} is at least $(k-1-m)$-times continuously differentiable. The derivative of N_{ik} satisfies

$$N_{ik}'(t) = (k-1) \left(\frac{N_{i,k-1}(t)}{\tau_{i+k-1} - \tau_i} - \frac{N_{i+k,k-1}(t)}{\tau_{i+k} - \tau_{i+1}} \right) .$$

Proof. The first statement follows from the fact that the divided difference $[\tau_i, \ldots, \tau_{i+k}]f$ contains at most the $(m-1)$-th derivative of the function f at the position τ_j. However, the truncated power $f(s) = (s - \tau_j)_+^{k-1}$ is $(k-2)$ times continuously differentiable. The second statement follows from

$$\begin{aligned} N_{ik}'(t) &= -(k-1)(\tau_{i+k} - \tau_i)[\tau_i, \ldots, \tau_{i+k}](\cdot - t)_+^{k-2} \\ &= -(k-1)(\tau_{i+k} - \tau_i) \\ &\quad \left(\frac{[\tau_{i+1}, \ldots, \tau_{i+k}](\cdot - t)_+^{k-2} - [\tau_i, \ldots, \tau_{i+k-1}](\cdot - t)_+^{k-2}}{\tau_{i+k} - \tau_i} \right) \\ &= (k-1) \left(\frac{N_{i,k-1}(t)}{\tau_{i+k-1} - \tau_i} - \frac{N_{i+1,k-1}(t)}{\tau_{i+k} - \tau_{i+1}} \right) . \end{aligned}$$

$\qquad\square$

We now return to the space $\mathbf{S}_{k,\Delta}$ of splines of order k with respect to the grid $\Delta = \{t_j\}_{j=0,\ldots,l+1}$:

$$\Delta : a = t_0 < t_1 < \cdots < t_{l+1} = b .$$

For the construction of the desired basis, to Δ we assign the following *extended sequence of knots* $T = \{\tau_j\}_{j=1,\ldots,n+k}$, where the boundary knots

$a = t_0$ and $b = t_{l+1}$ are counted k times.

$$\Delta: \quad a \quad = \quad t_0 \quad < \quad t_1 \quad < \cdots < \quad t_{l+1} \quad = \quad b$$

$$\| \qquad\qquad \| \qquad\qquad\qquad \|$$

$$T: \quad \tau_1 = \cdots \quad = \quad \tau_k \quad < \quad \tau_{k+1} \quad < \cdots < \quad \tau_{n+1} \quad = \quad \cdots = \tau_{n+k}$$

Here $n = l + k = \dim \mathbf{S}_{k,\Delta}$ is the dimension of the spline space $\mathbf{S}_{k,\Delta}$. Consider the n B-splines N_{ik} for $i = 1, \ldots, n$ which correspond to the extended sequence of knots $T = \{\tau_j\}$. In the following, we shall see that they form the desired basis of $\mathbf{S}_{k,\Delta}$ (see Figure 7.15). To begin with, it is clear from

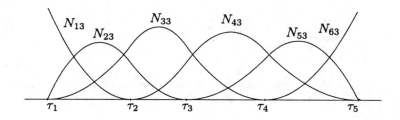

Figure 7.15: B-spline basis of order $k = 3$ (locally quadratic)

Corollary 7.52 that the B-splines N_{ik} are indeed splines of order k, i.e.

$$N_{ik} \in \mathbf{S}_{k,\Delta} \quad \text{for all} \quad i = 1, \ldots, n \ .$$

Since the number n coincides with the dimension $n = \dim \mathbf{S}_{k,\Delta}$, it only remains to show that they are linearly independent. For this, we need the following technical statement, which is also known as *Marsden identity*.

Lemma 7.53 *With the above notation, we have for all $t \in [a, b]$ and $s \in \mathbf{R}$ that*

$$(t - s)^{k-1} = \sum_{i=1}^{n} \varphi_{ik}(s) N_{ik}(t) \quad \text{with} \quad \varphi_{ik}(s) := \prod_{j=1}^{k-1} (\tau_{i+j} - s) \ .$$

Proof. Because of the recursive definition of the B-splines, the proof is by induction over k. The statement is clear for $k = 1$, because of $1 = \sum_{i=1}^{n} N_{i1}(t)$. Thus let $k > 1$, and suppose that the statement is true for all

$l \leq k - 1$. Insertion of the recurrence relation (7.37) on the right hand side yields

$$\sum_{i=1}^{n} \varphi_{ik}(s) N_{ik}(t)$$

$$= \sum_{i=2}^{n} \left(\frac{t - \tau_i}{\tau_{i+k-1} - \tau_i} \varphi_{ik}(s) + \frac{\tau_{i+k-1} - t}{\tau_{i+k-1} - \tau_i} \varphi_{i-1,k}(s) \right) N_{i,k-1}(t)$$

$$= \sum_{i=2}^{n} \prod_{j=1}^{k-2} (\tau_{i+j} - s) \cdot$$

$$\cdot \underbrace{\left(\frac{t - \tau_i}{\tau_{i+k-1} - \tau_i}(\tau_{i+k-1} - s) + \frac{\tau_{i+k-1} - t}{\tau_{i+k-1} - \tau_i}(\tau_i - s) \right)}_{= t - s} N_{i,k-1}(t)$$

$$= (t - s) \sum_{i=2}^{n} \varphi_{i,k-1}(s) N_{i,k-1}(t)$$

$$= (t - s)(t - s)^{k-2} = (t - s)^{k-1}.$$

Here note that the expression which is 'bracketed from below' is the linear interpolation of $t - s$, hence $t - s$ itself. $\quad\square$

Corollary 7.54 *The space* $\mathbf{P}_{k-1}[a,b]$ *of polynomials of degree* $\leq k - 1$ *over* $[a,b]$ *is contained in the space, which is spanned by the B-splines of order* k, *i.e.*

$$\mathbf{P}_{k-1}[a,b] \subset span\ (N_{1k}, \ldots, N_{nk}).$$

In particular,

$$1 = \sum_{i=1}^{n} N_{ik}(t)\ \ for\ all\ \ t \in [a,b],$$

i.e. the B-splines form a partition of unity on $[a,b]$.

Proof. For the l-th derivative of the function $f(s) := (t - s)^{k-1}$, it follows from the Marsden identity that

$$f^{(l)}(0) = (k - 1) \cdots (k - l)(-1)^l t^{k-l-1} = \sum_{i=1}^{n} \varphi_{ik}^{(l)}(0) N_{ik}(t)$$

and therefore, with $m = k - l - 1$,

$$t^m = \frac{(-1)^{k-m-1}}{(k - 1) \cdots (m + 1)} \sum_{i=1}^{n} \varphi_{ik}^{(k-m-1)}(0) N_{ik}(t).$$

The $(k-1)$-th derivative of ϕ_{ik} satisfies

$$\phi_{ik}^{k-1}(s) = \left(\prod_{j=1}^{k-1} (\tau_{i+j} - s) \right)^{k-1} = ((-1)^{k-1} s^{k-1} + \cdots)^{k-1} = (-1)^{k-1}(k-1)!$$

and the second statement thus also follows. □

After these preparations, we can now prove the linear independence of the B-splines. They are *locally* independent as the following theorem shows.

Theorem 7.55 *The B-splines N_{ik} are locally linear independent, i.e. if*

$$\sum_{i=1}^{n} c_i N_{ik}(t) = 0 \quad \text{for all} \ \ t \in]c, d[\subset [a, b]$$

and $]c, d[\cap]\tau_i, \tau_{i+k}[\neq \phi$, *then*

$$c_i = 0 \ .$$

Proof. Without loss of generality, we may assume that the open interval $]c, d[$ does not contain any knots (otherwise we decompose $]c, d[$ into sub-intervals). According to Corollary 7.54, each polynomial of degree $\leq k - 1$ over $]c, d[$ can be represented by the B-splines N_{ik}. However, only $k = \dim \mathbf{P}_{k-1}$ B-splines are different from zero on the interval $]c, d[$. They therefore have to be linearly independent. □

Let us summarize briefly what we have shown: The B-splines N_{ik} of order k with respect to the sequence of knots $T = \{\tau_j\}$ form a basis $\mathcal{B} := \{N_{1k}, \ldots, N_{nk}\}$ of the spline space $\mathbf{S}_{k,\Delta}$. They are locally linear independent, are locally supported, and form a positive partition of unity. Each spline $s \in \mathbf{S}_{k,\Delta}$ therefore has a unique representation as a linear combination of the form

$$s = \sum_{i=1}^{n} d_i N_{ik} \ .$$

The coefficients d_i are called *de Boor points* of s. The function values $s(t)$ are therefore convex combinations of the de Boor points d_i. For the evaluation, we can use the recursive definition of the B-Splines N_{ik}, and we can therefore also derive the recurrence relation for the linear combinations themselves which is given in Exercise 7.9, the *algorithm of de Boor*.

Remark 7.56 By employing the Marsden identity, one can explicitly give the dual basis $\mathcal{B}' = \{\nu_1, \ldots, \nu_n\}$ of the B-spline basis $\mathcal{B}$,

$$\nu_j : \mathbf{S}_{k,\Delta} \to \mathbf{R} \ \text{ linearly with } \ \nu_j(N_{ik}) = \delta_{ij} \ .$$

With this at hand, it can be shown that there is a constant D_k, which depends only on the order k, such that

$$D_k \max_{j=1,\ldots,n} |d_j| \leq \| \sum_{j=1}^{n} d_j N_{jk} \|_\infty \leq \max_{j=1,\ldots,n} |d_j| \ .$$

Here the second inequality follows from the fact that the B-splines form a positive partition of unity. Perturbations in the function values $s(t)$ of the spline $s = \sum_{i=1}^{n} c_i N_{ik}$ and the coefficients can therefore be estimated against each other. In particular, the evaluation of a spline in B-spline representation is *well conditioned*. Therefore, the basis is also called *well conditioned*.

7.4.2   Spline interpolation

We now turn our attention again towards the problem of interpolating a function f which is given pointwise on a grid $\Delta = \{t_0, \ldots, t_{l+1}\}$,

$$a = t_0 < t_1 < \cdots < t_{l+1} = b \ .$$

In the linear case $k = 2$, the number $l + 2$ of knots coincides with the dimension of the spline space $n = \dim \mathbf{S}_{2,\Delta} = l + k$. The linear B-splines N_{i2} with respect to the extended sequence of knots

$$T = \{\tau_1 = \tau_2 < \cdots < \tau_{n+1} = \tau_{n+2}\} \quad \text{with} \quad \tau_j = t_{j-2} \ \text{for} \ j = 2, \ldots, n$$

satisfy $N_{i2}(t_j) = \delta_{j+1,i}$. The piecewise linear spline $I_2 f \in \mathbf{S}_{2,\Delta}$, which interpolates f, is therefore uniquely determined with

$$I_2 f = \sum_{i=1}^{n} f(t_{i-1}) N_{i2} \ .$$

Besides this very simple case of linear spline interpolation, the case $k = 4$ of cubic splines plays the most important role in the applications. In this case, we are missing two conditions to uniquely characterize the interpolating cubic spline $s \in \mathbf{S}_{i,\Delta}$, because

$$\dim \mathbf{S}_{4,\Delta} - \text{number of knots} = l + k - l - 2 = 2 \ .$$

Now the starting idea for the construction of spline functions was to find interpolating curves which are as 'smooth' as possible; we could also say 'possibly least curved'. The *curvature* of a parametric curve (in the plane)

$$y : [a, b] \to \mathbf{R}, \quad y \in C^2[a, b]$$

at $t \in [a, b]$ is given by

$$\kappa(t) := \frac{y''(t)}{(1 + y'(t)^2)^{3/2}} \; .$$

The absolute value of the curvature is just the reciprocal $1/r$ of the radius r of the osculating circle to the curve at the point $(t, y(t))$ (see Figure 7.16), i. e. the curvature is zero iff the osculating circle has the radius ∞, hence

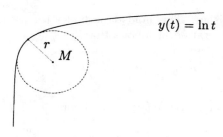

Figure 7.16: Osculating circle of the curve $(t, y(t))$.

the curve is straight. In order to simplify this, instead of the curvature, we consider for small $y'(t)$ the reasonable approximation $y''(t)$,

$$\frac{y''(t)}{(1 + y'(t)^2)^{3/2}} \approx y''(t) \; ,$$

and measure the curvature of the entire curve by the L_2-norm

$$\|y''\|_2 = \left(\int_a^b y''(t)^2 \, dt \right)^{\frac{1}{2}}$$

of this approximation with respect to $[a, b]$. The interpolating cubic splines, which satisfy the additional properties of Corollary 7.58, minimize this functional.

Theorem 7.57 *Let s be an interpolating cubic spline of f at the knots $a = t_0 < \cdots < t_{l+1} = b$, and let $y \in C^2[a, b]$ be an arbitrary interpolating function of f such that*

$$[s(t)''(y(t)' - s(t)')]_{t=a}^b = 0 \; . \tag{7.38}$$

Then

$$\|s''\|_2 \le \|y''\|_2 \; . \tag{7.39}$$

Proof. Trivially, $y'' = s'' + (y'' - s'')$, and, inserted into the right hand side of (7.39), it follows that

$$\int_a^b (y'')^2 \, dt = \int_a^b (s'')^2 \, dt + 2 \underbrace{\int_a^b s''(y'' - s'') \, dt}_{(*)} + \underbrace{\int_a^b (y'' - s'')^2 \, dt}_{\geq 0}$$

$$\geq \int_a^b (s'')^2 \, dt \,,$$

if the term (*) vanishes. This holds true under the assumption (7.38), because by partial integration, it follows that

$$\int_a^b s''(y'' - s'') \, dt = [s''(y' - s')]_a^b - \int_a^b s'''(y' - s') \, dt \,,$$

where s''' is in general discontinuous at the knots $t_1, \ldots, t_{n-1}$, and is constant

$$s'''(t) = s_i'''(t) = d_i \quad \text{for} \quad t \in (t_i, t_{i+1})$$

in the interior of the subintervals (the s_i are cubic polynomials). Therefore, under the assumption (7.38), it is true that

$$\begin{aligned}
\int_a^b s''(y'' - s'') \, dt &= -\sum_{i=1}^n \int_{t_{i-1}}^{t_i} d_i (y' - s_i') \, dt \\
&= -\sum_{i=1}^n d_i \int_{t_{i-1}}^{t_i} y' - s_i' \, dt \\
&= -\sum_{i=1}^n d_i [\underbrace{(y(t_i) - s(t_i))}_{=0} - \underbrace{(y(t_{i-1}) - s(t_{i-1}))}_{=0}] \\
&= 0.
\end{aligned}$$

$\square$

Corollary 7.58 *In addition to the interpolation conditions $s(t_i) = f(t_i)$, assume that the cubic spline $s \in \mathbf{S}_{4,\Delta}$ satisfies one of the following boundary conditions:*

i) $s'(a) = f'(a)$ *and* $s'(b) = f'(b)$,

ii) $s''(a) = s''(b) = 0$,

iii) $s'(a) = s'(b)$ *and* $s''(a) = s''(b)$ *(if f is periodic with period $b - a$).*

Then there exists a unique solution $s \in \mathbf{S}_{4,\Delta}$ which satisfies this boundary condition. An arbitrary interpolating function $y \in C^2[a,b]$, which satisfies the same boundary condition, furthermore satisfies

$$\|s''\|_2 \le \|y''\|_2 .$$

Proof. The requirements are linear in s, and their number coincides with the dimension $n = l + 4$ of the spline space $\mathbf{S}_{4,\Delta}$. It is therefore sufficient to show that the trivial spline $s \equiv 0$ is the only solution for the null-function $f \equiv 0$. Since $y \equiv 0$ satisfies all requirements, Theorem 7.57 implies that

$$\|s''\|_2 \le \|y''\|_2 = 0 . \tag{7.40}$$

Since s'' is continuous, this implies $s'' \equiv 0$, i.e. s is a continuously differentiable, piecewise linear function with $s(t_i) = 0$, and is therefore the null-function. $\square$

The three types i), ii) and iii) are called *complete*, *natural* and *periodic* cubic spline interpolation. The physical interpretation of the above minimization property (7.40) accounts for the name 'spline'. If $y(t)$ describes the position of a thin wooden beam, then

$$E = \int_a^b \left(\frac{y''(t)}{(1 + y'(t)^2)^{3/2}} \right)^2 dt$$

measures the 'deformation energy' of the beam. Because of Hamilton's principle, the beam takes a position so that this energy is minimized. For small deformations one has approximately

$$E \approx \int_a^b y''(t)^2\, dt = \|y''\|_2^2 .$$

The interpolating cubic spline $s \in \mathbf{S}_{4,\Delta}$ therefore describes approximately the position of a thin wooden beam which is fixed at the knots t_i. In the complete spline interpolation, we have clamped the beam at the boundary knots with an additional prescription of the slopes. The natural boundary conditions correspond to the situation when the beam is straight outside the interval $[a,b]$. Such thin wooden beams were in fact used as drawing tools and are called 'splines'.

Note that besides the function values, two additional pieces of information regarding the original function f at the knots enter in the complete spline interpolation. Thus their approximation properties (particularly at the boundary) are better than the ones of the other types ii) and iii). In fact, the complete interpolating spline $I_4 f \in \mathbf{S}_{4,\Delta}$ approximates a function $f \in C^4[a,b]$ of the order h^4, where

$$h := \max_{i=0,\dots,l} |t_{i+1} - t_i|$$

is the largest distance of the knots t_i. We state the following related result by HALL and MEYER [45] without proof.

Theorem 7.59 *Let $I_4 f \in \mathbf{S}_{4,\Delta}$ be the complete interpolating spline of a function $f \in C^4[a,b]$ with respect to the knots t_i with $h := \max_i |t_{i+1} - t_i|$. Then*

$$\|f - I_4 f\|_\infty \le \frac{5}{384} h^4 \|f^{(4)}\|_\infty .$$

Note that this estimate is independent of the position of the knots t_i.

7.4.3   Computation of cubic splines

In the following, we shall derive a system of equations for the cubic interpolating splines. For this purpose, we describe the spline $s \in \mathbf{S}_{4,\Delta}$ by employing the local Bézier representation

$$s(t) = s_i(t) = \sum_{j=0}^{3} b_{3i+j} B_j^3(t; t_i, t_{i+1}) \quad \text{for} \ \ t \in [t_i, t_{i+1}] \tag{7.41}$$

of the partial polynomials s_i with respect to the intervals $[t_i, t_{i+1}]$. Here

$$B_j^3(t; t_i, t_{i+1}) = B_j^3\left(\frac{t - t_i}{h_i}\right) \quad \text{with} \ \ h_i := t_{i+1} - t_i .$$

The continuity of s enters implicitly into the representation (7.41). By (7.34), the C^1-smoothness implies that

$$b_{3i} = \frac{h_i}{h_{i-1} + h_i} b_{3i-1} + \frac{h_{i-1}}{h_{i-1} + h_i} b_{3i+1} . \tag{7.42}$$

Furthermore, according to the C^2-smoothness of s, we have shown that there are de Boor points d_i such that

$$b_{3i+2} = -\frac{h_i}{h_{i-1}} d_i + \frac{h_{i-1} + h_i}{h_{i-1}} b_{3i+1} \tag{7.43}$$

$$b_{3i-2} = \frac{h_{i-1} + h_i}{h_i} b_{3i-1} - \frac{h_{i-1}}{h_i} d_i . \tag{7.44}$$

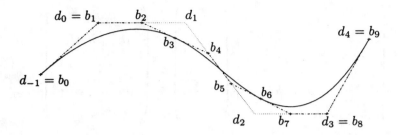

Figure 7.17: Cubic spline with de Boor points d_i and Bézier points b_i.

Graphically, this means that the straight line segment between b_{3i-2} and d_i respectively d_i and b_{3i+2} is partitioned at the ratio $h_{i-1} : h_i$ by the Bézier points b_{3i-1} respectively b_{3i+1}. The points d_i, b_{3i+1}, b_{3i+2} and d_{i+1} are therefore positioned as shown in figure 7.17. Taken together, this implies

$$b_{3i+1} = \frac{h_i + h_{i+1}}{h_{i-1} + h_i + h_{i+1}} d_i + \frac{h_{i-1}}{h_{i-1} + h_i + h_{i+1}} d_{i+1} \qquad (7.45)$$

$$b_{3i-1} = \frac{h_{i-2} + h_{i-1}}{h_{i-2} + h_{i-1} + h_i} d_i + \frac{h_i}{h_{i-2} + h_{i-1} + h_i} d_{i-1} \qquad (7.46)$$

If we define at the boundary $h_{-1} := h_{l+1} := 0$ and

$$d_0 := b_1, \quad d_{-1} := b_0, \quad d_{l+1} := b_{3l+2} \quad \text{and} \quad d_{l+2} := b_{3(l+1)}, \qquad (7.47)$$

then the Bézier coefficients b_{3i+j}, and thus also the spline s, are completely determined by the $l+4$ points d_{-1} to d_{l+2} and the equations (7.42) to (7.47).

By inserting the interpolation conditions

$$f_i = s(t_i) = b_{3i} \quad \text{for} \quad l = 0, \dots, l+1,$$

then it follows at the boundary that

$$d_{-1} := f_0 \quad \text{and} \quad d_{l+2} = f_{l+1}.$$

The remaining points $d_0, \ldots, d_{l+1}$ of the interpolating spline must solve the following system (proof as an exercise):

$$
\begin{bmatrix}
1 & & & & \\
\alpha_1 & \beta_1 & \gamma_1 & & \\
& \ddots & \ddots & \ddots & \\
& & \alpha_l & \beta_l & \gamma_l \\
& & & & 1
\end{bmatrix}
\begin{pmatrix}
d_0 \\
\vdots \\
\vdots \\
\vdots \\
d_{l+1}
\end{pmatrix}
=
\begin{pmatrix}
b_1 \\
(h_0 + h_1)f_1 \\
\vdots \\
(h_{l-1} + h_l)f_l \\
b_{3l+2}
\end{pmatrix}
\tag{7.48}
$$

with

$$
\begin{aligned}
\alpha_i &:= \frac{h_i^2}{h_{i-2} + h_{i-1} + h_i} \\
\beta_i &:= \frac{h_i(h_{i-2} + h_{i-1})}{h_{i-2} + h_{i-1} + h_i} + \frac{h_{i-1}(h_i + h_{i+1})}{h_{i-1} + h_i + h_{i+1}} \\
\gamma_i &:= \frac{h_{i-1}^2}{h_{i-1} + h_i + h_{i+1}} .
\end{aligned}
$$

We now only have to determine the Bézier points b_1 and b_{3l+2} from the boundary conditions. We confine ourselves to the first two types. For the *complete spline interpolation*, we obtain from

$$
f_0' = s'(a) = \frac{3}{h_0}(b_1 - b_0) \quad \text{and} \quad f_{l+1}' = s'(b) = \frac{3}{h_l}(b_{3(l+1)} - b_{3l+2})
$$

that we have to set

$$
b_1 = \frac{h_0}{3}f_0' + f_0 \quad \text{and} \quad b_{3l+2} = -\frac{h_l}{3}f_{l+1}' + f_{l+1}. \tag{7.49}
$$

For the *natural boundary conditions* we have to choose b_1 and b_{3l+2}, so that $s''(a) = s''(b) = 0$. This is satisfied for

$$
b_1 := b_0 = f_0 \quad \text{and} \quad b_{3l+2} := b_{3(l+1)} = f_{l+1} \tag{7.50}
$$

(see figure 7.18).

Remark 7.60 For an equidistant grid, i.e. $h_i = h$ for all i, we have

$$
\alpha_i = \gamma_i = \frac{h}{3} \quad \text{and} \quad \beta_i = \frac{4h}{3} \quad \text{for} \quad i = 2, \ldots, l - 1 .
$$

In this case (and also for an almost equidistant grid) the matrix is strictly diagonally dominant, and it can therefore be solved efficiently and in a stable manner by the Gaussian elimination without interchanging columns.

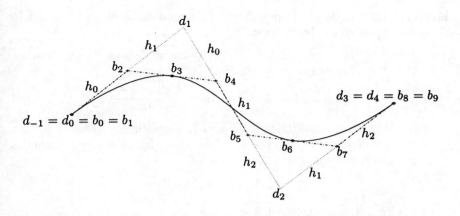

Figure 7.18: Cubic interpolating spline with natural boundary conditions

Remark 7.61 The de Boor points d_i are just the B-spline coefficients of the interpolating cubic splines, i.e.

$$s = \sum_{i=1}^{l+4} d_{i-2} N_{i4} \; ,$$

if, as above, N_{i4} are the B-splines for the extended sequence of knots $T = \{\tau_j\}$.

7.5   Exercises

Exercise 7.1 Let $\Lambda_n(K,I)$ denote the Lebesgue constant with respect to the set of knots K on the interval I.

a) Let $K = \{t_0, \ldots, t_n\} \subset I = [a,b]$ be pairwise distinct knots. Suppose that the affine transformation

$$\chi : I \to I_0 = [-1,1], \quad t \mapsto \frac{2t - a - b}{b - a}$$

of this interval onto the unit interval I_0 maps the set of knots K onto the set of knots $K_0 = \chi(K)$. Show that the Lebesgue constant is invariant under this transformation, i.e.

$$\Lambda_n(K,I) = \Lambda_n(K_0, I_0) .$$

b) Let $K = \{t_0, \ldots, t_n\}$ with $a \le t_0 < t_1 < \cdots < t_n \le b$ be knots in the interval $I = [a,b]$. Give the affine transformation

$$\chi : [t_0, t_n] \to I$$

on I, which satisfies the property that for $\bar{K} = \chi(K) = \{\bar{t}_0, \ldots, \bar{t}_n\}$:

$$a = \bar{t}_0 < \bar{t}_1 < \cdots < \bar{t}_n = b .$$

Show that

$$\Lambda_n(\bar{K}, I) \le \Lambda_n(K, I) ,$$

i.e. inclusion of the boundary knots improves the Lebesgue constant.

Exercise 7.2 Consider the class of functions

$$\mathcal{F} := \left\{ f \in C^{n+1}[-1,1] \mid \|f^{(n+1)}\|_\infty \le (n+1)! \right\} .$$

For $f \in \mathcal{F}$, let $p_n(f)$ denote the polynomial of degree n of the (Hermite) interpolation for the knots $K = \{t_0, \ldots, t_n\} \subset I_0 = [-1,1]$.

a) Show that

$$\varepsilon_n(K) := \sup_{f \in \mathcal{F}} \|f - p_n(f)\|_\infty = \|\omega_{n+1}\|_\infty ,$$

where $\omega_{n+1}(t) = (t - t_0) \cdots (t - t_n)$.

b) Show that $\varepsilon_n(K) \ge 2^{-n}$ and that equality holds iff K is the set of the Chebyshev knots, i.e.

$$t_j = \cos \frac{2j + 1}{2n + 2} \pi \quad \text{for } j = 0, \ldots, n .$$

Exercise 7.3 Count how many computations and how much storage space an economically written program requires for the evaluation of interpolation polynomials on the basis of the Lagrange representation.

Compare with the algorithms of Aitken-Neville and the representation over Newton's divided differences.

Exercise 7.4 Let $a = t_0 < t_1 < \cdots < t_{n-1} < t_n = b$ be a distribution of knots in the interval $I = [a, b]$. For a continuous function $g \in C(I)$, the *interpolating polygon* $\mathcal{I}g \in C(I)$ is defined by

a) $\mathcal{I}g(t_i) = g(t_i)$ for $i = 0, \ldots, n$,

b) $\mathcal{I}g|_{[t_i, t_{i+1}]}$ is a polynomial of degree one for $i = 0, \ldots, n-1$.

Show:

a) Any function $g \in C^2(I)$ satisfies

$$\|g - \mathcal{I}g\|_\infty \leq \frac{h^2}{8} \|g''\|_\infty ,$$

where $h = \max_{0 \leq i \leq n-1} (t_{i+1} - t_i)$ is the 'grid-width parameter'.

b) The absolute condition of the polygonal interpolation satisfies

$$\kappa_{\mathrm{abs}} = 1 .$$

Discuss and evaluate the difference between this and the polynomial interpolation.

Exercise 7.5 For the approximation of the first derivative of a pointwise given function f, one utilizes the first divided difference

$$(D_h f)(x) := [x, x + h]f .$$

a) Estimate the approximation error $|D_h f(x) - f'(x)|$ for $f \in C^3$. (Leading order in h for $h \to 0$ is sufficient.)

b) Instead of $D_h f(x)$, the floating point arithmetic computes $\hat{D}_h f(x)$. Estimate the error $|\hat{D}_h f(x) - D_h f(x)|$ in leading order.

c) Which h turns out to be optimal, i.e. minimizes the total error?

d) Test your prediction at $f(x) = e^x$ at the position $x = 1$ with

$$h = 10^{-1}, \ 5 \cdot 10^{-2}, \ 10^{-2}, \ldots, \ \mathrm{eps} .$$

Exercise 7.6 Show that the derivatives of a polynomial in Bézier representation with respect to the interval $[t_0, t_1]$

$$P(t) = \sum_{i=0}^{n} b_i B_i^n(\lambda), \quad \lambda := \frac{t - t_0}{t_1 - t_0},$$

are given by

$$\frac{d^k}{dt^k} P(t) = \frac{n!}{(n-k)! h^k} \sum_{i=0}^{n-k} \Delta^k b_i B_i^{n-k}(\lambda), \quad h := t_1 - t_0.$$

Exercise 7.7 Find the Bézier representation with respect to $[0, 1]$ of the Hermite polynomials H_i^3 for the knots t_0, t_1, and sketch the Hermite polynomials together with the Bézier polygons.

Exercise 7.8 We have learned three different bases for the space $\mathbf{P}_3$ of polynomials of degree ≤ 3: the monomial basis $\{1, t, t^2, t^3\}$, the Bernstein basis $\{B_0^3(t), B_1^3(t), B_2^3(t), B_3^3(t)\}$ with respect to the interval $[0, 1]$, and the Hermite basis $\{H_0^3(t), H_1^3(t), H_2^3(t), H_3^3(t)\}$ for the knots t_0, t_1. Determine the matrices for the basis changes.

Exercise 7.9 Show that a spline $s = \sum_{i=1}^{n} d_i N_{ik}$ in B-spline representation with respect to the knots $\{\tau_i\}$ satisfies the following recurrence relation:

$$s(t) = \sum_{i=l+1}^{n} d_i^l(t) N_{i,k-1}(t) .$$

Here the d_i^l are defined by $d_i^0(t) := d_i$ and

$$d_i^l(t) := \begin{cases} \dfrac{t - \tau_i}{\tau_{i+k-l} - \tau_i} d_i^{l-1}(t) + \dfrac{\tau_{i+k-l} - t}{\tau_{i+k-l} - \tau_i} d_{i-1}^{l-1}(t) & \text{if } \tau_{i+k-l} \neq \tau_i \\ \\ 0 & \text{else} \end{cases}$$

for $l > 0$. Show that $s(t) = d_i^{k-1}(t)$ for $t \in [\tau_i, \tau_{i+1}]$. Use this to derive a scheme for the computation of the spline $s(t)$ through continued convex combination of the coefficients d_i (algorithm of de Boor).

8 Large Symmetric Systems of Equations and Eigenvalue Problems

The previously described *direct* methods for the solution of a linear system $Ax = b$ (Gaussian elimination, Cholesky factorization, QR-factorization with Householder or Givens transformations) have two properties in common.

a) The methods start with arbitrary (for the Cholesky factorization symmetric) full (or dense) matrices $A \in \mathrm{Mat}_n(\mathbf{R})$.

b) The cost of solving the system is of the order $O(n^3)$ (multiplications).

However, there are many important cases of problems $Ax = b$, where

a) the matrix A is highly structured (see below) and most of the components are zero (i.e. A is *sparse*),

b) the dimension n of the problem is very large.

E.g. discretization of the Laplace equation in two space dimensions leads to *block-tridiagonal matrices*,

$$A = \begin{bmatrix} A_{11} & A_{12} & & & \\ A_{21} & A_{22} & A_{23} & & \\ & \ddots & \ddots & \ddots & \\ & & A_{q-1,q-2} & A_{q-1,q-1} & A_{q-1,q} \\ & & & A_{q,q-1} & A_{qq} \end{bmatrix} \tag{8.1}$$

with $A_{ij} \in \mathrm{Mat}_{n/q}(\mathbf{R})$, which, in addition, are symmetric, i.e. $A_{ij} = A_{ji}^T$. The direct methods are unsuitable for the treatment of such problems; they do not exploit the special structure, and they take far too long. There are essentially two approaches to develop new solution methods. The first consists of exploiting the special structure of the matrix in the direct methods, in particular its *sparsity pattern*, as much as possible. We have already discussed questions of this kind, when we compared the Givens and Householder transformations. The rotations operate only on two rows (from the left) or columns (from the right) of a matrix at a time, and they are therefore suited to largely maintain a sparsity pattern. In contrast, the Householder transformations are completely unsuitable for this purpose. Already in one step, they destroy any pattern of the starting matrix, so that from then on, the algorithm has to work with a full matrix. In general, the Gaussian elimination treats the sparsity pattern of matrices most sparingly. It is therefore the most commonly used starting basis for the construction of direct methods which utilize the structure of the matrix (*direct sparse solver*). Typically, column pivoting with possible row interchange and row pivoting with possible column interchange alternate with each other, depending on which strategy spares the most zero elements. In addition, the pivot rule is relaxed (*conditional pivoting*) in order to keep the number of additional non-zero elements (*fill-in elements*) small. In the last few years, the direct sparse solvers have developed into a sophisticated art form. Their description requires, in general, to resort to graphs which characterize the prevailing systems (see e.g. [37]). Their presentation is not suitable for this introduction.

The second approach to solve large systems which are rich in structure, is to develop *iterative methods* for the approximation of the solution x. This seems reasonable, also because we are generally only interested in the solution x up to a prescribed precision ε which depends on the precision of the input data (compare the evaluation of approximate solutions in Section 2.4.3). If, for example, the linear system was obtained by discretization of a differential equation, then the precision of the solution of the system only has to lie within the error bounds which are induced by the discretization. Any extra work would be a waste of time.

In the following sections, we shall be concerned with the most common iterative methods for the solution of large linear systems and eigenvalue problems for symmetric matrices. The goal is then always the construction of an iteration prescription $x_{k+1} = \phi(x_0, \ldots, x_k)$ such that

a) the sequence $\{x_k\}$ of iterates converges as fast as possible to the solution x,

b) x_{k+1} can be computed with as little cost as possible from $x_0, \ldots, x_k$.

In the second requirement, one usually asks that the evaluation of ϕ does not cost much more than a simple matrix-vector multiplication $(A, y) \mapsto Ay$. It is notable that the cost for sparse matrices is of the order $O(n)$ and not $O(n^2)$ (as with full matrices), because often the number of non-zero elements in a row is independent of the dimension n of the problem.

8.1 Classical Iteration Methods

In Chapter 4, we have solved nonlinear systems by using *fixed point methods*. This idea is also the basis of most classical iteration methods.

For a fixed point method $x_{k+1} = \phi(x_k)$ for the solution of a linear system $Ax = b$, we shall of course construct an iteration function ϕ, so that it has a unique fixed point x^* which is the exact solution $x^* = x$ of $Ax = b$. This is most easily achieved by transforming the equation $Ax = b$ into a fixed point equation,

$$
\begin{aligned}
Ax = b \quad &\Longleftrightarrow \quad Q^{-1}(b - Ax) = 0 \\
&\Longleftrightarrow \quad \phi(x) := \underbrace{(I - Q^{-1}A)}_{=:\,G}\,x + \underbrace{Q^{-1}b}_{=:\,c} = x \ ,
\end{aligned}
$$

where $Q \in \mathrm{GL}(n)$ is an arbitrary regular matrix. In order to obtain a reasonable iteration method, we have to take care that the fixed point method $x_{k+1} = \phi(x_k) = Gx_k + c$ converges.

Theorem 8.1 *The fixed point method $x_{k+1} = Gx_k + c$ with $G \in \mathrm{Mat}_n(\mathbf{R})$ converges for each starting value $x_0 \in \mathbf{R}^n$ iff*

$$
\rho(G) < 1 \ ,
$$

where $\rho(G) = \max\limits_{j} |\lambda_j(G)|$ is the spectral radius of G.

Beweis. We again restrict ourselves to the simple case of a symmetric matrix $G = G^T$, which is the only case that we need in the following. Then there is an orthogonal matrix $Q \in \mathbf{O}(n)$ such that

$$
QGQ^T = \Lambda = \mathrm{diag}(\lambda_1, \ldots, \lambda_n)
$$

is the diagonal matrix of the eigenvalues of G. Since $|\lambda_i| \leq \rho(G) < 1$ for all i and $D^k = \mathrm{diag}(\lambda_1^k, \ldots, \lambda_n^k)$, we have $\lim_{k\to\infty} D^k = 0$, and therefore also

$$
\lim_{k\to\infty} G^k = \lim_{k\to\infty} Q^T D^k Q = 0 \ . \qquad \square
$$

Since $\rho(G) \leq \|G\|$ for any corresponding matrix-norm, it follows that $\|G\| < 1$ is sufficient for $\rho(G) < 1$. In this case, we can estimate the errors $x_k - x = G^k(x_0 - x)$ by

$$
\|x_k - x\| \leq \|G\|^k \|x_0 - x\| \ .
$$

Besides the convergence, we require that $\phi(y) = Gy + c$ can be easily computed. For this purpose, the matrix Q has to be easily invertible. The matrix which is most easy to invert, is doubtless the identity $Q = I$. The method which thus arises for the iteration function $G = I - A$,

$$x_{k+1} = x_k - Ax_k + b ,$$

is the *Richardson method*. If we start with a spd-matrix A then we obtain for the spectral radius of G

$$\rho(G) = \rho(I - A) = \max\{|1 - \lambda_{\max}(A)|, |1 - \lambda_{\min}(A)|\} .$$

A necessary condition for the convergence of the Richardson iteration is therefore $\lambda_{\max}(A) < 2$. Taken by itself, this iteration is thus only rarely usable. However, we shall below discuss possibilities to improve the convergence. The next more complicated matrices are the diagonal matrices, so that the diagonal D of

$$A = L + D + R ,$$

is a candidate for a second possibility for Q. Here $D = \operatorname{diag}(a_{11}, \ldots, a_{nn})$ and

$$L := \begin{bmatrix} 0 & \cdots & \cdots & 0 \\ a_{21} & \ddots & & \vdots \\ \vdots & \ddots & \ddots & \vdots \\ a_{n1} & \cdots & a_{n,n-1} & 0 \end{bmatrix} , \quad R := \begin{bmatrix} 0 & a_{12} & \cdots & a_{11} \\ \vdots & \ddots & \ddots & \vdots \\ \vdots & \ddots & \ddots & a_{n-1,n} \\ 0 & \cdots & \cdots & 0 \end{bmatrix} .$$

The corresponding method

$$x_{k+1} = (I - D^{-1}A)x_k + D^{-1}b = -D^{-1}(L + R)x_k + D^{-1}b$$

is called *Jacobi method*. A sufficient condition for its convergence is, that A is strictly diagonally dominant.

Theorem 8.2 *The Jacobi iteration* $x_{k+1} = -D^{-1}(L + R)x_k + D^{-1}b$ *converges for any starting value* x_0 *to the solution* $x = A^{-1}b$, *if* A *is strictly diagonally dominant, i.e.*

$$|a_{ii}| > \sum_{i \neq j} |a_{ij}| \quad \text{for all} \ \ i = 1, \ldots, n .$$

Proof. The statement follows from Theorem 8.1 because

$$\rho(D^{-1}(L + R)) \leq \|D^{-1}(L + R)\|_\infty = \max_i \sum_{j \neq i} |\frac{a_{ij}}{a_{ii}}| . \qquad \square$$

In the first chapter, after the diagonal ones, the triangular systems have proven to be simply solvable. For full lower or upper triangular matrices, the cost is of the order $O(n^2)$ per solution, for sparse matrices the cost is often of the order $O(n)$, i.e. of an order which we consider acceptable. By taking Q as the lower triangular half $Q := D + L$, we obtain the *Gauss-Seidel method*

$$
\begin{aligned}
x_{k+1} &= (I - (D + L)^{-1}A)x_k + (D + L)^{-1}b \\
&= -(D + L)^{-1}Rx_k + (D + L)^{-1}b \ .
\end{aligned}
$$

It converges for any spd-matrix A. In order to prove this property, we derive a condition for the contraction property of $\rho(G) < 1$ of $G = I - Q^{-1}A$, which is easy to verify. For this, we note that every spd-Matrix A induces a scalar product $(x, y) := \langle x, Ay \rangle$ on $\mathbf{R}^n$. For any matrix $B \in \mathrm{Mat}_n(\mathbf{R})$, $B^* := A^{-1}B^T A$ is the *adjoint matrix* with respect to this scalar product, i.e.

$$
(Bx, y) = (x, B^*y) \quad \text{for all} \ \ x, y \in \mathbf{R}^n \ .
$$

A self adjoint matrix $B = B^*$ is called *positive* with respect to $(\cdot, \cdot)$, if

$$
(Bx, x) > 0 \quad \text{for all} \ \ x \neq 0 \ .
$$

Lemma 8.3 *Let $G \in \mathrm{Mat}_n(\mathbf{R})$, and let G^* be the adjoint matrix of G with respect to a scalar product $(\cdot, \cdot)$. Then, if $B := I - G^*G$ is a positive matrix with respect to $(\cdot, \cdot)$, it follows that $\rho(G) < 1$.*

Proof. Since B is positive, we have for all $x \neq 0$

$$
(Bx, x) = (x, x) - (G^*Gx, x) = (x, x) - (Gx, Gx) > 0 \ ;,
$$

which is derived from $(\cdot, \cdot)$ that $\|x\| > \|Gx\|$ for all $x \neq 0$. This implies

$$
\rho(G) \leq \|G\| := \sup_{\|x\|=1} \frac{\|Gx\|}{\|x\|} < 1 \ ,
$$

$\|x_0\| = 1$, because of the compactness of the sphere.

$\square$

Theorem 8.4 *The Gauss-Seidel method converges for any spd-matrix A.*

Proof. We have to show that $B := I - G^*G$, with $G = I - (D + L)^{-1}A$, is a positive matrix with respect to $(\cdot, \cdot) := \langle \cdot, A \cdot \rangle$. Because of $R^T = L$:

$$G^* = I - A^{-1}A^T(D + L)^{-T}A = I - (D + R)^{-1}A$$

$$B = I - G^*G = (D + R)^{-1}D(D + L)^{-1}A . \tag{8.2}$$

The trick in the last manipulation consists of inserting the equation

$$(D + M)^{-1} = (D + M)^{-1}(D + M)(D + M)^{-1}$$

for $M = R, L$, after carrying out the multiplications and then factoring. From (8.2) it follows for all $x \neq 0$ that

$$
\begin{aligned}
(Bx, x)_A &= \langle (D + R)^{-1}D(D + L)^{-1}Ax, Ax \rangle \\
&= \langle D^{1/2}(D + L)^{-1}Ax, D^{1/2}(D + L)^{-1}Ax \rangle > 0 ,
\end{aligned}
$$

i.e. B is positive and $\rho(G) < 1$. $\square$

The speed of convergence of a fixed point iteration $x_{k+1} = Gx_k + c$ depends strongly on the spectral radius $\rho(G)$. For any concretely chosen G, there is however a way to improve it, namely the *extrapolation* or better *relaxation method*. For this we consider convex combinations of the respectively 'old' and 'new' iterate

$$
\begin{aligned}
x_{k+1} &= \omega(Gx_k + c) + (1 - \omega)x_k \\
&= G_\omega x_k + \omega c \text{ with } G_\omega := \omega G + (1 - \omega)I ,
\end{aligned}
$$

where $\omega \in [0, 1]$ is a *damping parameter*. This way we obtain from a fixed point iteration $x_{k+1} = Gx_k + c$ an entire family of *relaxed fixed point iterations* with the iteration function which depends on ω

$$\phi_\omega(x) = \omega\phi(x) + (1 - \omega)x = G_\omega x + \omega c .$$

The art consists now of choosing the damping parameter ω so that $\rho(G_\omega)$ is as small as possible. In fact, for a class of fixed point iterations it is possible to even force convergence by a suitable choice of ω, despite the fact that the starting iteration in general does not converge.

Definition 8.5 A fixed point method $x_{k+1} = Gx_k + c$, $G = G(A)$, is called *symmetrizable*, if for any spd-matrix A, $I - G$ is equivalent (similar) to a spd-matrix, i.e. there is a regular matrix $W \in \mathrm{GL}(n)$ such that

$$W(I - G)W^{-1}$$

is a spd-matrix.

Example 8.6 The Richardson method $G = I - A$ is trivially symmetrizable. The same is also true for the Jacobi method $G = I - D^{-1}A$: With $W := D^{\frac{1}{2}}$ we have that

$$D^{\frac{1}{2}}(I - G)D^{-\frac{1}{2}} = D^{\frac{1}{2}}D^{-1}AD^{-\frac{1}{2}} = D^{-\frac{1}{2}}AD^{-\frac{1}{2}}$$

is a spd-matrix.

The iteration matrices G of symmetric fixed point methods have the following properties.

Lemma 8.7 *Let* $x_{k+1} = Gx_k + c$, $G = G(A)$ *be a symmetrizable fixed point method, and let A a spd-matrix. Then all eigenvalues of G are real and less than 1, i.e. the spectrum $\sigma(G)$ of G satisfies*

$$\sigma(G) \subset\,]-\infty, 1[\,.$$

Proof. Since $I - G$ is similar to a spd-matrix, the eigenvalues of $I - G$ are real and positive, and the eigenvalues of G are therefore real and < 1.   □

Now let $\lambda_{\min} \leq \lambda_{\max} < 1$ be the extreme eigenvalues of G. Then the eigenvalues of G_ω are just

$$\lambda_i(G_\omega) = \omega\lambda_i(G) + 1 - \omega = 1 - \omega(1 - \lambda_i(G)) < 1\,,$$

i.e.

$$\rho(G_\omega) = \max\left\{|1 - \omega(1 - \lambda_{\min}(G))|\,,\ |1 - \omega(1 - \lambda_{\max}(G))|\right\}\,.$$

Because of $0 < 1 - \lambda_{\max}(G) \leq 1 - \lambda_{\min}(G)$, the *optimal damping parameter* $\bar{\omega}$ with

$$\rho(G_{\bar{\omega}}) = \min_{0<\omega\leq 1} \rho(G_\omega) = 1 - \bar{\omega}(1 - \lambda_{\min}(G))$$

satisfies the equation

$$1 - \bar{\omega}(1 - \lambda_{\max}(G)) = -1 + \bar{\omega}(1 - \lambda_{\min}(G))$$

(see figure 8.1). We thus obtain the following result.

Lemma 8.8 *With the above notation,*

$$\bar{\omega} = \frac{2}{2 - \lambda_{\max}(G) - \lambda_{\min}(G)}$$

is the optimal damping parameter for the symmetrizable iteration method $x_{k+1} = Gx_k + c$. *The spectral radius of the iteration matrix of the relaxed method satisfies*

$$\rho(G_{\bar{\omega}}) = 1 - \bar{\omega}(1 - \lambda_{\min}(G)) < 1\,.$$

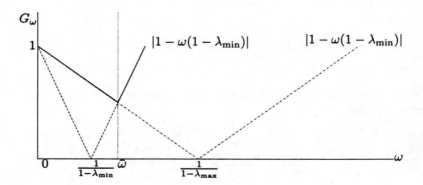

Figure 8.1: Spectral radius in dependence of the damping parameter ω

In other words: We have just shown that, for any symmetrizable iteration method, convergence can be forced by a suitable choice of the damping parameter. Here information about the matrix A enters in the determination of $\bar{\omega}$, so that extrapolating methods can only be given for certain classes of matrices.

Example 8.9 For the Richardson method with $G = I - A$, the relaxed iteration has a particularly simple form:

$$G_\omega = \omega G + (1 - \omega)I = I - \omega A .$$

A spd-matrix A satisfies $\lambda_{\min}(G) = 1 - \lambda_{\max}(A)$ and $\lambda_{\max}(G) = 1 - \lambda_{\min}(A)$, and therefore

$$\bar{\omega} = \frac{2}{2 - \lambda_{\max}(G) - \lambda_{\min}(G)} = \frac{2}{\lambda_{\max}(A) + \lambda_{\min}(A)} .$$

The spectral radius of the optimally damped Richardson method is therefore

$$\rho(G_{\bar{\omega}}) = \rho(I - \bar{\omega}A) = \frac{\lambda_{\max}(A) - \lambda_{\min}(A)}{\lambda_{\max}(A) + \lambda_{\min}(A)} = \frac{\kappa_2(A) - 1}{\kappa_2(A) + 1} < 1$$

with the condition $\kappa_2(A)$ of A with respect to the Euclidian norm.

For a more detailed account on optimal relaxation methods, we refer to the book by HAGEMAN and YOUNG [43]. We now leave this topic, because nowadays, the most important application of relaxation methods, namely the *multi-grid methods*, in fact do *not* use the above computed optimal damping parameter $\bar{\omega}$. Instead, the contraction behavior is analyzed in more detail, namely with respect to the frequency parts of the iterates. If the solutions of a linear system are interpreted as a superposition of 'standing waves', then it turns out that, with a suitable choice (with respect to the problem) of the relaxation parameter, the relaxed Jacobi, as well as the relaxed Gauss-Seidel method, prefer to damp the high frequency parts of the error. This property is called *smoothing property* and is used in the construction of multi-grid methods for the iterative solution of discretized partial differential equations. The fundamental idea is to damp the high frequency parts of the error with a relaxation method over a fine discretization grid and then to go to the next coarser grid, which serves to eliminate the low frequency parts. By recursive application of this rule, an iteration method emerges, which only requires the *direct* solution of a linear system on the *coarsest* grid, and thus with *few* unknowns. Since we do not address the question of discretization in this elementary introduction, we have to leave it here with a brief hint. For further details we recommend the survey article by WITTUM [82], or the textbook by HACKBUSCH [42].

8.2   Chebyshev Acceleration

Without any exception, the methods which were introduced in the last section are all fixed point methods, $x_{k+1} = \phi(x_k) = Gx_k + c$. For the computation of x_{k+1}, only information of the last iteration step is utilized, and the already computed values $x_0, \ldots, x_{k-1}$ are not taken into consideration. In the following, we shall present a technique to improve a given fixed point method, by constructing a linear combination

$$y_k = \sum_{j=0}^{k} v_{kj} x_j \tag{8.3}$$

from all values $x_0, \ldots, x_k$. With a suitable choice of the coefficients v_{kj}, the replacement sequence $\{y_0, y_1, \ldots\}$ will then converge faster than the original sequence $\{x_0, x_1, \ldots\}$. How should the v_{kj} be determined? If $x_0 = \cdots = x_k = x$ is already the solution, then we require that $y_k = x$ also solves, from which it follows that

$$\sum_{j=0}^{k} v_{kj} = 1 .$$

Thus the error $y_k - x$ in particular satisfies

$$y_k - x = \sum_{j=0}^{k} v_{kj}(x_j - x) \,. \tag{8.4}$$

We therefore seek a vector y_k in the affine sub-space

$$V_k := \left\{ y_k = \sum_{j=0}^{k} v_{kj}x_j \ \middle| \ v_{kj} \in \mathbf{R} \ \text{ with } \ \sum_{j=0}^{k} v_{kj} = 1 \right\} \subset \mathbf{R}^n \,,$$

which approximates the exact solution as well as possible, i.e.

$$\|y_k - x\| = \min_{y \in V_k} \|y - x\| \tag{8.5}$$

with a suitable norm $\| \cdot \|$. According to Remark 3.6, y_k is the (affine) orthogonal projection of x onto V_k with respect to the Euclidean norm $\|y\| = \sqrt{\langle y, y \rangle}$, and the minimization problem is equivalent to the *variational problem*

$$\langle y_k - x, y - x_0 \rangle = 0 \ \text{ for all } \ y \in V_k \,.$$

Now, if $q_1, \ldots, q_k$ is an orthogonal basis of the linear sub-space U_k which is parallel to V_k, i.e. $V_k = x_0 + U_k$, then we can give explicitly the affine orthogonal projection $Q_k : \mathbf{R}^n \longrightarrow V_k$:

$$y_k = Q_k x = x_0 + \sum_{j=1}^{k} \frac{\langle q_j, x - x_0 \rangle}{\langle q_j, q_j \rangle} q_j \,. \tag{8.6}$$

(8.6), i.e. the formula cannot be evaluated, and the minimization problem (8.5) is thus not solvable yet. There are two ways to escape this situation: One possibility is to replace the Euclidean scalar product with a different one which better suits the problem at hand. We shall pursue this approach in Section 8.3. The second possibility, which is considered here, is to construct a *solvable substitute problem* instead of the minimization problem (8.5).

The iterates x_k of the given fixed point iteration are $x_k = \phi^k(x_0)$, where $\phi(y) = Gy + c$ is the iteration rule. Therefore,

$$y_k = \sum_{j=0}^{k} v_{kj}x_j = P_k(\phi)x_0 \,,$$

where $P_k \in \mathbf{P}_k$ is a polynomial of degree k with

$$P_k(\lambda) = \sum_{j=0}^{k} v_{kj}\lambda^j \ \text{ and } \ P_k(1) = \sum_{j=0}^{k} v_{kj} = 1 \,.$$

According to (8.4), for the error $y_k - x$ we thus obtain:

$$y_k - x = P_k(G)(x_0 - x) .$$

In order to split off the initially unknown solution, we make the (generally rough) estimate

$$\|x_k - x\| \le \|P_k(G)\| \ \|x_0 - x\| .$$

Instead of the solution of the minimization problem (8.5), we now seek a polynomial P_k with $P_k(1) = 1$ such that

$$\|P_k(G)\| = \min .$$

For this purpose, we assume that the underlying fixed point iteration is symmetrizable and set

$$a := \lambda_{\min}(G) \ \text{ and } \ b := \lambda_{\max}(G) .$$

Thus the 2-norm of $P_k(G)$ satisfies

$$\|P_k(G)\|_2 = \max_i |P_k(\lambda_i)| \le \max_{\lambda \in [a,b]} |P_k(\lambda)| =: \bar{\rho}(P_k(G)) .$$

The value $\bar{\rho}(P_k(G))$ is also called the *virtual spectral radius* of G. This way we finally arrive at the *min-max problem*

$$\max_{\lambda \in [a,b]} |P_k(\lambda)| = \min \ \text{ with } \ \deg P_k = k \ \text{ and } \ P_k(1) = 1 .$$

We have already encountered and solved this min-max problem in Section 7.1.4. According to Theorem 7.21, the P_k turn out to be the specially normalized Chebyshev polynomials

$$P_k(\lambda) = \frac{T_k(t(\lambda))}{T_k(t(1))} \ \text{ with } \ t(\lambda) = 2\frac{\lambda - a}{b - a} - 1 .$$

For the computation of y_k we can utilize the three-term recurrence relation

$$T_k(t) = 2tT_{k-1}(t) - T_{k-2}(t), \ \ T_0(t) = 1 \ \ , T_1(t) = t. \tag{8.7}$$

Note that in the first improvement step, from

$$P_1(\lambda) = \frac{t(\lambda)}{t(1)} = \bar{\omega}\lambda + 1 - \bar{\omega} ,$$

we recover the relaxation method with the optimal damping parameter $\bar{\omega} = 2/(2 - b - a)$, which was described in Section 8.1. If we set

$$\bar{t} := t(1) = \frac{2 - b - a}{b - a} \ \text{ and } \ \rho_k := 2\bar{t}\frac{T_{k-1}(\bar{t})}{T_k(\bar{t})} ,$$

then it follows that

$$P_k(\lambda) = 2t\frac{T_{k-1}(\bar{t})}{T_k(\bar{t})}P_{k-1}(\lambda) - \frac{T_{k-2}(\bar{t})}{T_k(\bar{t})}P_{k-2}(\lambda)$$
$$= \rho_k(1 - \bar{\omega} + \bar{\omega}\lambda)P_{k-1}(\lambda) + (1 - \rho_k)P_{k-2}(\lambda) . \qquad (8.8)$$

Here note that

$$-\frac{T_{k-2}(\bar{t})}{T_k(\bar{t})} = \frac{T_k(\bar{t}) - 2\bar{t}T_{k-1}(\bar{t})}{T_k(\bar{t})} = 1 - \rho_k$$

and

$$2t\frac{T_{k-1}(\bar{t})}{T_k(\bar{t})} = \rho_k\frac{t}{\bar{t}} = \rho_k\frac{2\lambda - b - a}{2 - b - a} = \rho_k(1 - \bar{\omega} + \bar{\omega}\lambda) .$$

If we insert (8.8) into $y_k = P_k(\phi)x_0$ for the fixed point method $\phi(y) = Gy+c$, then we obtain the recurrence relation

$$y_k = P_k(\phi)x_0$$
$$= \Big(\rho_k\big((1 - \bar{\omega})P_{k-1}(\phi) + \bar{\omega}\phi P_{k-1}(\phi)\big) + (1 - \rho_k)P_{k-2}(\phi)\Big)x_0$$
$$= \rho_k((1 - \bar{\omega})y_{k-1} + \bar{\omega}(Gy_{k-1} + c)) + (1 - \rho_k)y_{k-2} .$$

For a fixed point method of the form $G = I - Q^{-1}A$ and $c = Q^{-1}b$ we have in particular

$$y_k = \rho_k(y_{k-1} - y_{k-2} + \bar{\omega}Q^{-1}(b - Ay_{k-1})) + y_{k-2} .$$

This iteration for the y_k is called *Chebyshev iteration* or *Chebyshev acceleration* for the fixed point iteration $x_{k+1} = Gx_k + c$ with $G = I - Q^{-1}A$ and $c = Q^{-1}b$.

Algorithm 8.10 *Chebyshev iteration* (for the starting value $y_0 = x_0$ with a prescribed relative precision tol).

$$\bar{t} := \frac{2 - \lambda_{\max}(G) - \lambda_{\min}(G)}{\lambda_{\max}(G) - \lambda_{\min}(G)}; \quad \bar{\omega} := \frac{2}{2 - \lambda_{\max}(G) - \lambda_{\min}(G)};$$
$$T_0 := 1, T_1 := \bar{t};$$
$$y_1 := \bar{\omega}(Gy_0 + c) + (1 - \bar{\omega})y_0;$$

for $k := 2$ **to** $k_{\max}$ **do**

$$T_k := 2\bar{t}T_{k-1} - T_{k-2};$$
$$\rho_k := 2\bar{t}\frac{T_{k-1}}{T_k};$$

solve the system $Qz = b - Ay_{k-1}$ in z;

$$y_k := \rho_k(y_{k-1} - y_{k-2} + \bar{\omega}z) + y_{k-2};$$

if $\|y_k - y_{k-1}\| \leq \text{tol}\|y_k\|$ **then exit;**

end for

symmetrizable fixed point iteration can be estimated as follows:

Theorem 8.11 *Let $G = G(A)$ be the iteration matrix of a symmetrizable fixed point iteration $x_{k+1} = \phi(x_k) = Gx_k + c$ for the spd-matrix A, and let $x_0 \in \mathbf{R}^n$ be an arbitrary starting value. Then the corresponding Chebyshev iteration $y_k = P_k(\phi)x_0$ satisfies*

$$\|y_k - x\| \leq \frac{1}{|T_k(\bar{t})|}\|x_0 - x\| \quad \text{with} \quad \bar{t} = \frac{2 - \lambda_{\max}(G) - \lambda_{\min}(G)}{\lambda_{\max}(G) - \lambda_{\min}(G)} .$$

Proof. According to the construction of the iteration, we have

$$\|y_k - x\| \leq \|P_k(G)\| \, \|x_0 - x\| \leq \max_{\lambda \in [a,b]} \frac{|T_k(t(\lambda))|}{|T_k(t(1))|}\|x_0 - x\| ,$$

where, as above, $t(\lambda)$ is the transformation

$$t(\lambda) = 2\,\frac{\lambda - \lambda_{\min}(G)}{\lambda_{\max}(G) - \lambda_{\min}(G)} - 1 = \frac{2\,\lambda - \lambda_{\max}(G) - \lambda_{\min}(G)}{\lambda_{\max}(G) - \lambda_{\min}(G)}.$$

The statement follows, because $|T_k(t)| \leq 1$ for $t \in [-1, 1]$. $\qquad\square$

Example 8.12 For the Chebyshev acceleration of the Richardson iteration, we have in particular:

$$\bar{t} = \frac{2 - \lambda_{\max}(G) - \lambda_{\min}(G)}{\lambda_{\max}(G) - \lambda_{\min}(G)} = \frac{\lambda_{\max}(A) + \lambda_{\min}(A)}{\lambda_{\max}(A) - \lambda_{\min}(A)} = \frac{\kappa_2(A) + 1}{\kappa_2(A) - 1} > 1 .$$

Lemma 8.13 *For the Chebyshev polynomials T_k and $\kappa > 1$, we have the estimate*

$$\left|T_k\left(\frac{\kappa + 1}{\kappa - 1}\right)\right| \geq \frac{1}{2}\left(\frac{\sqrt{\kappa} + 1}{\sqrt{\kappa} - 1}\right)^k .$$

Proof. One easily computes that for $z := (\kappa + 1)/(\kappa - 1)$, it follows that

$$z \pm \sqrt{z^2 - 1} = \frac{\sqrt{\kappa} \pm 1}{\sqrt{\kappa} \mp 1} ,$$

$$T_k\left(\frac{\kappa+1}{\kappa-1}\right) = \frac{1}{2}\left[\left(\frac{\sqrt{\kappa}+1}{\sqrt{\kappa}-1}\right)^k + \left(\frac{\sqrt{\kappa}-1}{\sqrt{\kappa}+1}\right)^k\right] \geq \frac{1}{2}\left(\frac{\sqrt{\kappa}+1}{\sqrt{\kappa}-1}\right)^k ,$$

$\square$

According to Theorem 8.11 and Lemma 8.13, we therefore have

$$\|y_k - x\| \leq 2\left(\frac{\sqrt{\kappa(A)}-1}{\sqrt{\kappa(A)}+1}\right)^k \|x_0 - x\| .$$

Chebyshev iteration requires a very good knowledge of the limits $\lambda_{\min}$ and $\lambda_{\max}$ of the (real) spectrum of A. Modern methods therefore combine these methods with a vector iteration, which in a few steps generally produces usable estimates of $\lambda_{\min}$ and $\lambda_{\max}$.

Remark 8.14 The idea of the Chebyshev acceleration which we presented here, can also be carried over to non-symmetric matrices. Instead of the intervals, which in the symmetric case enclose the real spectrum, one has ellipses, which in general enclose the complex spectrum. Details can be found in the article by MANTEUFFEL [54].

8.3   Method of Conjugate Gradients

In the last section, we tried to approximate the solution of the linear problem $Ax = b$ by vectors y_k in an affine subspace V_k. The use of the Euclidian norm $\|y\| = \sqrt{\langle y, y\rangle}$ first lead us to an impasse, which we overcame by passing to a solvable substitute problem. Here we want to pursue further our original idea by passing to a scalar product which is adapted to the problem at hand. In fact, any spd-matrix A defines in a natural way a scalar product

$$(x, y) := \langle x, Ay\rangle$$

with the corresponding norm

$$\|y\|_A = \sqrt{(y, y)} ,$$

the *energy norm* of A. We encountered both in Section 8.1. Now we repeat the chain of reasoning of Section 8.2, only with the energy norm instead of the Euclidian norm of the iteration error. Let $V_k = x_0 + U_k \subset \mathbf{R}^n$ be a k-dimensional affine subspace, where U_k is the linear subspace which is parallel to V_k. The solution x_k of the minimization problem

$$\|x_k - x\|_A = \min_{y \in V_k} \|y - x\|_A , \tag{8.9}$$

$$\|x_k - x\|_A = \min_{V_k} ,$$

is also called *Ritz-Galerkin approximation* of x in V_k. According to Theorem 3.4, x_k is the orthogonal projection of x onto V_k with respect to $(\cdot,\cdot)$, i.e. the minimization problem $\|x_k - x\|_A = \min$ is equivalent to the variational problem

$$(x - x_k, u) = 0 \quad \text{for all} \quad u \in U_k . \tag{8.10}$$

Instead of 'orthogonal with respect to $(\cdot,\cdot) = \langle \cdot, A\cdot \rangle$', we shall in the following simply say 'A-orthogonal' (historically also A-*conjugate*), i.e., according to (8.10), the errors $x - x_k$ must be A-orthogonal to U_k. If we denote the residues again by $r_k := b - Ax_k$, then we have

$$(x - x_k, u) = \langle A(x - x_k), u \rangle = \langle r_k, u \rangle .$$

The variational problem (8.10) is therefore equivalent to the requirement that the residues r_k must be orthogonal (with respect to the Euclidian scalar product) to U_k, i.e.

$$\langle r_k, u \rangle = 0 \quad \text{for all} \quad u \in U_k . \tag{8.11}$$

Now let $p_1, \ldots, p_k$ be an A-orthogonal basis of U_k, i.e.

$$(p_k, p_j) = \delta_{kj}(p_k, p_k) .$$

Then, for the A-orthogonal projection $P_k : \mathbf{R}^n \to V_k$, it follows that

$$x_k = P_k x = x_0 + \sum_{j=1}^{k} \frac{(p_j, x - x_0)}{(p_j, p_j)} p_j = x_0 + \sum_{j=1}^{k} \frac{\langle p_j, Ax - Ax_0 \rangle}{(p_j, p_j)} p_j$$

$$= x_0 + \sum_{j=1}^{k} \underbrace{\frac{\langle p_j, r_0 \rangle}{(p_j, p_j)}}_{:= \alpha_j} p_j . \tag{8.12}$$

We note that, in contrast to Section 8.2, the initially unknown solution x does not appear on the right hand side, i.e. we can explicitly compute the A-orthogonal projection x_k of x onto V_k, without knowing x. From (8.12), for x_k and r_k, we immediately obtain the recurrence relations

$$x_k = x_{k-1} + \alpha_k p_k \quad \text{and} \quad r_k = r_{k-1} - \alpha_k A p_k , \tag{8.13}$$

because

$$r_k = A(x - x_k) = A(x - x_{k-1} - \alpha_k p_k) = r_{k-1} - A\alpha_k p_k .$$

For the construction of an approximation method, we only lack suitable subspaces $V_k \subset \mathbf{R}^n$, for which an A-orthogonal basis $p_1, \ldots, p_k$ can be easily computed. By the Cayley-Hamilton theorem (see e.g. [8]), there is a polynomial $P_{n-1} \in \mathbf{P}_{n-1}$ such that

$$A^{-1} = P_{n-1}(A) ,$$

and therefore

$$x - x_0 = A^{-1}r_0 = P_{n-1}(A)r_0 \in \mathrm{span}\{r_0, Ar_0, \ldots, A^{n-1}r_0\} .$$

If we chose $V_k = x_0 + U_k$ with $U_0 := \{0\}$ for the approximation spaces, and

$$U_k := \mathrm{span}\{r_0, Ar_0, \ldots, A^{k-1}r_0\} \quad \text{for} \quad k = 1, \ldots, n ,$$

then $x \in V_n$, i.e. in the worst case, the n-th approximation x_n is the solution itself. The spaces $U_k = U_k(A, x_0)$ are called *Krylov spaces*. They are also automatically obtained from our requirement to essentially only carry out a single matrix-vector multiplication $(A, y) \mapsto Ay$ in each iteration step. If we furthermore recall Theorem 6.4, then we see that we can construct an A-orthogonal basis $p_1, \ldots, p_k$ of U_k with the three-term recurrence relation (6.5). However, we can compute the p_k directly from the residues.

Lemma 8.15 *Let $r_k \neq 0$. Then the residues $r_0, \ldots, r_k$ are pairwise orthogonal, i.e.*

$$\langle r_i, r_j \rangle = \delta_{ij} \langle r_i, r_i \rangle \quad \text{for} \quad i, j = 0, \ldots, k ,$$

and they span U_{k+1}, i.e.

$$U_{k+1} = \mathrm{span}\{r_0, \ldots, r_k\} .$$

Proof. The proof is by induction over k. The case $k = 0$ is trivial because of $U_1 = \mathrm{span}(r_0)$. Suppose the statement is true for $k - 1$. From (8.13), it follows immediately that $r_k \in U_{k+1}$. Furthermore, we have seen in (8.11) that r_k is orthogonal to U_k. By induction assumption we have

$$U_k = \mathrm{span}\{r_0, \ldots, r_{k-1}\} ,$$

and thus $\langle r_k, r_j \rangle = 0$ for $j < k$. Finally, $r_k \neq 0$ implies that $U_{k+1} = \mathrm{span}\{r_0, \ldots, r_k\}$. $\square$

We therefore construct the A-orthogonal basis vectors p_k as follows: If $r_0 \neq 0$ (otherwise x_0 is the solution), then we set $p_1 := r_0$. Lemma 8.15 now states for $k > 1$ that r_k either vanishes, or that the vectors $p_1, \ldots, p_{k-1}$ and r_k are linearly independent, and span U_{k+1}. In the first case, we have $x = x_k$, and we are done. In the second case, by choosing

$$p_{k+1} = r_k - \sum_{j=1}^{k} \frac{(r_k, p_j)}{(p_j, p_j)} p_j = r_k - \underbrace{\frac{(r_k, p_k)}{(p_k, p_k)}}_{=: \ \beta_{k+1}} p_k \ , \tag{8.14}$$

we obtain an orthogonal basis of U_{k+1}. By inserting (8.14), the evaluation of α_k and β_k can be further simplified. Since

$$(x - x_0, p_k) = (x - x_{k-1}, p_k) = \langle r_{k-1}, p_k \rangle = \langle r_{k-1}, r_{k-1} \rangle \ ,$$

it follows that

$$\alpha_k = \frac{\langle r_{k-1}, r_{k-1} \rangle}{(p_k, p_k)} \ ,$$

and because of

$$-\alpha_k(r_k, p_k) = \langle -\alpha_k A p_k, r_k \rangle = \langle r_k - r_{k-1}, r_k \rangle = \langle r_k, r_k \rangle$$

we have

$$\beta_{k+1} = \frac{\langle r_k, r_k \rangle}{\langle r_{k-1}, r_{k-1} \rangle} \ .$$

Together we obtain the *method of conjugate gradients* or briefly *cg-method* which was introduced by HESTENES and STIEFEL [47] in 1952.

Algorithm 8.16 *cg-method* (for the starting value x_0).

$$p_1 := r_0 := b - Ax_0;$$

for $k := 1$ **to** $k_{\max}$ **do**

$$\alpha_k \quad := \frac{\langle r_{k-1}, r_{k-1} \rangle}{(p_k, p_k)} = \frac{\langle r_{k-1}, r_{k-1} \rangle}{\langle p_k, A p_k \rangle};$$
$$x_k \quad := x_{k-1} + \alpha_k p_k;$$

if accurate **then exit**;

$$r_k \quad := r_{k-1} - \alpha_k A p_k;$$
$$\beta_{k+1} := \frac{\langle r_k, r_k \rangle}{\langle r_{k-1}, r_{k-1} \rangle};$$
$$p_{k+1} := r_k + \beta_{k+1} p_k;$$

end for

Note that in fact essentially only one matrix-vector multiplication has to be carried out per iteration step, namely Ap_k, and the method thus fully satisfies our requirements with respect to cost of computation. In the above, we have not further specified the termination criterion 'accurate'. Of course, one would like to have a criterion of the kind

$$\|x - x_k\| \quad \text{'sufficiently small'} , \qquad (8.15)$$

which is however not feasible in this form. Because of this, (8.15) is in practice replaced by requiring

$$\|r_k\|_2 = \|x - x_k\|_{A^2} \quad \text{'sufficiently small'}. \qquad (8.16)$$

As we have already explained in great detail in Section 2.4.3, the residue norm is not a suitable measure for the convergence: For ill-conditioned systems, i.e. for $\kappa(A) \gg 1$, the iterates can improve drastically, even though the residue norms grow. We shall return to this topic again in Section 8.4. The question regarding convergence properties remains to be answered. Since the cg-method produces the solution $x_n = x$ in finitely many steps, we only lack a statement regarding convergence speed.

Theorem 8.17 *The approximation error $x - x_k$ of the cg-method can be estimated in the energy norm $\|y\|_A = \sqrt{\langle y, Ay \rangle}$ by*

$$\|x - x_k\|_A \leq 2 \left(\frac{\sqrt{\kappa_2(A)} - 1}{\sqrt{\kappa_2(A)} + 1} \right)^k \|x - x_0\|_A , \qquad (8.17)$$

where $\kappa_2(A)$ is the condition of A with respect to the Euclidian norm.

Proof. Since x_k is the solution of the minimization problem $\|x - x_k\|_A = \min_{V_k}$, we have

$$(x - x_k, x - x_k) \leq (x - y, x - y) \quad \text{for all} \ \ y \in V_k .$$

The elements of V_k are of the form

$$y = x_0 + P_{k-1}(A)r_0 = x_0 + P_{k-1}(A)(b - Ax_0) = x_0 + AP_{k-1}(A)(x - x_0) ,$$

with a polynomial $P_{k-1} \in \mathbf{P}_{k-1}$ such that

$$x - y = x - x_0 - AP_{k-1}(A)(x - x_0) = \underbrace{(I - AP_{k-1}(A))}_{=: \ Q_k(A)}(x - x_0) ,$$

where $Q_k \in \mathbf{P}_k$ is a polynomial with $Q_k(0) = 1$. By inserting into the minimization condition, it follows that

$$
\begin{aligned}
\|x - x_k\|_A &\leq \min_{Q_k(0)=1} \|Q_k(A)(x - x_0)\|_A \\
&\leq \min_{Q_k(0)=1} \max_{\lambda \in \sigma(A)} |Q_k(\lambda)| \ \|x - x_0\|_A \ .
\end{aligned}
$$

Here we have used the fact that

$$
\|Q_k(A)\|_A = \max_{\lambda \in \sigma(A)} |Q_k(\lambda)|,
$$

which follows from

$$
\|Q_k(A)\|_A = \sup_{z \neq 0} \frac{\|Q_k(A)z\|_A}{\|z\|_A}
$$

by inserting $z = A^{\frac{1}{2}}w$. It remains to be shown that the solution of the min-max problem can be estimated by

$$
\alpha := \min_{Q_k(0)=1} \max_{\lambda \in \sigma(A)} |Q_k(\lambda)| \leq 2 \left(\frac{\sqrt{\kappa_2(A)} - 1}{\sqrt{\kappa_2(A)} + 1} \right)^k .
$$

In order to prove this, let

$$
0 < a = \lambda_1 \leq \lambda_2 \leq \cdots \leq \lambda_n = b
$$

be the eigenvalues of the spd-matrix A. Theorem 7.21 is applicable, since $0 \notin [a, b]$; and we therefore have

$$
\alpha \leq \min_{Q_k(0)=1} \max_{\lambda \in [a,b]} |Q_k(\lambda)| \leq \frac{1}{c} ,
$$

where $c := |T_k(\lambda(0))|$ is the maximal absolute value of the modified Chebyshev polynomial on $[a, b]$. Here

$$
\lambda(0) = 2\frac{0 - a}{b - a} - 1 = -\frac{b + a}{b - a} = -\frac{\kappa_2(A) + 1}{\kappa_2(A) - 1} ,
$$

because $\kappa_2(A) = \lambda_n/\lambda_1 = b/a$, so that we obtain the statement from Lemma 8.13 as in Example 8.12. $\qquad \square$

Corollary 8.18 *In order to reduce the error in the energy-norm by a factor ε, i.e.*

$$
\|x - x_k\|_A \leq \varepsilon \|x - x_0\|_A ,
$$

at most k cg-iterations are needed, where k is the smallest integer such that

$$
k \geq \frac{1}{2}\sqrt{\kappa_2(A)} \ \ln(2/\varepsilon) .
$$

Proof. According to Theorem 8.17, we have to show that

$$2 \left(\frac{\sqrt{\kappa_2(A)} - 1}{\sqrt{\kappa_2(A)} + 1} \right)^k \leq \varepsilon \,,$$

or, equivalently,

$$\theta^k \leq \frac{2}{\varepsilon} \quad \text{with} \quad \theta := \left(\frac{\sqrt{\kappa_2(A)} + 1}{\sqrt{\kappa_2(A)} - 1} \right) > 1 \,,$$

Thus the reduction factor is achieved if

$$k \geq \log_\theta(2/\varepsilon) = \frac{\ln(2/\varepsilon)}{\ln \theta} \,.$$

Now the natural logarithm satisfies

$$\ln \left(\frac{a+1}{a-1} \right) > \frac{2}{a} \quad \text{for } a > 1 \,.$$

(By differentiating both sides with respect to a, one sees that their difference is strictly decreasing for $a > 1$. In the limit case $a \to \infty$ both sides vanish.) By assumption we thus have

$$k \;\geq\; \frac{1}{2} \sqrt{\kappa_2(A)} \ln(2/\varepsilon) \;\geq\; \frac{\ln(2/\varepsilon)}{\ln \theta}$$

and therefore the statement. $\square$

Remark 8.19 Because of the striking properties of the cg-method for spd-matrices, the natural question to ask is, which properties can be carried over to non-symmetric matrices. First one needs to point out the fact that an arbitrary, only invertible matrix, A does *not*, in general, induce a scalar product. Two principle possibilities have been pursued so far:

If one interprets the cg-method for spd-matrices as an orthogonal similarity transformation to tridiagonal form (compare Chapter 6.1.1), then one would have to transform an arbitrary, not necessarily symmetric matrix to *Hessenberg form* (compare Remark 5.13). This means that a k-term recurrence relation with growing k will replace a three-term recurrence relation. This variant is also called *Arnoldi method* (compare [2]). Besides the fact that it uses more storage space, it is not particularly robust.

If one insists on maintaining the three-term recurrence relation as a structural element (it uses little storage space), then one generally passes to the normal equations

$$A^T A x = A^T b$$

and realizes a cg-method for the spd-matrix $A^T A$. However, because of $\kappa_2(A^T A) = \kappa_2(A)^2$ (see Lemma 3.10), in this method, in the estimate (8.17) of the convergence speed, the factor $\sqrt{\kappa_2(A)}$ has to be replaced by $\kappa_2(A)$. This is in general a significant change for the worse. A nice overview on this scientific area is given by the extensive presentation by STOER [73]. Regarding algorithms on this basis, two variants have so far essentially established themselves, namely the *cgs-method* or *conjugate gradient squared method* by SONNEVELD [71] and the *bi-cg-method*, which was originally proposed by FLETCHER [29]. Furthermore, the cgs-method avoids the additional evaluation of the mapping $y \mapsto A^T y$, which in certain applications is costly to program.

Remark 8.20 In the derivation of the cg-method, we did not follow the historical development but have chosen the more lucid way using Galerkin approximations, which at the same time was supposed to illustrate a more general concept. Here the meaning of the notion 'conjugate gradients' at first remained mysterious. The original approach starts with a different iteration method, the *method of steepest descent*. Here one tries to successively approximate the solution x of the minimization problem

$$\phi(x) = \frac{1}{2}\langle x, Ax \rangle - \langle x, b \rangle = \min$$

by minimizing ϕ in the direction of the steepest descent

$$-\Delta\phi(x_k) = b - Ax_k = r_k \ .$$

We divide the minimization problem $\phi(x) = \min$ into a sequence of one-dimensional minimization problems

$$\phi_k(\alpha_{k+1}) = \phi(x_k + \alpha_{k+1} r_k) = \min_{\alpha_{k+1}} \ ,$$

whose solution, the *optimal line search* is given by

$$\alpha_{k+1} = \frac{\langle r_k, r_k \rangle}{\langle r_k, Ar_k \rangle} \ .$$

We now expect that the sequence of the thus constructed approximations $x_{k+1} = x_k + \alpha_{k+1} r_k$ converges to x. This is indeed the case, because

$$|\phi(x_k) - \phi(x)| \le \left(1 - \frac{1}{\kappa_2(A)}\right) |\phi(x_{k-1}) - \phi(x)| \ ,$$

which, however, is very slow for a large condition $\kappa_2(A) = \lambda_n(A)/\lambda_1(A)$. From geometric intuition, it is clear that there exists the problem that the

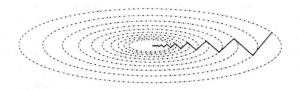

Figure 8.2: Method of steepest descent for large $\kappa_2(A)$

level surfaces $\{x \mid \phi(x) = c\}$ for $c \geq 0$ of ϕ are ellipsoids with very differing axis, where $\lambda_1(A)$ and $\lambda_n(A)$ are just the lengths of the smallest respectively largest semi-axis of $\{\phi(x) = 1\}$ (see Figure 8.2). The quantity $\kappa_2(A)$ thus describes the geometric 'distortion' of the ellipsoids as compared with spheres. However, the method of steepest descent converges best, when the level surfaces are approximately spheres. An improvement can be reached by replacing the directions of search r_k by other 'gradients'. Here, the A-orthogonal (or A-conjugate) vectors p_k with $p_1 = r_0 = b$ have particularly good properties, which explains the historical naming.

Finally, we wish to mention that the cg-method, as opposed to the Chebyshev method, does not require adjustment of parameters.

8.4   Preconditioning

The estimates of the convergence speed for the Chebyshev acceleration, as well as the estimates for the cg-method depend monotonically on the condition $\kappa_2(A)$ with respect to the Euclidian norm. Our next question is therefore the following: How can one make the condition of the matrix A smaller? Or, more precisely: How can the problem $Ax = b$ be transformed, so that the condition of the resulting matrix is as small as possible? This question is the topic of the *preconditioning*. Geometrically speaking, this means: We want to transform the problem, such that the level surfaces, which in general are ellipsoids, become as close as possible to spheres.

Instead of the equation $Ax = b$ with an spd-matrix $A \in \text{Mat}_n(\mathbf{R})$, we can solve also for any invertible matrix $B \in \text{GL}(n)$ the equivalent problem

$$\bar{A}\bar{x} = b \quad \text{with } \bar{x} := B^{-1}x \text{ and } \bar{A} := AB.$$

Here we have to take care that the symmetry of the problem does not get destroyed, so that our iteration methods remain applicable. If B is also symmetric and positive definite, then the matrix $\bar{A} = AB$ is not any more self-adjoint with respect to the Euclidian scalar-product $\langle \cdot, \cdot \rangle$, but with respect to the product which is induced by B,

$$(\cdot, \cdot)_B := \langle \cdot, B \cdot \rangle \, ,$$

because

$$(x, ABy)_B = \langle x, BABy \rangle = \langle ABx, By \rangle = (ABx, y)_B \, .$$

The cg-method is therefore again applicable, if we change the scalar products accordingly: $(\cdot, \cdot)_B$ takes on the role of the Euclidian product $\langle \cdot, \cdot \rangle$, and the corresponding 'energy product'

$$(\cdot, \cdot)_{AB} = (AB \cdot, \cdot)_B = \langle AB \cdot, B \cdot \rangle$$

of $\bar{A} = AB$ the role of $(\cdot, \cdot)$. This immediately yields the following iteration $\bar{x}_0, \bar{x}_1, \ldots$ for the solution of $\bar{A}\bar{x} = b$:

$p_1 := r_0 := b - AB\bar{x}_0;$

for $k := 1$ **to** $k_{\max}$ **do**

$$\alpha_k \quad := \frac{(r_{k-1}, r_{k-1})_B}{(p_k, p_k)_{AB}} = \frac{\langle r_{k-1}, Br_{k-1} \rangle}{\langle ABp_k, Bp_k \rangle};$$

$\bar{x}_k \quad := \bar{x}_{k-1} + \alpha_k p_k;$

if accurate **then exit**;

$r_k \quad := r_{k-1} - \alpha_k ABp_k;$

$$\beta_{k+1} := \frac{(r_k, r_k)_B}{(r_{k-1}, r_{k-1})_B} = \frac{\langle r_k, Br_k \rangle}{\langle r_{k-1}, Br_{k-1} \rangle};$$

$p_{k+1} := r_k + \beta_{k+1} p_k;$

end for

We are of course interested in an iteration for the actual solution $x = B\bar{x}$, and we thus replace the row for the $\bar{x}_k$ by

$$x_k = x_{k-1} + \alpha_k Bp_k \, .$$

It strikes us that the p_k now only occur explicitly in the last row. If for this reason, we introduce the (A-orthogonal) vectors $q_k := Bp_k$, then this yields the following economical version of the method, the *preconditioned cg-method* or briefly *pcg-method*.

Algorithm 8.21 *pcg-method* for the starting value x_0.

$$r_0 := b - Ax_0;$$

$$q_1 := Br_0;$$

for $k := 1$ **to** $k_{\max}$ **do**

$$\alpha_k \quad := \frac{\langle r_{k-1}, Br_{k-1} \rangle}{\langle q_k, Aq_k \rangle};$$

$$x_k \quad := x_{k-1} + \alpha_k q_k;$$

if accurate **then exit**;

$$r_k \quad := r_{k-1} - \alpha_k Aq_k;$$

$$\beta_{k+1} := \frac{\langle r_k, Br_k \rangle}{\langle r_{k-1}, Br_{k-1} \rangle};$$

$$q_{k+1} := Br_k + \beta_{k+1} q_k;$$

end for

Per iteration step, each time we only need to carry out one multiplication by the matrix A (for Aq_k), respectively by B (for Br_k), thus, as compared with the original cg-method, only one more multiplication by B.

Let us turn to the error $x - x_k$ of the pcg-method. According to Theorem 8.17, for the error $\|\bar{x} - \bar{x}_k\|_{AB}$ of the transformed iterate $\bar{x}_k$ in the 'new' energy norm

$$\|\bar{y}\|_{AB} := \sqrt{(\bar{y}, \bar{y})_{AB}} = \sqrt{\langle AB\bar{y}, B\bar{y} \rangle} \,,$$

we have the estimate

$$\|\bar{x} - \bar{x}_k\|_{AB} \leq 2 \left(\frac{\sqrt{\kappa_B(AB)} - 1}{\sqrt{\kappa_B(AB)} + 1} \right)^k \|\bar{x} - \bar{x}_0\|_{AB} \,.$$

Here $\kappa_B(AB)$ is the condition of AB with respect to the energy norm $\| \cdot \|_B$. However, the condition

$$\kappa_B(AB) = \frac{\lambda_{\max}(AB)}{\lambda_{\min}(AB)} = \kappa_2(AB)$$

is now independent of the underlying scalar product, and, because of

$$\|\bar{y}\|_{AB} = \sqrt{\langle AB\bar{y}, B\bar{y} \rangle} = \sqrt{\langle Ay, y \rangle} = \|y\|_A \,,$$

the norm $\| \cdot \|_{AB}$ is nothing else than the transformed energy norm $\| \cdot \|_A$. We therefore obtain the following analogue to Theorem 8.17.

Theorem 8.22 *Consider the approximation error $x - x_k$ of the cg-method 8.21, which is preconditioned with the spd-matrix B. $x - x_k$ can be estimated in the energy norm $\|y\|_A = \sqrt{\langle y, Ay \rangle}$ by*

$$\|x - x_k\|_A \leq 2 \left(\frac{\sqrt{\kappa_2(AB)} - 1}{\sqrt{\kappa_2(AB)} + 1} \right)^k \|x - x_0\|_A . \qquad (8.18)$$

We therefore seek a spd-matrix B, a *preconditioner*, with the properties

a) The mapping $(B, y) \mapsto By$ is 'simple' to carry out

b) The condition $\kappa_2(AB)$ of AB is 'small',

where, for the time being, we have to leave it with the vague expressions 'simple' and 'small'. The ideal matrix to satisfy b), $B = A^{-1}$, unfortunately has the disadvantage that the evaluation of the mapping $y \mapsto By = A^{-1}y$ possesses the complexity of the entire problem and contradicts therefore the requirement a). However, the following lemma says that it is sufficient, if the energy norms $\| \cdot \|_B$ and $\| \cdot \|_{A^{-1}}$, which are induced by B and A^{-1}, can be estimated (as sharply as possible) from above and below (compare [84]).

Lemma 8.23 *Suppose that for two positive constants $\mu_0, \mu_1 > 0$, one of the following three equivalent conditions is satisfied*

i) $\quad \mu_0 \langle A^{-1}y, y \rangle \; \leq \; \langle By, y \rangle \; \leq \; \mu_1 \langle A^{-1}y, y \rangle \quad$ *for all* $y \in \mathbf{R}^n$

ii) $\quad \mu_0 \langle By, y \rangle \; \leq \; \langle BABy, y \rangle \; \leq \; \mu_1 \langle By, y \rangle \quad$ *for all* $y \in \mathbf{R}^n$

iii) $\quad \lambda_{\min}(AB) \geq \mu_0 \;$ *and* $\; \lambda_{\max}(AB) \leq \mu_1.$ $\qquad (8.19)$

Then the condition of AB satisfies

$$\kappa_2(AB) \leq \frac{\mu_1}{\mu_0} .$$

Proof. The equivalence of i) and ii) follows by inserting $y = Au$ into i), because

$$\langle A^{-1}y, y \rangle = \langle Au, u \rangle \;\; \text{and} \;\; \langle By, y \rangle = \langle BAu, Au \rangle = \langle ABAu, u \rangle .$$

Because of

$$\lambda_{\min}(AB) = \min_{y \neq 0} \frac{(ABy, y)_B}{(y, y)_B} = \min_{y \neq 0} \frac{\langle BABy, y \rangle}{\langle By, y \rangle}$$

and

$$\lambda_{\max}(AB) = \max_{y \neq 0} \frac{(ABy, y)_B}{(y, y)_B} = \max_{y \neq 0} \frac{\langle BABy, y \rangle}{\langle By, y \rangle} ,$$

the latter condition is equivalent to iii) (compare Lemma 8.29), from which the statement

$$\kappa_2(AB) = \frac{\lambda_{\max}(AB)}{\lambda_{\min}(AB)} \leq \frac{\mu_1}{\mu_0}$$

follows immediately.   □

If both norms $\| \cdot \|_B$ and $\| \cdot \|_{A^{-1}}$ are approximately equal, i.e. $\mu_0 \approx \mu_1$, then B and A^{-1} are called *spectrally equivalent*, or briefly also $B \overset{\sim}{\sim} A^{-1}$. In this case, according to Lemma 8.23, we have that $\kappa_2(AB) \approx 1$.

Remark 8.24 The three conditions of Lemma 8.23 are in fact symmetric in A and B. One can see this most easily in the condition for the eigenvalues, since

$$\lambda_{\min}(AB) = \lambda_{\min}(BA) \quad \text{and} \quad \lambda_{\max}(AB) = \lambda_{\max}(BA) \ .$$

(This follows for example from $\langle ABy, y \rangle = \langle BAy, y \rangle$.) If we assume that the vague relation $\approx$ is transitive, then one can call the spectrally equivalence with full right an 'equivalence' in the sense of equivalence relations.

An important consequence of Lemma 8.23 concerns the termination criterium.

Lemma 8.25 *Suppose that the assumptions of Lemma 8.23 are satisfied, and let* $\| \cdot \|_A := \sqrt{\langle A\cdot, \cdot \rangle}$ *be the energy norm and* $\| \cdot \|_B := \sqrt{\langle B\cdot, \cdot \rangle}$. *Then*

$$\frac{1}{\sqrt{\mu_1}} \|r_k\|_B \ \leq \ \|x - x_k\|_A \ \leq \ \frac{1}{\sqrt{\mu_0}} \|r_k\|_B \ ,$$

i.e., if B *and* A^{-1} *are spectrally equivalent, then the norm* $\|r_k\|_B$ *of the residue, which can be evaluated, estimates very well the energy norm* $\|x - x_k\|_A$ *of the error.*

Proof. According to Lemma 8.23 and Remark 8.24, (8.19), with A and B exchanged, implies that

$$\mu_0 \langle Ay, y \rangle \ \leq \ \langle ABAy, y \rangle \ \leq \ \mu_1 \langle Ay, y \rangle \ ,$$

or, equivalently

$$\frac{1}{\mu_1} \langle ABAy, y \rangle \ \leq \ \langle Ay, y \rangle \ \leq \ \frac{1}{\mu_0} \langle ABAy, y \rangle \ .$$

The residue $r_k = b - Ax_k = A(x - x_k)$ however satisfies

$$\langle Br_k, r_k \rangle = \langle BA(x - x_k), A(x - x_k) \rangle = \langle ABA(x - x_k), x - x_k \rangle \ ,$$

and therefore, as claimed,

$$\frac{1}{\sqrt{\mu_1}}\|r_k\|_B \leq \|x - x_k\|_A \leq \frac{1}{\sqrt{\mu_0}}\|r_k\|_B .$$

$\square$

As a termination criterion for the pcg-method, instead of (8.16), we therefore utilize the much more sensible condition

$$\|r_k\|_B = \sqrt{\langle Br_k, r_k \rangle} \text{ 'sufficiently small' .}$$

Example 8.26 A very simple, but often already effective preconditioning is the inverse $B := D^{-1}$ of the diagonal D of A, called the *diagonal preconditioning*. Variants of this utilize *block diagonal matrices* (see (8.1)), where the blocks have to be invertible.

Example 8.27 If one applies the Cholesky factorization $A = LL^T$ from Section 1.4 to a symmetric sparse matrix, then one observes that outside the sparsing pattern of A, usually only 'relatively small' elements l_{ij} are produced. This observation leads to the idea of the *incomplete Cholesky factorization* (IC): It consists of simply omitting these elements. If by

$$P(A) := \{(i,j) \mid a_{ij} \neq 0\}$$

we denote the index set of the non-vanishing elements of a matrix A, then, instead of L, we construct a matrix $\tilde{L}$ with

$$P(\tilde{L}) \subset P(A) ,$$

by proceeding as in the Cholesky factorization and setting $\tilde{l}_{ij} := 0$ for all $(i,j) \notin P(A)$. Here we expect that

$$A \approx \tilde{A} := \tilde{L}\tilde{L}^T .$$

In [55], a proof of the existence and numerical stability was given, in the case that A is a *M-Matrix*, i.e.

$$a_{ii} > 0, \quad a_{ij} \leq 0 \text{ for } i \neq j \text{ and } A^{-1} \geq 0 \text{ (element-wise) .}$$

Such matrices occur when discretizing simple partial differential equations (see [78]). In this case, an M-matrix is produced in each elimination step. For B, we then set

$$B := \tilde{A}^{-1} = \tilde{L}^{-T}\tilde{L}^{-1} .$$

This way, in many cases one obtains a drastic acceleration of the cg-method, far beyond the class of M-matrices, which is accessible to proofs.

Remark 8.28 For systems which originate from the discretization of partial differential equations, this additional knowledge about the origin of the system allows for a much more refined and effective construction of preconditioners. For examples, we refer to the articles by YSERENTANT [85], XU [84], and (for time-dependent partial differential equations) to the dissertation of BORNEMANN [6]. As a matter of fact, in solving partial differential equations by discretization, one has to deal not only with *a single* linear system of fixed, though high dimension. An adequate description is by a nested *sequence* of linear systems, whose dimension grows when the discretization is successively refined. This sequence is solved by a cascade of cg-methods of growing dimension. Methods of this type are genuine alternatives to classical multi-grid methods — for details see e.g. the fundamental paper by DEUFLHARD, LEINEN and YSERENTANT [22]. In these methods, each linear system is only solved up to the precision of the corresponding discretization. In addition, it allows for a simultaneous construction of discretization grids, which fit the problem under consideration. We shall explain this aspect in Section 9.7 in the simple example of numerical quadrature. In Exercise 8.4, we illustrate an aspect of the cascade principle, which is suitable for this introductory text.

8.5   Lanczos Methods

In Section 5.3, we have computed the eigenvalues of a symmetric matrix by first transforming it to tridiagonal form and then applying the QR-algorithm. In this section, we shall again turn our attention to the eigenvalue problem

$$Ax = \lambda x \tag{8.20}$$

with a real symmetric matrix A. We shall be concerned with large sparse matrices A, as they occur in most applications. For these problems, the methods which were presented in Chapter 5 are too expensive. In the following, we shall therefore develop iterative methods for the approximation of the eigenvalues of a symmetric matrix, which essentially only require one matrix-vector multiplication per iteration step. The idea for these methods goes back to the Hungarian mathematician LANCZOS in a work from the year 1950 [53]. As in the derivation of the cg-method in Section 8.3, we pose the eigenvalue problem (8.20) first as an extremum problem.

Lemma 8.29 *Suppose $\lambda_{\min}$ and $\lambda_{\max}$ are the smallest respectively largest eigenvalues of the real symmetric matrix $A \in \mathrm{Mat}_n(\mathbf{R})$. Then*

$$\lambda_{\min} = \min_{x \neq 0} \frac{\langle x, Ax \rangle}{\langle x, x \rangle} \quad and \quad \lambda_{\max} = \max_{x \neq 0} \frac{\langle x, Ax \rangle}{\langle x, x \rangle} \ .$$

Proof. Since A is symmetric, there exists an orthogonal matrix $Q \in \mathbf{O}(n)$ such that

$$QAQ^T = \Lambda = \mathrm{diag}(\lambda_1, \ldots, \lambda_n) \ .$$

With $y := Qx$, we have $\langle x, x \rangle = \langle y, y \rangle$, and

$$\langle x, Ax \rangle = \langle Q^T y, AQ^T y \rangle = \langle y, QAQ^T y \rangle = \langle y, \Lambda y \rangle \ .$$

Thus the statement is reduced to the case of a diagonal matrix, for which the statement is obvious. $\qquad \Box$

The function $\mu : \mathbf{R}^n \to \mathbf{R}$,

$$\mu(x) := \frac{\langle x, Ax \rangle}{\langle x, x \rangle}$$

is called *Rayleigh quotient* of A.

Corollary 8.30 *Let $\lambda_1 \leq \cdots \leq \lambda_n$ be the eigenvalues of the symmetric matrix $A \in \mathrm{Mat}_n(\mathbf{R})$, and let $\eta_1, \ldots, \eta_n$ be the corresponding eigenvectors. Then*

$$\lambda_i = \min_{\substack{x \in \mathrm{span}(\eta_i, \ldots, \eta_n) \\ x \neq 0}} \mu(x) = \max_{\substack{x \in \mathrm{span}(\eta_1, \ldots, \eta_i) \\ x \neq 0}} \mu(x) \ .$$

If we compare the formulation of the eigenvalue problem as an extremum problem in Lemma 8.29 with the derivation of the cg-method in Section 8.3, then this suggests, to approximate the eigenvalues $\lambda_{\min}$ and $\lambda_{\max}$, by solving the corresponding extremum problem on a sequence of subspaces $V_1 \subset V_2 \subset \cdots \subset \mathbf{R}^n$.

Since the Krylov spaces proved useful in Section 8.3, we choose as subspaces

$$V_k(x) := \mathrm{span}\{x, Ax, \ldots, A^{k-1}x\}$$

for a starting value $x \neq 0$ and $k = 1, 2, \ldots$. We expect that the extreme values

$$\lambda_{\min}^{(k)} := \min_{\substack{y \in V_k(x) \\ y \neq 0}} \frac{\langle y, Ay \rangle}{\langle y, y \rangle} \geq \lambda_{\min} \ , \quad \lambda_{\max}^{(k)} := \max_{\substack{y \in V_k(x) \\ y \neq 0}} \frac{\langle y, Ay \rangle}{\langle y, y \rangle} \leq \lambda_{\max}$$

to approximate the eigenvalues $\lambda_{\min}$ and $\lambda_{\max}$ well for growing k. According to Theorem 6.4, we can construct an orthonormal basis $v_1, \ldots, v_k$ of $V_k(x)$ by the following three-term recurrence relation:

$$
\begin{aligned}
v_0 &:= 0, \quad v_1 := \frac{x}{\|x\|_2} \\
\alpha_k &:= \langle v_k, Av_k \rangle \\
w_{k+1} &:= Av_k - \alpha_k v_k - \beta_k v_{k-1} \\
\beta_{k+1} &:= \|w_{k+1}\|_2 \\
v_{k+1} &:= \frac{w_{k+1}}{\beta_{k+1}} \text{ falls } \beta_{k+1} \neq 0 .
\end{aligned}
\tag{8.21}
$$

This iteration is called *Lanczos algorithm*. Thus $Q_k := [v_1, \ldots, v_k]$ is a column-orthonormal matrix, and

$$
Q_k^T A Q_k =
\begin{bmatrix}
\alpha_1 & \beta_2 & & & & \\
\beta_2 & \alpha_2 & \beta_3 & & & \\
& \ddots & \ddots & \ddots & & \\
& & \beta_{k-1} & \alpha_{k-1} & \beta_k \\
& & & \beta_k & \alpha_k
\end{bmatrix}
=: T_k
$$

is a symmetric tridiagonal matrix. Now set $y = Q_k v$. Then, for a $v \in \mathbf{R}^k$, we have $\langle y, y \rangle = \langle v, v \rangle$, and

$$
\langle y, Ay \rangle = \langle Q_k v, AQ_k v \rangle = \langle v, Q_k^T A Q_k v \rangle = \langle v, T_k v \rangle .
$$

Hence it follows that

$$
\lambda_{\min}^{(k)} = \min_{\substack{y \in V_k(x) \\ y \neq 0}} \frac{\langle y, Ay \rangle}{\langle y, y \rangle} = \min_{\substack{v \in \mathbf{R}^k \\ v \neq 0}} \frac{\langle v, T_k v \rangle}{\langle v, v \rangle} = \lambda_{\min}(T_k) ,
$$

and similarly that $\lambda_{\max}^{(k)} = \lambda_{\max}(T_k)$. Because of $V_{k+1} \supset V_k$, the minimal property yields immediately that

$$
\lambda_{\min}^{(k+1)} \leq \lambda_{\min}^{(k)} \quad \text{and} \quad \lambda_{\max}^{(k+1)} \geq \lambda_{\max}^{(k)} .
$$

The approximations $\lambda_{\min}^{(k)}$ and $\lambda_{\max}^{(k)}$ are therefore the extreme eigenvalues of the symmetric tridiagonal matrix T_k, and, as such, can be easily computed. However, in contrast to the cg-method, it is not guaranteed that $\lambda_{\min}^{(n)} = \lambda_{\min}$, since in general $V_n(x) \neq \mathbf{R}^n$. This shows in the three-term recurrence relation (8.21) in a vanishing of β_{k+1} for a $k < n$. In this case, the computation has to be started again with an $\tilde{x} \in V_k(x)^\perp$.

The convergence speed of the method can again be estimated by utilizing the Chebyshev polynomials.

Theorem 8.31 *Let A be a symmetric matrix with the eigenvalues $\lambda_1^\cdot \leq \cdots \leq \lambda_n$ and corresponding orthonormal eigenvectors $\eta_1, \ldots, \eta_n$. Furthermore, let $\mu_1 \leq \cdots \leq \mu_k$ be the eigenvalues of the tridiagonal matrix T_k of the Lanczos method for the starting value $x \neq 0$, and with the orthonormal basis $v_1, \ldots, v_k$ of $V_k(x)$ as in (8.21). Then*

$$\lambda_n \geq \mu_k \geq \lambda_n - \frac{(\lambda_n - \lambda_1)\tan^2(\measuredangle\,(v_1, \eta_n))}{T_{k-1}^2(1 + 2\rho_n)} ,$$

where $\rho_n := (\lambda_n - \lambda_{n-1})/(\lambda_{n-1} - \lambda_1)$.

Proof. Because of $V_k(x) = \{P(A)x \mid P \in \mathbf{P}_{k-1}\}$, we have

$$\mu_k = \max_{\substack{y \in V_k(x) \\ y \neq 0}} \frac{\langle y, Ay \rangle}{\langle y, y \rangle} = \max_{P \in \mathbf{P}_{k-1}} \frac{\langle P(A)v_1, AP(A)v_1 \rangle}{\langle P(A)v_1, P(A)v_1 \rangle} .$$

By representing v_1 with respect to the orthonormal basis $\eta_1, \ldots, \eta_n$, $v_1 = \sum_{j=1}^n \xi_j \eta_j$ with $\xi_j = \langle v_1, \eta_j \rangle = \cos(\measuredangle\,(v_1, \eta_j))$, it then follows that

$$\langle P(A)v_1, AP(A)v_1 \rangle = \sum_{j=1}^n \xi_j^2 P^2(\lambda_j)\lambda_j$$

$$\langle P(A)v_1, P(A)v_1 \rangle = \sum_{j=1}^n \xi_j^2 P^2(\lambda_j) .$$

We thus obtain

$$\frac{\langle P(A)v_1, AP(A)v_1 \rangle}{\langle P(A)v_1, P(A)v_1 \rangle} = \lambda_n + \frac{\sum_{j=1}^{n-1} \xi_j^2 P^2(\lambda_j)(\lambda_j - \lambda_n)}{\sum_{j=1}^n \xi_j^2 P^2(\lambda_j)}$$

$$\geq \lambda_n + (\lambda_1 - \lambda_n)\frac{\sum_{j=1}^{n-1} \xi_j^2 P^2(\lambda_j)}{\xi_n^2 P^2(\lambda_n) + \sum_{j=1}^{n-1} \xi_j^2 P^2(\lambda_j)} .$$

In order to obtain as sharp an estimate as possible, we need to insert a polynomial $P \in \mathbf{P}_{n-1}$, which is as small as possible in the interval $[\lambda_1, \lambda_{n-1}]$. According to Theorem 7.19, one should take the transformed Chebyshev polynomial

$$P(\lambda) := T_{k-1}(t(\lambda)) \quad \text{with} \quad t(\lambda) = 2\frac{\lambda - \lambda_1}{\lambda_{n-1} - \lambda_1} - 1 = 1 + 2\frac{\lambda - \lambda_{n-1}}{\lambda_{n-1} - \lambda_1}$$

with the property $|P(\lambda_j)| \le 1$ for $j = 1, \ldots, n-1$. Because of $\sum_{j=1}^{n} \xi_j^2 = \|v_1\|_2^2 = 1$, it thus follows that

$$\mu_k \ge \lambda_n - (\lambda_n - \lambda_1)\frac{1-\xi_n^2}{\xi_n^2}\frac{1}{T_{k-1}^2(1+2\rho_n)} \,,$$

and the statement follows from the fact that

$$\frac{1-\xi_n^2}{\xi_n^2} = \tan^2(\measuredangle\,(v_1, \eta_n)) \,.$$

$\square$

In many applications (e.g. in structural mechanics) one encounters the *generalized symmetric eigenvalue problem*,

$$Ax = \lambda Bx \,, \tag{8.22}$$

where the matrices $A, B \in \mathrm{Mat}_n(\mathbf{R})$ are both symmetric and B is in addition positive definite. If we insert the Cholesky factorization $B = LL^T$ of B into (8.22), then we have

$$Ax = \lambda Bx \iff Ax = \lambda LL^T x \iff \underbrace{(L^{-1}AL^{-T})}_{=:\,\bar{A}}\underbrace{L^T x}_{=:\,\bar{x}} = \lambda L^T x \,.$$

Since $\bar{A} = L^{-1}AL^{-T}$ is again symmetric, it follows that the generalized eigenvalue problem $Ax = \lambda Bx$ is equivalent to the symmetric eigenvalue problem $\bar{A}\bar{x} = \lambda\bar{x}$. Thus all eigenvalues λ_i are real. Furthermore, there is an orthonormal basis $\eta_1, \ldots, \eta_n$ of generalized eigenvectors $A\eta_i = \lambda_i B\eta_i$. If we therefore define the *generalized Rayleigh quotient* of (A, B) by

$$\mu(x) := \frac{\langle x, Ax\rangle}{\langle x, Bx\rangle} \,,$$

then we obtain the following statement, which is an analogue to Lemma 8.29.

Lemma 8.32 *Let $\lambda_{\min}$ and $\lambda_{\max}$ be the smallest respectively largest eigenvalue of the generalized eigenvalue problem $Ax = \lambda Bx$, where the matrices $A, B \in \mathrm{Mat}_n(\mathbf{R})$ are symmetric and B is positive definite. Then*

$$\lambda_{\min} = \min_{x\ne 0}\frac{\langle x, Ax\rangle}{\langle x, Bx\rangle} \quad and \quad \lambda_{\max} = \max_{x\ne 0}\frac{\langle x, Ax\rangle}{\langle x, Bx\rangle} \,.$$

Proof. With the above notation, we have

$$\langle x, Bx \rangle = \langle x, LL^T x \rangle = \langle \bar{x}, \bar{x} \rangle$$

and

$$\langle x, Ax \rangle = \langle L^{-T}\bar{x}, AL^{-T}\bar{x} \rangle = \langle \bar{x}, \bar{A}\bar{x} \rangle \ .$$

The statement follows from $Ax = \lambda Bx \Leftrightarrow \bar{A}\bar{x} = \lambda \bar{x}$ and Lemma 8.29. $\square$

Similarly, the Lanczos algorithm (8.21) carries over, by maintaining $\langle x, Ax \rangle$, but replacing $\langle x, x \rangle$ by $\langle x, Bx \rangle$ and $\|x\|_2$ by $\langle x, Bx \rangle^{1/2}$. A detailed presentation can be found in the book by CULLUM and WILLOUGHBY [11].

Finally, we want to discuss the case that not only the extremal eigenvalues, but eigenvalues in a given interval, or even all eigenvalues are to be found. In this case, we go back to the *fundamental idea of the inverse vector iteration*. Let $\bar{\lambda}$ be a given estimate in the neighborhood of the unknown eigenvalues. Then (for the generalized eigenvalue problem) the matrix

$$C := (\bar{A} - \bar{\lambda}I)^{-1} = L^T(A - \bar{\lambda}B)^{-1}L$$

has the eigenvalues

$$\nu_i := (\lambda_i - \bar{\lambda})^{-1} \quad \text{for} \quad i = 1, 2, \dots \tag{8.23}$$

The application of the Lanczos method to the matrix C then produces just the dominant eigenvalues ν_i, thus, because of (8.23), eigenvalues λ_i in a neighborhood of $\bar{\lambda}$. By variation of the *shift parameter* $\bar{\lambda}$ in a given interval, one can therefore obtain all eigenvalues. This variant of the algorithm is called *spectral-Lanczos* and was described in 1980 by ERICSSON and RUHE [27].

8.6   Exercises

Exercise 8.1 Show that the coefficients $\rho_k = 2\bar{t}T_{k-1}(\bar{t})/T_k(\bar{t})$, which occur in the Chebyshev acceleration, satisfy the two-term recurrence relation

$$\rho_1 = 2, \quad \rho_{k+1} = \frac{1}{1 - \frac{1}{4}\sigma^2 \rho_k},$$

where $\sigma := 1/\bar{t}$. Furthermore, show that the limit of the sequence $\{\rho_k\}$ satisfies

$$\lim_{k \to \infty} \rho_k =: \rho = \frac{2}{1 + \sqrt{1 - \sigma^2}} \ .$$

Exercise 8.2 Given are sparse matrices of the following structure (band matrices, block diagonal matrices, arrow matrices, block cyclical matrices).

$$
\begin{bmatrix}
* & * & * & * & & \\
* & \ddots & \ddots & \ddots & \ddots & \\
* & \ddots & \ddots & \ddots & \ddots & * \\
 & \ddots & \ddots & \ddots & \ddots & * \\
 & & \ddots & \ddots & \ddots & * \\
 & & & * & * & *
\end{bmatrix}
\begin{bmatrix}
* & * & * & & & \\
* & * & * & & & \\
* & * & * & & & \\
 & & & * & & \\
 & & & & * & * \\
 & & & & * & *
\end{bmatrix}
$$

$$
\begin{bmatrix}
* & * & * & * & * & * \\
* & * & & & & \\
* & & * & & & \\
* & & & * & & \\
* & & & & * & \\
* & & & & & *
\end{bmatrix}
\begin{bmatrix}
* & * & * & * & & \\
* & * & * & * & & \\
 & & * & * & * & * \\
 & & * & * & * & * \\
* & * & & & * & * \\
* & * & & & * & *
\end{bmatrix}
$$

Estimate the required storage space and cost of computation (number of operations) for

a) LU-factorization with Gaussian elimination without pivoting respectively with column pivot search and exchange of rows.

b) QR-factorization with Householder transformations without respectively with exchange of columns.

c) QR-factorization with Givens transformations.

Exercise 8.3 Let $B \stackrel{.}{\sim} A^{-1}$ be a spectrally equivalent preconditioning matrix. In the special case that B is of the form $B = CC^T$, a preconditioned cg-method can also be derived differently. For this, from the system $Ax = b$, one passes formally to the equivalent system

$$
\hat{A}\hat{x} = \hat{b} \quad \text{with} \quad \hat{A} = CAC^T, \quad \hat{x} = C^{-T}x, \quad \text{and} \quad \hat{b} = Cb .
$$

Here $\hat{A}$ is again a spd-matrix. One applies the classical cg-method to this transformed system. Using this idea, derive a convergence result, which is based on the energy norm of the error (which, as is well known, is not directly approachable). Use this approach to derive two different effective variants of the pcg-method, one of which coincides with our algorithm 8.21. Derive both variants, and consider, which would be preferable under which conditions. Implement both variants in the special case of the incomplete Cholesky factorization, and carry out computational comparisons.

Exercise 8.4 *Short introduction to the cascade principle.* We consider a sequence of linear systems

$$A_j x_j = b_j, \quad j = 1, \dots, m$$

of dimension n_j, as it could come up by successively finer (uniform) discretization of an (elliptic) partial differential equation. The dimension of the systems is assumed to grow geometrically, i.e.

$$n_{j+1} = K n_j \text{ for a } K > 1.$$

We seek an approximate solution $\tilde{x}_m$ of the largest system (corresponding to the finest discretization), such that the error in the energy norm satisfies

$$\|\tilde{x}_m - x_m\|_{A_m} \leq \varepsilon^m \delta$$

for given $\varepsilon, \delta > 0$. Assume that the connection of the linear systems with the discretizations is shown by the following properties:

i) The matrices A_j are symmetric positive definite, and their conditions are uniformly bounded by

$$\kappa_2(A_j) \leq C \text{ for } j = 1, \dots, m.$$

(This is only true after a suitable preconditioning.)

ii) If $\tilde{x}_j$ is an approximation of x_j with

$$\|\tilde{x}_j - x_j\|_{A_j} \leq \varepsilon^j \delta,$$

then the vector $\tilde{x}_{j+1}^0 := (\tilde{x}_j, 0)$, which is augmented by zeros, is an approximation of x_{j+1} with

$$\|\tilde{x}_{j+1}^0 - x_{j+1}\|_{A_{j+1}} \leq \varepsilon^j \delta,$$

iii) The cost of a cg-iteration for A_j is

$$C_j = \sigma n_j \text{ for a } \sigma > 0.$$

We compare two algorithms for the computation of $\tilde{x}_m$:

a) *Standard cg-method:*   How many cg-iterations are necessary to compute $\tilde{x}_m$, beginning with a starting solution x_m^0 with

$$\|x_m^0 - x_m\|_{A_m} \leq \delta ?$$

What is the cost?

b) *Cascade cg-method:*   Let x_1^0 be an approximation of x_1 with

$$\|x_1^0 - x_1\|_{A_1} \leq \delta \ .$$

From this, we compute successive approximations $\tilde{x}_1, \dots, \tilde{x}_m$ with

$$\|\tilde{x}_j - x_j\|_{A_j} \leq \varepsilon^j \delta \ ,$$

by always taking the approximate solutions of the previous system, which are augmented by zeros, as a starting solution

$$\tilde{x}_j^0 := (\tilde{x}_{j-1}, 0) \quad \text{for } j = 2, \dots, m$$

in the cg-method for the system $A_j x_j = b_j$. Show that the approximate solution $\tilde{x}_m$, which is computed this way, has the required precision. How many cg-iterations are necessary for each j, and what is their cost? What is the total cost of the computation of $\tilde{x}_m$?

The method b) even opens the possibility to compute the starting value x_1^0 by a direct method, e.g. the Gaussian elimination, if the dimension n_1 is sufficiently small.

Exercise 8.5 Carry the spectral Lanczos algorithm out in detail. In particular, replace the explicit computation of $\bar{A}$ by the product representation $L^{-1}AL^{-T}$.

9 Definite Integrals

A relatively common problem is the computation of the Riemann integral

$$I(f) := I_a^b(f) := \int_a^b f(t) \, dt \ .$$

Here f is a piecewise continuous function on the interval $[a, b]$, which, however, is usually piecewise smooth in the applications. (A piecewise continuous, but not piecewise smooth function cannot be implemented on a
computer.) In analysis one learns many techniques to 'solve' such an inte-

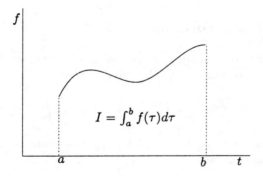

$$I = \int_a^b f(\tau)d\tau$$

Figure 9.1: Problem of quadrature

gral in the sense that a "simpler" expression is found for $I(f)$. If this is
possible, then one says that the integral is *analytically* solvable, and that
the solution can be expressed *in closed form*. However, these two notions do
not mean much more, than that the "analytic expressions" are better known
mathematically, and one therefore prefers these to the original integral expression. However, in many cases, a closed analytic representation of the
integral does not exist. Then only the purely numerical evaluation remains.
In many cases, this is even preferable, when a closed form solution is known.

One can very easily convince oneself of this fact by looking into an integral table (e.g. [41]).

The numerical computation of $I(f)$ is also called *numerical quadrature*. This notion reminds us of the *quadrature of the circle*, the problem of only using compass and ruler to construct a square which has the area of the unit circle, which is unsolvable because of the transcendence of π. Todays meaning of the term quadrature emerged from this construction of the square of equal area. More often, one finds the notion of *numerical integration*, which however, and more generally, at the same time describes the solution of differential equations. In fact, the computation of the integral $I(f)$ corresponds formally to the solution of the *initial value problem*

$$y'(t) = f(t), \quad y(a) = 0 , \quad t \in [a, b] ,$$

since $y(b) = I(f)$. We will be coming back to this formulation in the course of this chapter.

9.1 Quadrature Formulas

First, we shall list some analytic properties of the Riemann integral. In the following we assume that $b > a$. The definite integral

$$I_a^b = I : C[a, b] \longrightarrow \mathbf{R}, \quad f \longmapsto I(f) = \int_a^b f(x) \, dx$$

is a *positive linear form* on the space $C[a, b]$ of continuous functions on the interval $[a, b]$. In other words:

a) I is *linear*, i.e. for all continuous functions f, g and $\alpha, \beta \in \mathbf{R}$, we have

$$I(\alpha f + \beta g) = \alpha I(f) + \beta I(g) .$$

b) I is *positive*, i.e. if f is non-negative, then so is the integral $I(f)$,

$$f \geq 0 \implies I(f) \geq 0 .$$

In addition, the integral is *additive* with respect to a partition of the integration interval, i.e. for all $\tau \in [a, b]$, we have

$$I_a^\tau + I_\tau^b = I_a^b .$$

Before starting with the computation, we should ask the question of the condition of the integral. In order to do this, we first have to clarify, how to measure perturbations δf of the integrand. If we choose the supremum-norm (as the standard norm on C^0), then, on an infinite integration interval, the perturbations can become infinite. Because of this, we decide here for the so-called L^1-norm

$$\|f\|_1 := \int_a^b |f(t)|\,dt = I(|f|) .$$

Lemma 9.1 *The absolute and relative condition of the quadrature problem* (I, f), $I(f) = \int_a^b f$, *with respect to the* L^1*-norm* $\|\cdot\|_1$ *are*

$$\kappa_{\mathrm{abs}} = 1 \quad resp. \quad \kappa_{\mathrm{rel}} = \frac{I(|f|)}{|I(f)|} .$$

Proof. For any perturbation $\delta f \in L^1[a, b]$, the perturbation of the integral can be estimated by

$$\left| \int_a^b (f + \delta f) - \int_a^b f \right| = \left| \int_a^b \delta f \right| \leq \int_a^b |\delta f| = \|\delta f\|_1 ,$$

where equality holds for a positive perturbation. □

In the absolute concept, perturbation is therefore a harmless problem. In our relative view, however, we expect difficulties when the ratio of the integral of the absolute value and the absolute value of the integral is very large, and the problem is thus ill-conditioned. Obviously, this is completely analogous to the condition of addition (resp. subtraction) (compare Example 2.3). A danger for the relative condition comes from *highly oscillatory* integrands, which do occur in many applications. Already the integration of a single oscillation of the sine-function over a single period is ill-conditioned with respect to the relative error concept.

Of course, we expect that a method to compute the integral preserves its structural properties. The goal of the numerical quadrature is therefore the construction of *positive linear forms*

$$\hat{I} : C[a, b] \longrightarrow \mathbf{R}, \quad f \longmapsto \hat{I}(f) ,$$

which approximate the integral I as well as possible, i.e.

$$\hat{I}(f) - I(f) \quad \text{is "small".}$$

Example 9.2 The first, and, because of the definition of the Riemann integral, most obvious method for the computation of $I_a^b(f)$, is the so-called *trapezoidal sum* (see Figure 9.2). We partition the interval into n sub-

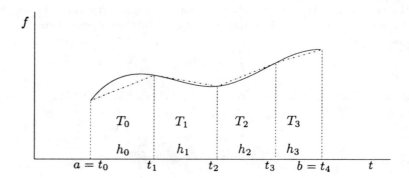

Figure 9.2: Trapezoidal sum with equidistant nodes

intervals $[t_{i-1}, t_i]$ $(i = 1, \ldots, n)$ of length $h_i := t_i - t_{i-1}$

$$a = t_0 < t_1 < \cdots < t_n = b \,,$$

and we approximate the integral $I(f)$ by the sum of the areas of the trapezoids

$$T^{(n)} := \sum_{i=1}^{n} T_i, \quad T_i := \frac{h_i}{2}(f(t_{i-1}) + f(t_i)) \,.$$

The trapezoidal sum $T^{(n)}$ is obviously a positive linear form. We also interpret it as an application of the *trapezoidal rule* (see Section 9.2)

$$T := \frac{b - a}{2}(f(a) + f(b)) \tag{9.1}$$

to the sub-intervals $[t_{i-1}, t_i]$. By comparing the trapezoidal sum $T^{(n)}$ with the Riemann lower resp. upper sums,

$$R_{\min}^{(n)} = \sum_{i=1}^{n} h_i \min_{t \in [t_{i-1}, t_i]} f(t) \quad \text{and} \quad R_{\max}^{(n)} = \sum_{i=1}^{n} h_i \max_{t \in [t_{i-1}, t_i]} f(t) \,,$$

it is obvious that

$$R_{\min}^{(n)} \leq T^{(n)} \leq R_{\max}^{(n)} \,.$$

For continuous $f \in C^0[a, b]$, the convergence of the Riemann sums therefore implies the convergence of the trapezoidal sums:

$$
\begin{array}{ccccc}
R_{\min}^{(n)} & \leq & T^{(n)} & \leq & R_{\max}^{(n)} \\
\downarrow & & \downarrow & & \downarrow \qquad \text{for } n \to \infty, \quad h_i \leq h \to 0 . \\
I(f) & & I(f) & & I(f)
\end{array}
$$

Below (see Lemma 9.8), we shall give the approximation error in more detail.

The trapezoidal sum is a simple example for a *quadrature formula*, which we define as follows.

Definition 9.3 A *quadrature formula* $\hat{I}$ for the computation of the definite integral is a sum

$$
\hat{I}(f) = (b - a) \sum_{i=0}^{n} \lambda_i f(t_i) , \tag{9.2}
$$

with *knots* $t_0, \ldots, t_n$, and *weights* $\lambda_0, \ldots, \lambda_n$, such that

$$
\sum_{i=0}^{n} \lambda_i = 1 . \tag{9.3}
$$

The condition (9.3) concerning the weights guarantees that a quadrature formula integrates constant functions exactly, i.e. $\hat{I}(1) = I(1) = b - a$. It is furthermore obvious that a quadrature formula is positive, if all weights are, i.e.

$$
\hat{I} \text{ positive} \iff \lambda_k \geq 0 \text{ for all } k = 0, \ldots, n.
$$

This means that for all practical purposes, only quadrature formulas with positive weights are interesting. If this is not the case, then the sum of the absolute values of the weights

$$
\sum_{i=0}^{n} |\lambda_i| \geq 1 \tag{9.4}
$$

measures, by how much the quadrature formula deviates from the positivity requirement. Because of the results for the scalar product, we do not have to worry about the stability of the evaluation of a quadrature formula, (see Lemma 2.30).

9.2   Newton-Cotes Formulas

The idea of the trapezoidal sum consists of replacing the function f by an approximation $\hat{f}$, here the linear interpolation, for which the quadrature can be easily carried out, and to view $I(\hat{f})$ as an approximation of $I(f)$, i.e.,

$$\hat{I}(f) := I(\hat{f}) \, .$$

Here, of course, not only linear approximations can be used, as is the case in the trapezoidal rule (9.1), but arbitrary approximations, like the ones which were introduced in the last chapter. In particular, for given knots $t_0, \ldots, t_n$, the function

$$\hat{f}(t) := P(f \,|\, t_0, \ldots, t_n) = \sum_{i=0}^{n} f(t_i) L_{in}(t),$$

is the interpolation polynomial of f, where $L_{in} \in \mathbf{P}_n$ is the i-th Lagrange polynomial for the knots t_j, i.e. $L_{in}(t_j) = \delta_{ij}$. This approach yields the quadrature formulas

$$\hat{I}(f) := I(P(f \,|\, t_0, \ldots, t_n)) = (b - a) \sum_{i=0}^{n} \lambda_{in} f(t_i),$$

where the weights

$$\lambda_{in} := \frac{1}{b - a} \int_{a}^{b} L_{in}(t) \, dt$$

only depend on the choice of the knots $t_0, \ldots, t_n$.

It is clear from the construction that the quadrature formulas, which are defined this way, are exact for polynomials $P \in \mathbf{P}_n$ of degree less than or equal to n,

$$\hat{I}(P) = I(P_n(P)) = I(P) \quad \text{for } P \in \mathbf{P}_n.$$

For given knots t_i, the quadrature formula is already uniquely determined by this property.

Lemma 9.4 *For $n + 1$ pairwise distinct knots $t_0, \ldots, t_n$, there exists one and only one quadrature formula*

$$\hat{I}(f) = (b - a) \sum_{i=0} \lambda_i f(t_i) \, ,$$

which is exact for all polynomials $P \in \mathbf{P}_n$ of degree less than or equal to n.

Proof.   We insert the Lagrange polynomials $L_{in} \in \mathbf{P}_n$, which belong to the knots t_i, and which by assumption are integrated exactly, into the quadrature formula

$$I(L_{in}) = \hat{I}(L_{in}) = (b-a)\sum_{j=0}^{n}\lambda_{jn}L_{in}(t_j) = (b-a)\sum_{j=0}^{n}\lambda_j\delta_{ij} = (b-a)\lambda_i$$

and thus, in a unique way, obtain back the weights $\lambda_i = (b-a)^{-1}I(L_{in}) = \lambda_{in}$. $\square$

In the special case of *equidistant knots*

$$h_i = h = \frac{b-a}{n}, \quad t_i = a+ih, \quad i = 0,\ldots,n\,,$$

the constructed quadrature formulas are called *Newton-Cotes formulas*. The term for the corresponding *Newton-Cotes weights* λ_{in} simplifies via the substitution $s := (t-a)/h$:

$$\lambda_{in} = \frac{1}{b-a}\int_a^b \prod_{\substack{j=0\\j\neq i}}^{n}\frac{t-t_j}{t_i-t_j}\,dt = \frac{1}{n}\int_0^n \prod_{\substack{j=0\\j\neq i}}^{n}\frac{s-j}{i-j}\,ds\,.$$

The weights λ_{in}, which are independent of the interval boundaries, only have to be computed once, respectively given once. In Table 9.1, we have listed them up to order $n = 4$. The weights, and therefore also the quadrature formula, are always positive for the orders $n = 1,\ldots,7$. Higher orders are less attractive, since starting with $n = 8$, negative weights may occur. In this case, the Lebesgue constant is the characteristic quantity (9.4), up to the normalization factor $(b-a)^{-1}$. Note that we have already encountered the Lebesgue constant in Section 7.1 as the condition number of the polynomial interpolation.

Table 9.1: Newton-Cotes weights λ_{in} for $n = 1,\ldots,4$

n	$\lambda_{0n},\ldots,\lambda_{nn}$					Error	Name
1		$\frac{1}{2}$	$\frac{1}{2}$			$\frac{h^3}{12}f''(\tau)$	Trapezoidal rule
2		$\frac{1}{6}$	$\frac{4}{6}$	$\frac{1}{6}$		$\frac{h^5}{90}f^{(4)}(\tau)$	Simpson's rule, Kepler's barrel rule
3		$\frac{1}{8}$	$\frac{3}{8}$	$\frac{3}{8}$	$\frac{1}{8}$	$\frac{3h^5}{80}f^{(4)}(\tau)$	Newton's 3/8-rule
4	$\frac{7}{90}$	$\frac{32}{90}$	$\frac{12}{90}$	$\frac{32}{90}$	$\frac{7}{90}$	$\frac{8h^7}{945}f^{(6)}(\tau)$	Milne's rule

In Table 9.1, we have already assigned the respective approximation errors to the Newton-Cotes formulas (for sufficiently smooth integrands), expressed as a power of the stepsize h and a derivative at an intermediate position $\tau \in [a, b]$. Observe that for the even orders $n = 2, 4$, the power of h and the degree of the derivative always jump by 2. In the following we shall verify these estimates for the first two formulas, by which one can already see the principle for odd n (see Exercise 9.3).

Before starting, we recall a not so obvious variant of the mean value theorem, which we shall encounter repeatedly in the proof of the approximation statements.

Lemma 9.5 *Let $g, h \in C[a, b]$ be continuous functions on $[a, b]$, where g has only one sign, i.e. either $g(t) \geq 0$ or $g(t) \leq 0$ for all $t \in [a, b]$. Then*

$$\int_a^b h(t)g(t)\, dt = h(\tau) \int_a^b g(t)\, dt$$

for a $\tau \in [a, b]$.

Proof. Assume without loss of generality that g is non-negative. Then

$$\min_{t \in [a,b]} h(t) \int_a^b g(s)\, ds \leq \int_a^b h(s)g(s)\, ds \leq \max_{t \in [a,b]} h(t) \int_a^b g(s)\, ds\,.$$

Therefore, for the continuous function

$$F(t) := \int_a^b h(s)g(s)\, ds - h(t) \int_a^b g(s)\, ds$$

there exist $t_0, t_1 \in [a, b]$ with $F(t_0) \geq 0$ and $F(t_1) \leq 0$, and thus, because of the mean value theorem, there exits also a $\tau \in [a, b]$ such that $F(\tau) = 0$, or, in other words

$$\int_a^b h(t)g(t)\, dt = h(\tau) \int_a^b g(t)\, dt\,,$$

as required.   □

Lemma 9.6 *Let $f \in C^2([a, b])$ be a twice continuously differentiable function. Then the approximation error of the trapezoidal rule*

$$T = \frac{b-a}{2}(f(a) + f(b))$$

with stepsize $h := b - a$ can be expressed by

$$T - \int_a^b f = \frac{h^3}{12} f''(\tau)$$

for a $\tau \in [a, b]$.

Proof. According to Theorem 7.10, and by using Newton's remainder, the linear interpolation $P = P_1(f)$ satisfies

$$f(t) = P(t) + [t, a, b]f \cdot (t - a)(t - b) \, ,$$

where, according to Corollary 7.13, the second divided difference can be expressed by

$$[t, a, b]f = \frac{f''(\tau)}{2}$$

with a $\tau = \tau(t) \in [a, b]$, which in independent of t. Inserted into the quadrature formula, Lemma 9.5 implies that

$$\int_a^b f = \int_a^b P(t) \, dt + \int_a^b \underbrace{[t, a, b]f \cdot (t - a)(t - b)}_{\leq 0} \, dt$$

$$= T + \frac{f''(\tau)}{2} \underbrace{\int_a^b (t - a)(t - b) \, dt}_{= -\frac{(b-a)^3}{6}}$$

for a $\tau \in [a, b]$, hence

$$T - \int_a^b f = \frac{h^3}{12} f''(\tau) \, .$$

$\square$

Lemma 9.7 *Kepler's barrel rule*

$$S = \frac{b - a}{6} \left(f(a) + 4f\left(\frac{a + b}{2}\right) + f(b) \right)$$

is also exact for polynomials $Q \in \mathbf{P}_3$ of degree 3. For $f \in C^4([a, b])$, the approximation error of stepsize $h := \frac{b-a}{2}$ can be expressed by

$$S - \int_a^b f = \frac{f^{(4)}(\tau)}{90} h^5$$

with a $\tau \in [a, b]$.

Proof. Let $Q \in \mathbf{P}_3$. Then, according to Newton's remainder formula, the quadratic interpolation $P = P_2(Q)$ at the knots a, b and $(a + b)/2$ satisfies

$$Q(t) = P(t) + \underbrace{\gamma (t - a)\left(t - \frac{a + b}{2}\right)(t - b)}_{= \omega_3(t)} \, ,$$

where $\gamma = Q'''(t)/6 \in \mathbf{R}$ is a constant. This implies for the integral that

$$\int_a^b Q = \int_a^b P + \gamma \int_a^b \omega_3(t)\, dt = S \, ,$$

because the integral $\int_a^b \omega_k$ of Newton's basis functions vanishes for odd k. Kepler's barrel rule is therefore also exact for polynomials of degree 3. For $f \in C^4([a,b])$, we now form the cubic Hermite interpolation $Q = P_3(f) \in \mathbf{P}_3$ with respect to the four knots

$$t_0 = a, \quad t_1 = \frac{a+b}{2}, \quad t_2 = \frac{a+b}{2}, \quad t_3 = b \, .$$

For the description of the approximation error of Q, we again utilize Newton's remainder,

$$f(t) = Q(t) + \left[t, a, \frac{a+b}{2}, \frac{a+b}{2}, b\right] f \underbrace{(t-a)\left(x - \frac{a+b}{2}\right)^2 (t-b)}_{= \; \omega_4(t) \, \leq \, 0} \, .$$

Estimation of the remainder according to Corollary 7.13, insertion into the integral and application of the mean value theorem

$$\int_a^b f = S + \frac{f^{(4)}(\tau)}{4!} \underbrace{\int_a^b \omega_4(t)\, dt}_{= \; -\frac{4}{15} h^5} = S - \frac{f^{(4)}(\tau)}{90} h^5$$

yield again the statement.   □

As we already saw in the introductory Example 9.2 of the trapezoidal sum, further quadrature formulas can be constructed by partitioning the interval and applying a quadrature formula on the respective sub-intervals. We partition the interval again into n sub-intervals $[t_{i-1}, t_i]$ with $i = 1, \dots, n$,

$$a = t_0 < t_1 < \cdots < t_n = b \, ,$$

so that, according to the additivity of the integral,

$$\int_a^b f = \sum_{i=1}^n \int_{t_{i-1}}^{t_i} f \, .$$

Thus

$$\hat{I}(f) := \sum_{i=1}^n \hat{I}_{t_{i-1}}^{t_i}(f)$$

is a (possibly better) approximation of the integral, where $\hat{I}_{t_i}^{t_{i+1}}$ denotes an arbitrary quadrature formula on the interval $[t_i, t_{i+1}]$. In the following, we shall derive the approximation error for the trapezoidal sum from the approximation error of the trapezoidal rule.

Lemma 9.8 *Let $h := (b-a)/n$, and let t_i be the equidistant knots $t_i = a+ih$ for $i = 0, \ldots, n$. Furthermore, let T_i denote the trapezoidal rule*

$$T_i := \frac{h}{2}(f(t_{i-1}) + f(t_i))$$

on the interval $[t_{i-1}, t_i]$, where $i = 1, \ldots, n$. Then, for $f \in C^2([a,b])$, the approximation error of the trapezoidal sum

$$T(h) := \sum_{i=1}^{n} T_i = h\left(\frac{1}{2}(f(a) + f(b)) + \sum_{i=1}^{n-1} f(a + ih)\right)$$

can be expressed in the form

$$T(h) - \int_a^b f = \frac{(b-a)h^2}{12} f''(\tau)$$

with a $\tau \in [a,b]$.

Proof. According to Lemma 9.6, there exists a $\tau_i \in [t_{i-1}, t_i]$ such that

$$T_i - \int_{t_{i-1}}^{t_i} f = \frac{h^3}{12} f''(\tau_i) \,,$$

and therefore

$$T(h) - \int_a^b f = \sum_{i=1}^{n}(T_i - \int_{t_{i-1}}^{t_i} f) = \sum_{i=1}^{n} \frac{h^3}{12} f''(\tau_i) = \frac{(b-a)h^2}{12} \frac{1}{n} \sum_{i=1}^{n} f''(\tau_i) \,.$$

Since

$$\min_{t\in[a,b]} f''(t) \leq \frac{1}{n} \sum_{i=1}^{n} f''(\tau_i) \leq \max_{t\in[a,b]} f''(t) \,,$$

and according the intermediate value theorem, there exists a $\tau \in [a,b]$ such that

$$\frac{1}{n} \sum_{i=1}^{n} f''(\tau_i) = f''(\tau) \,,$$

and therefore

$$T(h) - \int_a^b f = \frac{(b-a)h^2}{12} f''(\tau) \,,$$

as claimed. $\qquad\square$

9.3   Gauss-Christoffel Quadrature

In the construction of the Newton-Cotes formulas, starting with $n+1$ given integration knots t_i, we have determined the weights λ_i so that the quadrature formula integrates exactly polynomials up to degree n. Can we possibly achieve more by also putting the knots at our disposal?

In this section, we want to answer this question for the more general problem of *weighted* integrals

$$I(f) := \int_a^b \omega(t)\, f(t)\, dt\ ,$$

which we have already encountered, when we introduced orthogonal polynomials in Chapter 6.1.1. Again, let ω be a positive weight function, $\omega(t) > 0$ for all $t \in\,]a, b[$, so that the norms

$$\|P\| := (P, P)^{\frac{1}{2}} = \left(\int_a^b \omega(t) P(t)^2\, dt \right)^{\frac{1}{2}} < \infty$$

are well defined and finite for all polynomials $P \in \mathbf{P}_k$ and all $k \in \mathbf{N}$. In contrast to Section 9.2, the interval may be infinite here. It is only important that the corresponding moments

$$\mu_k := \int_a^b t^k \omega(t)\, dt$$

are bounded. For the definition of the absolute condition, we measure the perturbations δf in a natural way with respect to the weighted L^1-norm

$$\|f\|_1 = \int_a^b \omega(t)\, |f(t)|\, dt = I(|f|)\ .$$

This way, the results of Lemma 9.1 remain valid also for weighted integrals, only the interpretation of $I(f)$ is changed.

In Table 9.2, the most common weight functions are listed, together with the corresponding intervals.

9.3.1   Construction of the quadrature formula

Our goal is the construction of quadrature formulas of the form

$$\hat{I}_n(f) := \sum_{i=0}^n \lambda_{in} f(\tau_{in})\ ,$$

Table 9.2: Typical weight functions

$\omega(t)$	Interval $[a,b]$
$\frac{1}{\sqrt{1-t^2}}$	$[-1,1]$
e^{-t}	$[0,\infty]$
e^{-t^2}	$[-\infty,\infty]$
1	$[-1,1]$

which approximate the integral $I(f)$ as good as possible. More precisely, for a given n, we seek the $n+1$ *knots* $\tau_{0n},\ldots,\tau_{nn}$ and $n+1$ *weights* $\lambda_{0n},\ldots,\lambda_{nn}$, so that polynomials up to as high degree N as possible can be integrated exactly, i.e.

$$\hat{I}_n(P) = I(P) \quad \text{for all} \quad P \in \mathbf{P}_N .$$

If we first try to estimate, which degree N can be achieved for a given n, then we observe that we have $2n+2$ parameters (respectively $n+1$ knots and weights) at our disposal, as opposed to $N+1$ coefficients of a polynomial of degree N. The best that we can expect, is therefore that polynomials up to a degree $N \le 2n+1$ can be integrated exactly. Since the integration knots enter the quadrature formula in a nonlinear way, it is not enough to simply count the degrees of freedom. We instead try to draw conclusions from our wishful thinking, which may be helpful in the solution of the problem.

Lemma 9.9 *If $\hat{I}_n$ is exact for all polynomials $P \in \mathbf{P}_{2n+1}$, then the polynomials $\{P_j\}$, which are defined by their root representation*

$$P_{j+1}(t) := (t - \tau_{0j}) \cdots (t - \tau_{jj}) \in \mathbf{P}_{n+1} ,$$

are orthogonal with respect to the scalar product, which is induced by ω

$$(f,g) = \int_a^b \omega(t) f(t) g(t)\, dt .$$

Proof. For $j < n+1$, we have $P_{n+1}P_j \in \mathbf{P}_{2n+1}$; so that

$$(P_j, P_{n+1}) = \int_a^b \omega P_j P_{n+1} = \hat{I}_n(P_j P_{n+1}) = \sum_{i=0}^n \lambda_{in} P_j(\tau_{in}) \underbrace{P_{n+1}(\tau_{in})}_{=\,0} = 0 .$$

$\square$

Therefore the knots τ_{in}, which we seek, have to be roots of pairwise orthogonal polynomials $\{P_j\}$ of degree $\deg P_j = j$. Such orthogonal polynomials are not unknown to us. According to Theorem 6.2, there exists one and only one family $\{P_j\}$ of polynomials $P_j \in \mathbf{P}_j$ with leading coefficients one, i.e. $P_j(t) = t^j + \cdots$ so that

$$(P_k, P_j) = \delta_{kj}(P_k, P_k) .$$

From Theorem 6.5, we already know that the roots of these orthogonal polynomials are real and have to lie in the interval $[a, b]$. We have therefore constructed, in a unique way, candidates for the integration knots τ_{in} of the quadrature formula $\hat{I}_n$: t he roots of the orthogonal polynomial P_{n+1}. Once the knots are determined, there is no choice for the weights: In order that at least polynomials $P \in \mathbf{P}_n$ up to degree n are integrated exactly, according to Lemma 9.4, the weights

$$\lambda_{in} := \frac{1}{b-a} \int_a^b L_{in}(t) \, dt$$

have to be chosen with the Lagrange polynomials $L_{in}(\tau_{jn}) = \delta_{ij}$. This, at first, only guarantees exactness for polynomials up to degree n, which is in fact enough.

Lemma 9.10 *Let $\tau_{0n}, \ldots, \tau_{nn}$ be the roots of the $(n+1)$-st orthogonal polynomial P_{n+1}. Then any quadrature formula $\hat{I}_n(f) = \sum_{i=0}^n \lambda_i f(\tau_{in})$ satisfies:*

$$\hat{I}_n \text{ exact on } \mathbf{P}_n \quad \Longleftrightarrow \quad \hat{I}_n \text{ exact on } \mathbf{P}_{2n+1} .$$

Proof. Suppose that $\hat{I}_n$ is exact on $\mathbf{P}_n$ and $P \in \mathbf{P}_{2n+1}$. Then there exist polynomials $Q, R \in \mathbf{P}_n$ (Euclidean algorithm), such that

$$P = QP_{n+1} + R .$$

Since P_{n+1} is orthogonal to $\mathbf{P}_n$, it follows for the weighted integral that

$$\int_a^b \omega P = \underbrace{\int_a^b \omega Q P_{n+1}}_{= 0} + \int_a^b \omega R = \int_a^b \omega R = \hat{I}_n(R) .$$

On the other hand,

$$\hat{I}_n(R) = \sum_{i=0}^n \lambda_{in} R(\tau_{in}) = \sum_{i=0}^n \lambda_{in} \big(Q(\tau_{in}) \underbrace{P_{n+1}(\tau_{in})}_{= 0} + R(\tau_{in})\big) = \hat{I}_n(P) ,$$

i.e. $\hat{I}$ is exact on $\mathbf{P}_{2n+1}$.   □

We collect our results in the following theorem.

Theorem 9.11 *There exist uniquely determined knots* $\tau_{0n}, \ldots, \tau_{nn}$ *and weights* $\lambda_{0n}, \ldots, \lambda_{nn}$, *such that the quadrature formula*

$$\hat{I}_n(f) = \sum_{i=0}^n \lambda_{in} f(\tau_{in})$$

integrates exactly all polynomials of degree less than or equal to $2n + 1$, *i.e.*

$$\hat{I}_n(P) = \int_a^b \omega P \quad \text{for } P \in \mathbf{P}_{2n+1}.$$

The knots τ_{in} *are the roots of the* $(n+1)$-st *orthogonal polynomial* P_{n+1} *with respect to the weight function* ω *and the weights*

$$\lambda_{in} := \frac{1}{b - a} \int_a^b L_{in}(t)\, dt$$

with the Lagrange polynomials $L_{in}(\tau_{jn}) = \delta_{ij}$. *Furthermore, the weights are all positive,* $\lambda_{in} > 0$, *i.e.* $\hat{I}_n$ *is a positive linear form, and they satisfy the equation*

$$\lambda_{in} = \frac{1}{P'_{n+1}(\tau_{in})P_n(\tau_{in})}(P_n, P_n) \,. \tag{9.5}$$

Proof.   We only have to verify the positivity of the weights and their representation (9.5). Suppose $Q \in \mathbf{P}_{2n+1}$ is a polynomial, such that τ_{kn} is the only knot at which it does not vanish, i.e. $Q(\tau_{in}) = 0$ for $i \neq k$ and $Q(\tau_{kn}) \neq 0$. Then, obviously,

$$\int_a^b \omega Q = \lambda_{kn} Q(\tau_{kn}), \quad \text{hence} \quad \lambda_{kn} = \frac{1}{Q(\tau_{kn})} \int_a^b \omega Q \,.$$

If we set, e.g.,

$$Q(t) := \left(\frac{P_{n+1}(t)}{(t - \tau_{kn})} \right)^2 ,$$

then $Q \in \mathbf{P}_{2n}$ has the required properties, where $Q(\tau_{kn}) = P'_{n+1}(\tau_{kn})^2$. Thus the weights satisfy

$$\lambda_{kn} = \frac{1}{Q(\tau_{kn})} \int_a^b \omega Q = \int_a^b \omega \left(\frac{P_{n+1}(t)}{P'_{n+1}(\tau_{kn})(t - \tau_{kn})} \right)^2 dt > 0 \,,$$

i.e. all weights are positive. In order to verify formula (9.5), we put

$$Q(t) := \frac{P_{n+1}(t)}{t - \tau_{kn}} P_n(t) .$$

Again, $Q \in \mathbf{P}_{2n}$ has the required properties, and it follows that

$$\lambda_{kn} = \frac{1}{P'_{n+1}(\tau_{kn})P_n(\tau_{kn})} \int_a^b \omega(t) \frac{P_{n+1}(t)}{t - \tau_{kn}} P_n(t)\, dt .$$

The polynomial $P_{n+1}(t)/(t - \tau_{kn})$ again has leading coefficient 1, so that

$$\frac{P_{n+1}(t)}{t - \tau_{kn}} = P_n(t) + Q_{n-1}(t)$$

with a $Q_{n-1} \in \mathbf{P}_{n-1}$. Since P_n is orthogonal to $\mathbf{P}_{n-1}$, the statement finally follows

$$\lambda_{kn} = \frac{1}{P'_{n+1}(\tau_{kn})P_n(\tau_{kn})} (P_n, P_n) .$$

$\square$

These quadrature formulas $\hat{I}_n$ are the *Gauss-Christoffel formulas* for the weight function ω. As is the case for the Newton-Cotes formulas, it is easy to deduce the approximation error from the exactness for a certain polynomial degree.

Theorem 9.12 *For any function $f \in C^{2n+2}$, the approximation error of the Gauss-Christoffel quadrature can be expressed in the form*

$$\int_a^b \omega f - \hat{I}_n(f) = \frac{f^{(2n+2)}(\tau)}{(2n+2)!} (P_{n+1}, P_{n+1})$$

with a $\tau \in [a, b]$.

Proof. As expected, we employ the Newton remainder for the Hermite interpolation $P \in \mathbf{P}_{2n+1}$ for the $2n + 2$ knots $\tau_{0n}, \tau_{0n}, \ldots, \tau_{nn}, \tau_{nn}$:

$$f(t) = P(t) + [t, \tau_{0n}, \tau_{0n}, \ldots, \tau_{nn}, \tau_{nn}]f \cdot \underbrace{(t - \tau_{0n})^2 \cdots (t - \tau_{nn})^2}_{= P_{n+1}(t)^2 \geq 0}$$

Since $\hat{I}_n$ integrates the interpolation P exactly, it follows that

$$\int_a^b \omega f = \int_a^b \omega P + \frac{f^{(2n+2)}(\tau)}{(2n+2)!} \int_a^b \omega P_{n+1}^2$$

$$= \sum_{i=0}^n \lambda_{in} \underbrace{P(\tau_{in})}_{= f(\tau_{in})} + \frac{f^{(2n+2)}(\tau)}{(2n+2)!} (P_{n+1}, P_{n+1}) .$$

$\square$

Example 9.13 *Gauss-Chebyshev quadrature.* Consider the weight function $\omega(t) = 1/\sqrt{1-t^2}$ on the interval $[-1, 1]$. Then the Chebyshev polynomials T_k, with which we are so well acquainted, are orthogonal, because

$$\int_{-1}^{1} \frac{T_k(t)T_j(t)}{\sqrt{1-t^2}}\, dt = \begin{cases} \pi & \text{, if } k = j = 0 \\ \pi/2 & \text{, if } k = j > 0 \\ 0 & \text{, if } k \neq j. \end{cases}$$

The orthogonal polynomials with leading coefficient 1 are therefore $P_n(t) = 2^{1-n}T_n(t)$. The roots of P_{n+1} (respectively T_{n+1}) are the Chebyshev knots

$$\tau_{in} = \cos\frac{2i+1}{2n+2}\pi, \quad \text{for } i = 0, \ldots, n\,.$$

By employing (9.5), one easily calculates that the weights for $n > 0$ are given by

$$\begin{aligned} \lambda_{in} &= \frac{1}{2^{-n}T'_{n+1}(\tau_{in})2^{1-n}T_n(\tau_{in})} \int_{-1}^{1} \frac{2^{2-2n}T_n^2}{\sqrt{1-t^2}}\, dt \\ &= \frac{2\pi/2}{T'_{n+1}(\tau_{in})T_n(\tau_{in})} = \frac{\pi}{n+1}\,. \end{aligned}$$

The Gauss-Chebyshev quadrature has therefore the simple form

$$\hat{I}_n(f) = \frac{\pi}{n+1}\sum_{i=0}^{n} f(\tau_{in}) \quad \text{with} \quad \tau_{in} = \cos\frac{2i+1}{2n+2}\pi\,.$$

According to Theorem 9.12, and because of $(T_{n+1}, T_{n+1}) = \pi/2$, its approximation error satisfies

$$\int_{-1}^{1} \frac{f(t)}{\sqrt{1-t^2}}\, dt - \hat{I}_n(f) = \frac{\pi}{2^{2n+1}(2n+2)!} f^{(2n+2)}(\tau)$$

for a $\tau \in [-1, 1]$.

In Table 9.3.1, we have listed some names of classes of orthogonal polynomials, together with their associated weight functions. The corresponding quadrature methods always carry the name "Gauss", hyphenated with the name of the respective polynomial class. The Gauss-Legendre quadrature ($\omega \equiv 1$) is only used in special applications. For general integrands, the trapezoidal sum extrapolation, about which we shall learn in the next section, is superior. However, the weight function of the Gauss-Chebyshev quadrature is weakly singular at $t = \pm 1$, so that the trapezoidal rule is

Table 9.3: Commonly occurring classes of orthogonal polynomials

$\omega(t)$	Interval $I = [a, b]$	Class of orthogonal polynomials
$\frac{1}{\sqrt{1-t^2}}$	$[-1, 1]$	Chebyshev polynomials T_n
e^{-t}	$[0, \infty]$	Laguerre polynomials L_n
e^{-t^2}	$[-\infty, \infty]$	Hermite polynomials H_n
1	$[-1, 1]$	Legendre polynomials P_n

not applicable. Of particular interest are the Gauss-Hermite and the Gauss-Legendre quadrature, which allow the approximation of integrals over infinite intervals (and even solve exactly for polynomials $P \in \mathbf{P}_{2n+1}$).

Let us finally note an essential property of the Gauss quadrature for weight functions $\omega \not\equiv 1$: The quality of the approximation can only be improved by increasing the order. A partitioning into sub-intervals, however, is only possible for the Gauss-Legendre quadrature (resp. the Gauss-Lobatto quadrature, compare Exercise 9.11).

9.3.2   Computation of knots and weights

For the effective computation of the weights λ_{in}, we need another representation. For this purpose, let $\{\bar{P}_k\}$ be a family of *orthonormal* polynomials $\bar{P}_k \in \mathbf{P}_k$, i.e.

$$(\bar{P}_i, \bar{P}_j) = \delta_{ij} .$$

These satisfy the so-called *Christoffel-Darboux formula* (see e.g. [74] or [59]).

Lemma 9.14 *Suppose that k_n are the leading coefficients of the orthonormal polynomials $\bar{P}_n(t) = k_n t^n + O(t^{n-1})$. Then, for all $s, t \in \mathbf{R}$,*

$$\frac{k_n}{k_{n+1}} \left(\frac{\bar{P}_{n+1}(t)\bar{P}_n(s) - \bar{P}_n(t)\bar{P}_{n+1}(s)}{t - s} \right) = \sum_{j=0}^{n} \bar{P}_j(t)\bar{P}_j(s) .$$

The following formula for the weighs λ_{in} can be derived from this formula.

Lemma 9.15 *The weights λ_{in} satisfy*

$$\lambda_{in} = \left(\sum_{j=0}^{n} \bar{P}_j^2(\tau_{in}) \right)^{-1} . \tag{9.6}$$

Proof. Let $s = \tau_{in}$. Then the Christoffel-Darboux formula implies that

$$\frac{k_n}{k_{n+1}} \frac{\bar{P}_{n+1}(t)}{t - \tau_{in}} \bar{P}_n(\tau_{in}) = \sum_{j=0}^{n} \bar{P}_j(t)\bar{P}_j(\tau_{in}) , \qquad (9.7)$$

and, in the limit case $t \to \tau_{in}$,

$$\frac{k_n}{k_{n+1}} \bar{P}'_{n+1}(\tau_{in})\bar{P}_n(\tau_{in}) = \sum_{j=0}^{n} \bar{P}_j(\tau_{in})^2 . \qquad (9.8)$$

If we insert this into Formula (9.5) for the weights λ_{in}, then it follows that

$$
\begin{aligned}
\lambda_{in} &= \int_a^b \omega(t) \frac{P_{n+1}(t)}{(t - \tau_{in})P'_{n+1}(\tau_{in})} \, dt \\
&\stackrel{(9.7)}{=} \frac{k_{n+1}}{k_n \bar{P}_n(\tau_{in})\bar{P}'_{n+1}(\tau_{in})} \sum_{j=0}^{n} \bar{P}_j(\tau_{in}) \underbrace{\int_a^b \omega(t)\bar{P}_j(t) \, dt}_{= 0 \text{ for } j \neq 0} \\
&= \frac{k_{n+1}}{k_n \bar{P}_n(\tau_{in})\bar{P}'_{n+1}(\tau_{in})} \underbrace{\int_a^b \omega(t)\bar{P}_0(\tau_{in})\bar{P}_0(t) \, dt}_{= (\bar{P}_0, \bar{P}_0) = 1} \\
&\stackrel{(9.8)}{=} \left(\sum_{j=0}^{n} \bar{P}_j^2(\tau_{in}) \right)^{-1} .
\end{aligned}
$$

$\square$

The actual determination of the weights λ_{in} and knots τ_{in} is based on techniques, which are based on the contents of Chapter 6. For this recall that, according to Theorem 6.2, the orthogonal polynomials P_k with respect to the weight function ω satisfy a three-term recurrence relation

$$P_k(t) = (t - \beta_k)P_{k-1}(t) - \gamma_k^2 P_{k-2}(t) , \qquad (9.9)$$

where

$$\beta_k = \frac{(tP_{k-1}, P_{k-1})}{(P_{k-1}, P_{k-1})} , \quad \gamma_k^2 = \frac{(P_{k-1}, P_{k-1})}{(P_{k-2}, P_{k-2})} .$$

We therefore assume that the orthogonal polynomials are given by their three-term recurrence relation (9.9), which, for $k = 0, \ldots, n$, we can write as a linear system $Tp = tp + r$ with

$$
T := \begin{bmatrix}
\beta_1 & 1 & & & & \\
\gamma_2^2 & \beta_2 & 1 & & & \\
& \ddots & \ddots & \ddots & & \\
& & \gamma_n^2 & \beta_n & 1 & \\
& & & \gamma_{n+1}^2 & \beta_{n+1} &
\end{bmatrix}
$$

and

$$
p := (P_0(t), \ldots, P_n(t))^T, \quad r := (0, \ldots, 0, -P_{n+1}(t))^T \ .
$$

Thus

$$
P_{n+1}(t) = 0 \iff Tp = tp \ ,
$$

i.e. the roots of P_{n+1} are just the eigenvalues of T, where the eigenvector $p(\tau)$ corresponds to an eigenvalue τ .

Because the roots τ_{in} of P_{n+1} are all real, one could have the idea that the eigenvalue problem $Tp = tp$ can be transformed into a symmetric eigenvalue problem. The simplest possibility would be to scale with a diagonal matrix $D = \mathrm{diag}(d_0, \ldots, d_n)$ to obtain

$$
\hat{T}\hat{p} = t\hat{p} \text{ with } \hat{p} = Dp \ , \ \hat{T} = DTD^{-1} \ ,
$$

with the hope to achieve $\hat{T} = \hat{T}^T$. More explicitly, diagonal scaling, as applied to a matrix $A \in \mathrm{Mat}_{n+1}(\mathbf{R})$, satisfies

$$
A \mapsto \hat{A} := DAD^{-1} \text{ with } \hat{a}_{ij} = \frac{d_i}{d_j} a_{ij} \ .
$$

For $\hat{T}$ to be symmetric, it is necessary that

$$
\gamma_i^2 \frac{d_i}{d_{i-1}} = \frac{d_{i-1}}{d_i} \ , \quad \text{i.e. } d_i^2 = d_{i-1}^2 / \gamma_i^2 \ ,
$$

which we can satisfy without any problems, e.g. by

$$
d_0 := 1 \text{ and } d_i := (\gamma_2 \cdots \gamma_{i+1})^{-1} \text{ for } i = 1, \ldots, n \ .
$$

With this choice of D, $\hat{T}$ is the symmetric tridiagonal matrix

$$
\hat{T} = \begin{bmatrix}
\beta_1 & \gamma_2 & & & \\
\gamma_2 & \beta_2 & \gamma_3 & & \\
& \ddots & \ddots & \ddots & \\
& & \gamma_1 & \beta_n & \gamma_{n+1} \\
& & & \gamma_{n+1} & \beta_n
\end{bmatrix} = \hat{T}^T \ ,
$$

whose eigenvalues $\tau_{0n}, \ldots, \tau_{nn}$ we can compute by employing the QR-algorithm (compare Section 5.3). The weights λ_{in} can also be computed from (9.6) via the three-term recurrence relation (9.9). The Gauss quadrature can be carried out as soon as the $(\lambda_{in}, \tau_{in})$ are at hand (see also [40]).

9.4 Classical Romberg Quadrature

In the following, we want to learn about a different kind of integration method, which is based on the trapezoidal sum. The quadrature formulas, which were discussed so far, are all based on a single fixed grid of knots $t_0, \ldots, t_n$, at which the function was evaluated. In contrast to this, for the Romberg quadrature, we employ a sequence of grids, and we try to construct a better approximation of the integral from the corresponding trapezoidal sums.

9.4.1 Asymptotic expansion of the trapezoidal sum

Before we can describe the method, we have to analyse in more detail the structure of the approximation error of the trapezoidal rule. For a *stepsize* $h = (b - a)/n$, $n = 1, 2, \ldots$, let $T(h)$ denote the trapezoidal sum

$$T(h) := T^n := h \left(\frac{1}{2}(f(a) + f(b)) + \sum_{i=1}^{n-1} f(a + ih) \right)$$

for the equidistant knots $t_i = a + ih$. The following theorem shows that the trapezoidal sum $T(h)$, when viewed as a function of h, can be developed into an *asymptotic expansion in terms of* h^2.

Theorem 9.16 *Let* $f \in C^{2m+1}[a, b]$, *and let* $h = \frac{b-a}{n}$ *for an* $n \in \mathbf{N} \setminus \{0\}$. *Then the approximation error of the trapezoidal sum* $T(h)$ *has the following asymptotic expansion:*

$$T(h) = \int_a^b f(t)\, dt + \tau_2 h^2 + \tau_4 h^4 + \cdots + \tau_{2m} h^{2m} + R_{2m+2}(h) h^{2m+2} \quad (9.10)$$

with coefficients

$$\tau_{2k} = \frac{B_{2k}}{(2k)!} (f^{(2k-1)}(b) - f^{(2k-1)}(a)) ,$$

where B_{2k} are the Bernoulli numbers, and with the remainder

$$R_{2m+2}(h) = -\int_a^b K_{2m+2}(t,h)f^{(2m)}(t)\,dt\;.$$

The remainder R_{2m+2} is uniformly bounded in h, i.e. there is a constant $C_{2m+2} \geq 0$, which is independent of h, such that

$$|R_{2m+2}(h)| \leq C_{2m+2}|b-a| \quad \text{for all } h = (b-a)/n\;.$$

The proof of this classical theorem is based on the so-called *Euler formula* for the sums, for which we refer to [51]. The h^2-expansion also comes up in the general context of solving initial value problems for ordinary differential equations, however, in a much simpler way (see [44]).

Remark 9.17 The functions K_{2m+2}, which occur in the remainder, are closely related to the Bernoulli functions B_{2m+2}.

For periodic functions f of period $b-a$, all τ_{2k} vanish, i.e. the entire error stays in the remainder. In this case, no improvement can be achieved with the Romberg integration, as we shall describe below. In fact, in this case, the simple trapezoidal rule already yields the result of the trigonometric interpolation (compare Section 7.2).

For large k, the Bernoulli numbers B_{2k} satisfy

$$B_{2k} \approx (2k)!\;,$$

so that the series (9.10) in general also *diverges* with $m \to \infty$ for analytic functions $f \in C^\omega[a,b]$, i.e. in contrast to the series expansion, which we know from analysis (Taylor and Fourier series) in Theorem 9.16 the function is developed into a *divergent series*. At first, this does not seem to make a lot of sense; in practice, however, the finite partial sums can often be used to compute the function value with sufficient accuracy, even though the corresponding series diverges. In order to illustrate the fact that such an expansion into a divergent series can be numerically useful, we consider the following example (compare [51]).

Example 9.18 Let $f(h)$ be a function with an asymptotic expansion in h, such that for all $h \in \mathbf{R}$ and $n \in \mathbf{N}$

$$f(h) = \sum_{k=0}^n (-1)^k k! \cdot h^k + \theta(-1)^{n+1}(n+1)!\,h^{n+1} \quad \text{for a } 0 < \theta = \theta(h) < 1\;.$$

The series $\sum(-1)^k k!\,h^k$ diverges for all $h \neq 0$. When considering the sequence of partial sums

$$s_n(h) := \sum_{k=0}^n (-1)^k k!\,h^k$$

for small h, $0 \neq h \ll 1$, then it appears at first that they converge, because the terms $(-1)^k k! \, h^k$ of the series at first decay very much. However, starting from a certain index, the factor $k!$ dominates, the terms get arbitrarily large, and the sequence of partial sums diverges. Because

$$|f(h) - s_n(h)| = |(-1)^{n+1}\theta(n+1)! \, h^{n+1}| < |(n+1)! \, h^{n+1}| \, ,$$

the error made by approximating f by s_n is always smaller than the first term which we drop. In order to determine $f(h)$ to an (absolute) precision of tol, we have to find a n so that

$$|(n+1)! \, h^{n+1}| < \text{tol} \, .$$

We actually obtain $f(10^{-3})$ with ten precise decimal positions for $n = 3$ by

$$f(10^{-3}) \approx s_3(10^{-3}) = 1 - 10^{-3} + 2 \cdot 10^{-6} - 6 \cdot 10^{-9} \, .$$

Because of their "almost convergent" behavior, Legendre called such series also *semi-convergent*. Euler made his life easier by using the same notation for series, whether they converged or not.

9.4.2 Idea of extrapolation

In Section 9.2, we have approximated the integral

$$I(f) := \int_a^b f(t) \, dt$$

by the trapezoidal sum

$$T(h) = T^{(n)} = h\left(\frac{1}{2}(f(a) + f(b)) + \sum_{i=1}^{n-1} f(a + ih)\right) \quad \text{with} \ \ h = \frac{b-a}{n},$$

where the quality of the approximation depends on the stepsize h. For $h \to 0$, the expression $T(h)$ converges to the integral $I(f)$; more precisely we should say "for $n \to \infty$ and $h = (b-a)/n$", because $T(h)$ is only defined for discrete values $h = (b-a)/n$, $n = 1, 2, \ldots$. We then write

$$\lim_{h \to 0} T(h) := \lim_{n \to \infty} T^{(n)} = I(f) \, . \tag{9.11}$$

In order to illustrate the basic idea of the Romberg quadrature, we first start by assuming that we have computed $T(h)$ for two stepsizes

$$h_i := \frac{b-a}{n_i} \, , \quad i = 1, 2 \, ,$$

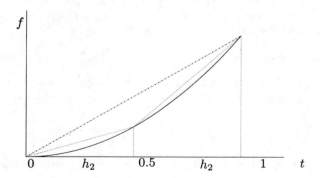

Figure 9.3: Trapezoidal sums $T(h_1)$ and $T(h_2)$ for $f(t) = t^2$

and we consider the most simple function $f(t) = t^2$ over the unit interval $[a, b] = [0, 1]$ and $n_i = i$ (see Figure 9.3). Because the second derivative $f^{(2)}$ is constant 2, we have $R_4(h) = 0$, and therefore

$$T(h) = I(f) + \tau_2 h^2 + R_4(h)h^4 = I(f) + \tau_2 h^2 . \qquad (9.12)$$

We can determine the coefficient τ_2 from the trapezoidal sums

$$
\begin{aligned}
T(h_1) &= T(1) = I(f) + \tau_2 h_1^2 = 1/2 \text{ and} \\
T(h_2) &= T(1/2) = I(f) + \tau_2 h_2^2 = 3/8 :
\end{aligned}
$$

$$\tau_2 = \frac{T(h_1) - T(h_2)}{h_1^2 - h_2^2} = \frac{1}{6} .$$

Again inserted into (9.12), we obtain the integral

$$I(f) = T(h_1) - \tau_2 h_1^2 = T(h_1) - \frac{T(h_2) - T(h_1)}{h_2^2 - h_1^2} h_1^2 = \frac{1}{3} \qquad (9.13)$$

from the two trapezoidal sums (see Figure 9.4). We can also explain formula (9.13) as follows: Based on the asymptotic expansion of the trapezoidal rule, we determine the integration polynomial in h^2 for the points $(h_1^2, T(h_1))$ and $(h_2^2, T(h_2)$,

$$P(T(h) \,|\, h_1^2, h_2^2)(h^2) = T(h_1) + \frac{T(h_2) - T(h_1)}{h_2^2 - h_1^2}(h^2 - h_1^2) ,$$

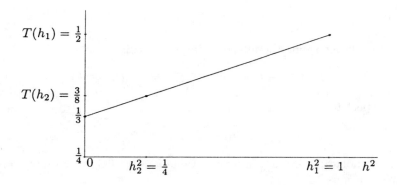

Figure 9.4: (Linear) extrapolation

and we extrapolate for $h^2 = 0$, i.e.

$$P(T(h) \,|\, h_1^2, h_2^2)(0) = T(h_1) - \frac{T(h_2) - T(h_1)}{h_2^2 - h_1^2} h_1^2 \; .$$

We expect that the extrapolated value $P(T(h) \,|\, h_1^2, h_2^2)(0)$ for $h^2 = 0$ to be a better approximation of $I(f)$.

This basic idea carries over to higher orders in a natural way, by respective repeated evaluation of $T(h)$ for successively smaller $h = h_i$. In particular, it can always be used in a more general context, if a method allows for an asymptotic expansion of the approximation error. This leads to the class of so-called *extrapolation methods*. In order to present the chain of reasoning, we start with a method $T(h)$, which, depending on of a "stepsize' h, computes the wanted value τ_0. Here we allow that $T(h)$ is only defined for discrete values h (see above). In addition, we require that the method converges to τ_0 for $h \to 0$, i.e.

$$\lim_{h \to 0} T(h) = \tau_0 \; . \tag{9.14}$$

Definition 9.19 The method $T(h)$ for the computation of τ_0 has an *asymptotic expansion in h^p up to the order pm*, if there exist constants $\tau_p, \tau_{2p}, \ldots,$ $\tau_{mp} \in \mathbf{R}$ such that

$$T(h) = \tau_0 + \tau_p h^p + \tau_{2p} h^{2p} + \cdots + \tau_{mp} h^{mp} + O(h^{(m+1)p}) \quad \text{for} \quad h \to 0 \; . \tag{9.15}$$

Remark 9.20 According to Theorem 9.16, the trapezoidal rule has an asymptotic expansion in h^2 up to order $2m$ for functions $f \in C^{2m+1}[a, b]$.

Once we have computed $T(h)$ for k different stepsizes

$$h = h_{i-k+1}, \ldots, h_i ,$$

then we can determine the interpolating polynomial in h^p

$$P_{ik}(h^p) = P(h^p; h^p_{i-k+1}, \ldots, h^p_i) \in \mathbf{P}_{k-1}(h^p)$$

with respect to the nodes

$$(h^p_{i-k+1}, T(h_{i-k+1})), \ldots, (h^p_i, T(h_i)) ,$$

and extrapolate by evaluating at the value $h = 0$. This way we obtain the approximations T_{ik},

$$T_{ik} := P_{ik}(0) \quad \text{for } 1 \leq k \leq i ,$$

of τ_0. According to Section 7.1.2, we of course employ the algorithm of Aitken and Neville for the computation of T_{ik}. The recurrence relation (7.4) is then transformed into the present situation as follows.

$$
\begin{aligned}
T_{i1} &:= T(h_i) \quad \text{for } i = 1, 2, \ldots \\
T_{ik} &:= T_{i,k-1} + \frac{T_{i,k-1} - T_{i-1,k-1}}{\left(\dfrac{h_{i-k+1}}{h_i}\right)^p - 1} \quad \text{for } 2 \leq k \leq i
\end{aligned}
\qquad (9.16)
$$

The Neville scheme turns into the so-called *extrapolation table*.

$$
\begin{array}{ccccccc}
T_{11} & & & & & & \\
& \searrow & & & & & \\
T_{21} & \rightarrow & T_{22} & & & & \\
\vdots & & & \ddots & & & \\
T_{k-1,1} & \rightarrow & \cdots & \rightarrow & T_{k-1,k-1} & & \\
& \searrow & & \searrow & & \searrow & \\
T_{k1} & \rightarrow & \cdots & \rightarrow & T_{k,k-1} & \rightarrow & T_{kk}
\end{array}
$$

Remark 9.21 In accordance with [19], we start to count with 1 in the extrapolation methods. As we shall see below, this leads to a more convenient connection between the order of approximation T_{kk} and the number k of computed values $T(h_1)$ to $T(h_k)$.

If we denote the approximation error of the approximations T_{ik} of τ_0, which were obtained by extrapolation, by

$$\varepsilon_{ik} := |T_{ik} - \tau_0| \quad \text{for} \ \ 1 \le k \le i,$$

then we can arrange these accordingly into an *error table*.

$$
\begin{array}{ccccccc}
\varepsilon_{11} & & & & & & \\
& \searrow & & & & & \\
\varepsilon_{21} & \rightarrow & \varepsilon_{22} & & & & \\
\vdots & & & \ddots & & & \\
\varepsilon_{k-1,1} & \rightarrow & \cdots & \rightarrow & \varepsilon_{k-1,k-1} & & \\
& \searrow & & \searrow & & \searrow & \\
\varepsilon_{k1} & \rightarrow & \cdots & \rightarrow & \varepsilon_{k,k-1} & \rightarrow & \varepsilon_{kk}
\end{array}
\qquad (9.17)
$$

The following theorem, which goes back to BULIRSCH [9], provides information about the behavior of these errors.

Theorem 9.22 *Let $T(h)$ be a method with an asymptotic expansion (9.15) in h^p up to the order pm, and let $h_1, \ldots, h_m$ be distinct stepsizes. Then the approximation error ε_{ik} of the extrapolation values T_{ik} satisfies*

$$\varepsilon_{ik} \ \dot{=} \ |\tau_{kp}| \underbrace{h^p_{i-k+1} \cdots h^p_i}_{k \text{ factors}} \quad \text{for} \ 1 \le k \le i \le m \text{ and } h_j \le h \to 0 \,.$$

More precisely,

$$\varepsilon_{ik} \ = \ |\tau_{kp}| \ h^p_{i-k+1} \cdots h^p_i + \sum_{j=i-k+1}^{i} O(h_j^{(k+1)p}) \quad \text{for } h_j \le h \to 0 \,.$$

This theorem says that, essentially, for each column of the extrapolation table, we can gain p orders. However, since we have to deal with asymptotic expansions, and not with series expansions, this viewpoint is too optimistic. The high order is of little use, if the remainders of the asymptotic expansion, which are hidden behind the $O(h_j^{(k+1)p})$ become very large. For the proof of the theorem we use the following auxiliary statement.

Lemma 9.23 *The Lagrange functions $L_0, \ldots, L_n$ with respect to the nodes $t_0, \ldots, t_n$ satisfy*

$$\sum_{j=0}^{n} L_j(0)t_j^m = \begin{cases} 1 & \text{for } m = 0 \\ 0 & \text{for } 1 \le m \le n \\ (-1)^n t_0 \cdots t_n & \text{for } m = n+1 \end{cases}.$$

Proof. For $0 \le m \le n$, $P(t) = t^m$ is the interpolating polynomial for the points (t_j, t_j^m) f—r $j = 0, \ldots, n$, and therefore

$$P(t) = t^m = \sum_{j=0}^{n} L_j(t)P(t_j) = \sum_{j=0}^{n} L_j(t)t_j^m.$$

If we set $t = 0$, then the statement follows in the first two cases. In the case $m = n+1$, we consider the polynomial

$$Q(t) := t^{n+1} - \sum_{j=0}^{n} L_j(t)t_j^{n+1}.$$

This is a polynomial of degree $n+1$ with leading coefficient 1 and the roots $t_0, \ldots, t_n$, thus

$$Q(t) = (t - t_0) \cdots (t - t_n),$$

and, in particular,

$$\sum_{j=0}^{n} L_j(0)t_j^{n+1} = -Q(0) = (-1)^n t_0 \cdots t_n.$$

$\square$

We now turn to the proof of Theorem 9.22.

Proof. Since $T(h)$ possesses an asymptotic expansion in h^p up to order pm, we have, for $1 \le k \le m$,

$$T_{j1} = T(h_j) = \tau_0 + \tau_p h_j^p + \cdots + \tau_{(k+1)p} h_j^{kp} + O(h_j^{(k+1)p}). \qquad (9.18)$$

It is sufficient, to show the statement for $i = k$ (the case $i \ne k$ follows then by shifting the indices of the h_j). Thus let

$$P_{kk} = P(h^p; h_1^p, \ldots, h_k^p)$$

be the interpolating polynomial in h^p with respect to the nodes $(h_j^p, T(h_j))$ for $j = 1, \ldots, k$, and let $L_1(h^p), \ldots, L_k(h^p)$ be the Lagrange polynomials for the nodes $h_1^p, \ldots, h_k^p$. Then

$$P_{kk}(h^p) = \sum_{j=1}^{k} L_j(h^p) T_{j1} .$$

Thus, according to (9.18) and Lemma 9.23,

$$
\begin{aligned}
T_{kk} &= P_{kk}(0) = \sum_{j=1}^{k} L_j(0) T_{j1} \\
&= \sum_{j=1}^{k} L_j(0) \left[\tau_0 + \tau_p h_j^p + \cdots + \tau_{kp} h_j^{kp} + O(h_j^{(k+1)p}) \right] \\
&= \tau_0 + \tau_{kp}(-1)^{k-1} h_1^p \cdots \cdots h_k^p + \sum_{j=1}^{k} O(h_j^{(k+1)p}) ,
\end{aligned}
$$

and therefore

$$\varepsilon_{kk} = |T_{kk} - \tau_0| = |\tau_{kp}| \, h_1^p \cdots \cdots h_k^0 + \sum_{j=1}^{k} O(h_j^{(k+1)p}) .$$

$\square$

The theory which we presented so far suggests, for a method $T(h)$ with an asymptotic expansion, the following *extrapolation algorithm*. We start with a *basic stepsize* H, and form the stepsizes h_i by dividing H, hence $h_i = H/n_i$ with $n_i \in \mathbf{N}$.

Algorithm 9.24 *Extrapolation method*

1. For a basic stepsize H, choose a sequence of stepsizes $h_1, h_2 \ldots$ with $h_j = H/n_j$, $n_{j+1} > n_j$ and set $i := 1$

2. Determine $T_{i1} = T(h_i)$

3. Compute T_{ik} for $k = 2, \ldots, i$ from Neville's scheme

$$T_{ik} = T_{i,k-1} + \frac{T_{i,k-1} - T_{i-1,k-1}}{\left(\dfrac{n_i}{n_{i-k+1}} \right)^p - 1}$$

4. If T_{ii} is precise enough, or if i is too large, then end the algorithm. Otherwise increase i by 1 and go back to 2.

This rough description leaves many questions open. It is obviously not clear, what is meant by "T_{ii} is precise enough" or "i is too large". It is also not clear, how the stepsizes h_j should be chosen. We shall discuss this in more detail in the context of the Romberg quadrature in the next sections.

9.4.3   Details of the algorithm

As we have seen in the above, the trapezoidal sum is a method with an asymptotic expansion in h^2, where the order up to which we can expand, depends on the smoothness of the integrand. We can therefore apply everything which we described in the previous section to the trapezoidal rule, and we obtain as an extrapolation method the classical *Romberg quadrature*, which was introduced by Romberg.

The *cost* A_i for the computation of T_{ii} can essentially be measured by the number of ncessary function evaluations of f,

A_i := number of f-evaluations used for the computation of T_{ii}.

These numbers of course depend on the chosen sequence $n_1, n_2, \ldots$. To each increasing sequence

$$\mathcal{F} = \{n_1, n_2, \ldots\}, \quad n_i \in \mathbf{N} \setminus \{0\}$$

we assign therefore the corresponding sequence of costs

$$\mathcal{A} = \{A_1, A_2, \ldots .\}$$

For the so-called *Romberg sequence*

$$\mathcal{F}_R = \{1, 2, 4, 8, 16, \ldots\}, \quad n_i = 2^{i-1},$$

we obtain

$$\mathcal{A}_R = \{2, 3, 5, 9, 17, \ldots\}, \quad A_i = n_i + 1 .$$

For this sequence, the recursive evaluation of the trapezoidal sums is particularly simple (see Figure 9.5). For $h = H/n$ we obtain

$$
\begin{aligned}
T(h/2) &= \frac{h}{4}\left(f(a) + 2\sum_{i=1}^{2n-1} f(a + ih/2) + f(b)\right) \\
&= \underbrace{\frac{h}{4}\left(f(a) + 2\sum_{k=1}^{n-1} f(a + kh) + f(b)\right)}_{= T(h)/2} + \frac{h}{2}\sum_{k=1}^{n} f\left(a + \frac{2k-1}{2}h\right)
\end{aligned}
$$

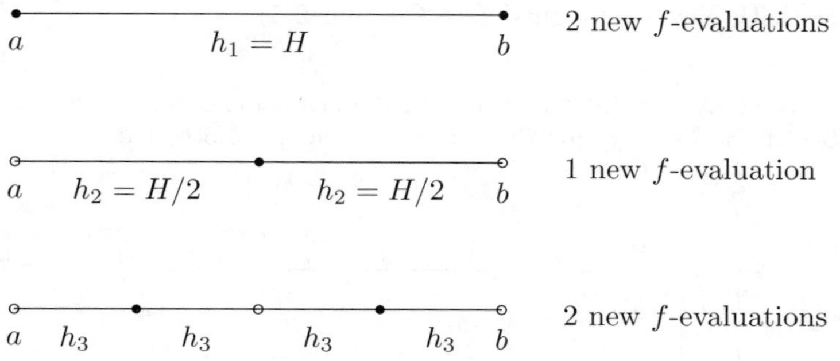

Figure 9.5: Computation of the trapezoidal sums for the Romberg sequence

and therefore for the Romberg sequence $h_i = H/2^{i-1}$

$$T_{i1} = \frac{1}{2}T_{i-1,1} + h_i \sum_{k=1}^{n_{i-1}} f(a + (2k-1)h_i) .$$

By representing the extrapolation values T_{ii} of the Romberg quadrature as a quadrature formula with weights λ_j,

$$T_{ii} = H \sum_{j=1}^{A_i} \lambda_j f_j,$$

where f_j is the j-th computed function value, it can be shown (see Exercise 9.6) that for the Romberg sequence $\mathcal{F}_R$ only positive weights λ_j result. Concerning the cost, Romberg [66] already proposed, in his original paper, an even more favorable sequence, which is now called the *Bulirsch sequence*:

$$\mathcal{F}_B := \{1, 2, 3, 4, 6, 8, 12, 16, 24, \ldots\}, \quad n_i = \begin{cases} 2^{k-1} & \text{if } i = 2k \\ 3 \cdot 2^k & \text{if } i = 2k + 1 \\ 1 & \text{if } i = 1 \end{cases} .$$

The corresponding sequence of costs is

$$\mathcal{A}_B = \{2, 3, 5, 7, 9, 13, 17, \ldots\}.$$

However, the sequence $\mathcal{F}_B$ has the disadvantage that the corresponding quadrature formula may now also contain negative weights (see Table 9.4).

In the context of a method of *variable* order, this property is not so dramatic, because such methods can, in critical regions, be switched back to lower order (with positive weights) (see Chapter 9.5).

Table 9.4: Weights λ_j for the diagonal and sub-diagonal elements of the extrapolation table at the knots t_j for the Bulirsch sequence (Note that $\lambda_j = \lambda_{n-j}$, $t_j = t_{n-j}$)

	0	$\frac{1}{6}$	$\frac{1}{4}$	$\frac{1}{3}$	$\frac{1}{2}$	$\sum |\lambda_i|$
$T_{1,1}$	$\frac{1}{2}$					1
$T_{2,1}$	$\frac{1}{4}$				$\frac{1}{2}$	1
$T_{2,2}$	$\frac{1}{6}$				$\frac{2}{3}$	1
$T_{3,2}$	$\frac{1}{10}$		$\frac{3}{5}$		$\frac{2}{3}$	1
$T_{3,3}$	$\frac{11}{120}$		$\frac{27}{40}$		$-\frac{8}{15}$	2.067
$T_{4,3}$	$\frac{13}{210}$	$\frac{16}{21}$	$-\frac{27}{35}$	$\frac{94}{105}$		4.086
$T_{4,4}$	$\frac{151}{2520}$	$\frac{256}{315}$	$-\frac{243}{280}$	$\frac{104}{105}$		4.471

Remark 9.25 In the solution of initial value problems for ordinary differential equations, the most simple *harmonic sequence*

$$\mathcal{F}_H = \{1, 2, 3, \ldots\}, \quad n_i = i$$

occurs. In our context of quadrature, it is however substantially less favorable than the Romberg sequence, because for one, the cost is larger,

$$\mathcal{A}_H = \{2, 3, 5, 7, 11, 13, 19, 23, 29, \ldots\},$$

and, for the other, the trapezoidal sums T_{i1} cannot be computed recursively.

The computation of the extrapolation table is carried out by *rows*. We stop, if sufficiently many digits "stand" or if no improvement of convergence is noticed.

Example 9.26 *Needle impulse*: Computation of the integral

$$\int_{-1}^{1} \frac{dt}{10^{-4} + x^2} \tag{9.19}$$

(compare Figure 9.10) with relative precision tol $= 10^{-8}$. The values T_{kk} of the extrapolation table are given in Table 9.5.

Table 9.5: Romberg quadrature for the needle impulse $f(t) = 1/(10^{-4} + x^2)$, ε_{kk}: relative precision, A_k: cost in terms of f-evaluations

k	T_{kk}	ε_{kk}	A_k
1	1.999800	$9.9 \cdot 10^{-1}$	2
2	13333.999933	$4.2 \cdot 10^{1}$	3
3	2672.664361	$7.6 \cdot 10^{0}$	5
4	1551.888793	$4.0 \cdot 10^{0}$	9
5	792.293096	$1.5 \cdot 10^{0}$	17
6	441.756664	$4.2 \cdot 10^{-1}$	33
7	307.642217	$1.4 \cdot 10^{-2}$	65
8	293.006708	$6.1 \cdot 10^{-2}$	129
9	309.850398	$7.4 \cdot 10^{-3}$	257
10	312.382805	$7.2 \cdot 10^{-4}$	513
11	312.160140	$2.6 \cdot 10^{-6}$	1025
12	312.159253	$2.5 \cdot 10^{-7}$	2049
13	312.159332	$1.1 \cdot 10^{-9}$	4097

9.5 Adaptive Romberg Quadrature

So far, we have taken the entire length of the interval $H = b - a$ as the basic stepsize. If we consider the needle impulse (9.19), then it strikes us that in this example, the essential contributions to the entire integral often come only from one or a more smaller sub-intervals. If we start with a basic stepsize $H = b - a$, then all regions of the basic interval $[a, b]$ are equal, and we apply the same method everywhere. This cannot be the best way to integrate functions. We should rather partition the integration interval so that we can choose in each sub-region a method which is tailor-made to the function and which thus, with as little effort as possible determines the integral with a given relative precision. Such methods, which control themselves in the course of the computation by adapting to the problem at hand, are called *adaptive methods*. Their essential advantage consists of the fact that a large class of problems can be handled with the same program, without the user having to make adaptions, i.e. without having to invest a-priori knowledge about the problem into the method. The program itself tries to adapt itself to the problem. In order to achieve this, the intermediate results which are computed in the course of the algorithm are constantly checked. This serves two purposes: On one hand, the algorithm can thus automatically choose an optimal solution strategy with respect to cost, and thus solve the posed problem *effectively*. On the other hand, this ensures that the program works more safely, and hopefully does not produce fictitious solutions, which, in reality do not have a lot to do with the posed problem. It should also be a goal that the program can recognize it's own limitations, and for instance determines that a prescribed precision cannot be achieved. This adaptive concept can in general only be carried out, if a reasonable estimate for the occurring approximation error is available, that can be computed at relatively little cost.

9.5.1   Principle of adaptivity

In quadrature the problem is more precisely formulated: Approximate the integral $I = \int_a^b f(t)\, dt$ up to a given relative precision tol, i.e. compute an approximation $\hat{I}$ of I so that

$$|\hat{I} - I| \le |I|\,\text{tol} . \tag{9.20}$$

Since we do not know I, we replace (9.20) by the requirement

$$|\hat{I} - I| \le I_{\text{scal}}\text{tol} \tag{9.21}$$

where I_{scal} ("scal" for "scaling") should be of the order of $|I|$. This value is either given by the user together with tol, or is obtained from the first approximations.

Whereas the classical Romberg quadrature merely adapts the order of the method in order to achieve a desired precision, in the adaptive Romberg quadrature the basic stepsize H is also adapted. There are two principal possibilities to attack the problem: The *initial value method* (in this section) and the *boundary value method* (two sections later).

The following considerations are based on [19] and [20]. We start with the formulation of the quadrature problem as an *initial value problem*

$$y'(t) = f(t), \quad y(a) = 0, \quad y(b) = \int_a^b f(t)\, dt$$

and try to compute the integral successively from the left to the right (see Figure 9.6). Here we partition the basic interval in suitable sub-intervals $[t_i, t_{i+1}]$ of length $H_i := t_{i+1} - t_i$, which are adapted to the function f and apply the Romberg quadrature to the thus obtained sub-problems

$$I_i = \int_{t_i}^{t_{i+1}} f(t)\, dt$$

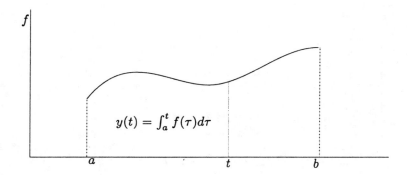

Figure 9.6: Quadrature as an initial value problem

up to a certain degree q_i.

Remark 9.27 In this initial value approach, however, the symmetry

$$I_a^b(f) = -I_b^a(f)$$

is destroyed, because we distinguish one direction (from left to right). This will not be the case in the boundary value approach (see Section 9.7).

Numerous questions arise from this first superficial description. Which stepsizes should the algorithm choose? Up to which order should the Romberg quadrature be carried out with respect to a sub-interval? How can the result be (locally) verified?

In the following, we construct a method, which, starting with input stepsize H and order q, computes the integral of the next sub-interval, and which makes proposals for the stepsize $\tilde{H}$ and the order $\tilde{q}$ of the following step. Should the computation of the sub-integral $I_t^{t+H}(f)$ not be possible with the required precision and with the given order, then the method is to choose a new order q and/or a reduced stepsize H. The method should stop after "too frequent" reduction of H.

9.5.2   Estimation of the approximation error

In order to realize the adaptive concept, which was sketched in the above, with respect to order and basic stepsize, we need a reasonable and cheap technique to estimate the approximation error. This is particularly simple

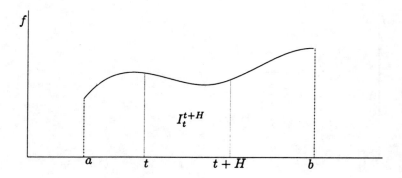

Figure 9.7: One step of the adaptive Romberg quadrature

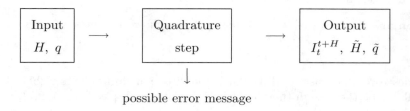

possible error message

Figure 9.8: Schematic description of a quadrature step

for extrapolation methods, since extrapolation tables include an array of different approximations. The estimation technique will be based on this table of approximations. In general, a quantity $\bar{\varepsilon}$ is called an *estimator* for an unapproachable approximation error ε, shortly $\bar{\varepsilon} = [\varepsilon]$, if $\bar{\varepsilon}$ can be estimated by ε both from above and below, i.e. if there exist constants $\kappa_1 \leq 1 \leq \kappa_2$, such that

$$\kappa_1 \varepsilon \;\leq\; \bar{\varepsilon} \;\leq\; \kappa_2 \varepsilon. \qquad (9.22)$$

The construction of an effective error estimator is one of the most difficult problems in the development of an adaptive algorithm. A common method consists of comparing an approximation of low order with one of higher order. All error estimators, which we shall encounter in this book, are based on this construction principle.

In our case, locally, with respect to a sub-interval $[t, t+H]$ of basic stepsize H, the approximation quality is described by the already earlier introduced error table (9.17) of the ε_{ik}. According to Theorem 9.22, we already know that

$$\varepsilon_{ik} = \left| T_{ik} - \int_t^{t+H} f(t)\, dt \right| \doteq |\tau_{2k}|\, h_{i-k+1}^2 \cdots h_i^2 \quad \text{for} \quad h_j \leq H \to 0 \,. \quad (9.23)$$

Besides, for $H \to 0$, the coefficients of the asymptotic expansion of the trapezoidal rule can be estimated by

$$\tau_{2k} = \frac{B_{2k}}{(2k)!}(f^{(2k-1)}(t+H) - f^{(2k-1)}(t)) \doteq \underbrace{\frac{B_{2k}}{(2k)!}f^{(2k)}(t)\, H}_{=:\, \bar{\tau}_{2k}} \,. \quad (9.24)$$

Here the constants $\bar{\tau}_{2k}$ depend on the integrand f, and thus on the problem. Inserted into (9.23), it follows that

$$\varepsilon_{ik} \doteq |\bar{\tau}_{2k}|\, h_{i-k+1}^2 \cdots h_i^2 \cdot H \doteq |\bar{\tau}_{2k}|\, \gamma_{ik}\, H^{2k+1} \quad \text{for} \quad H \to 0 \,, \quad (9.25)$$

where

$$\gamma_{ik} := (n_{i-k+1} \cdots n_i)^{-2} \,.$$

The order $2k + 1$ with respect to H only depends on the column index k of the extrapolation table. In particular, for two consecutive errors within column k, and independent from the problem, we have

$$\frac{\varepsilon_{i+1,k}}{\varepsilon_{ik}} \doteq \frac{\gamma_{i+1,k}}{\gamma_{ik}} = \left(\frac{n_{i-k+1}}{n_{i+1}}\right)^2 \ll 1 \,.$$

In other words, independent of the problem and independent of H, within each column k, the approximation errors decrease very fast with increasing row index i.

$$\varepsilon_{i+1,k} \ll \varepsilon_{ik} \quad (9.26)$$

For the relation between the columns, we need a further assumption, namely that *higher approximation orders yield smaller approximation errors*, i.e. for $1 \leq k < i$ we have

$$\varepsilon_{i,k+1} \ll \varepsilon_{ik} \,. \quad (9.27)$$

This assumption is plausible; however, it is not imperative: It is surely true for "sufficiently small" stepsizes H; however, it has to be verified in the program for each concrete H in a suitable way. In Section 9.5.3, we shall present one possibility to test whether our model coincides with the given situation. If we denote these relations in the error table by an arrow,

$$\varepsilon \to \delta \quad :\Longleftrightarrow \quad \varepsilon \ll \delta \,, \quad (9.28)$$

then, under the assumptions (9.26) and (9.27), the following picture emerges.

$$\varepsilon_{11}$$
$$\uparrow$$
$$\varepsilon_{21} \quad \leftarrow \quad \varepsilon_{22}$$
$$\uparrow \qquad\qquad \uparrow$$
$$\varepsilon_{31} \quad \leftarrow \quad \varepsilon_{32} \quad \leftarrow \quad \varepsilon_{33}$$
$$\uparrow \qquad\qquad \uparrow \qquad\qquad \uparrow$$
$$\vdots \qquad\quad\; \vdots \qquad\quad\; \vdots$$

The most precise approximation inside the row k is thus the diagonal element T_{kk}. It would therefore be ideal if we could estimate the error ε_{kk}. We however get into a dilemma here. In order to estimate the error of T_{kk}, we need a more precise approximation $\hat{I}$ of the integral, e.g. $T_{k+1,k}$. With such an approximation at hand, it would be possible to estimate ε_{kk} e.g. by

$$\bar{\varepsilon}_{kk} \; := \; |T_{k+1,k} - T_{k,k}| = [\varepsilon_{kk}] \; .$$

However, once we have computed $T_{k+1,k}$, then we can also directly produce the (better) approximation $T_{k+1,k+1}$. However, we do not have an estimate of the error $\varepsilon_{k+1,k+1}$, unless we again compute a more precise approximation. We escape this dilemma by the insight that the second-best solution may also be useful. The second best approximation, which is at our disposition, including the row k, is the *sub-diagonal element* $T_{k,k-1}$. The approximation error $\varepsilon_{k,k-1}$ can be estimated from known data up to this row as follows.

Lemma 9.28 *Under the assumption (9.27),*

$$\bar{\varepsilon}_{k,k-1} \; := \; |T_{k,k-1} - T_{kk}| = [\varepsilon_{k,k-1}]$$

is an error estimator for $\varepsilon_{k,k-1}$ in the sense of (9.22).

Proof. Let $I := \int_t^{t+H} f(t)\,dt$. Then

$$\bar{\varepsilon}_{k,k-1} = |(T_{k,k-1} - I) - (T_{k,k} - I)| \le \varepsilon_{k,k-1} + \varepsilon_{kk} \; ,$$

and, with the assumption (9.27),

$$\bar{\varepsilon}_{k,k-1} = |(T_{k,k-1} - I) - (T_{k,k} - I)| \ge \varepsilon_{k,k-1} - \varepsilon_{kk} \; ,$$

and therefore

$$\underbrace{\left(1 - \frac{\varepsilon_{kk}}{\varepsilon_{k,k-1}}\right)}_{\ll 1}\varepsilon_{k,k-1} \; \leq \; \bar{\varepsilon}_{k,k-1} \; \leq \; \underbrace{\left(1 + \frac{\varepsilon_{kk}}{\varepsilon_{k,k-1}}\right)}_{\ll 1}\varepsilon_{k,k-1} \; .$$

□

In order to simplify notation in the following, we assume that $I_{\mathrm{scal}} = 1$. Then we replace the termination criterion

$$|I - \hat{I}| \; \leq \mathrm{tol}$$

by the condition

$$\bar{\varepsilon}_{k,k-1} \leq \rho\mathrm{tol}, \tag{9.29}$$

which can be verified in the algorithm, where $\rho < 1$ (typically $\rho := 0.25$) is a safety factor. The diagonal element T_{kk} is thus accepted as a solution, iff the termination condition (9.29) is satisfied. This condition is also called the *sub-diagonal error criterion*.

Remark 9.29 For a long time, there had been intense discussions of whether one is allowed to consider the "best" solution (here, the diagonal element T_{kk}) as an approximation, even though only the error of the "second-best" solution (here, the sub-diagonal element $T_{k,k-1}$) was estimated. In fact, the employed error estimator is useful for the solution $T_{k,k-1}$ only if T_{kk} is the "best" solution, so that it would be inconsistent to skip this more precise solution.

9.5.3   Derivation of the algorithm

It is the goal of the adaptive algorithm, to approximate the integral up to a desired precision at as little cost as possible. At our disposition are two parameters of the algorithm for the adaption to the problem, namely the basic stepsize H, and the order $p = 2k$, i.e. the maximal used column k of the extrapolation table. We first start with a method $\hat{I}$ with a given fixed order p, i.e.

$$\varepsilon = \varepsilon(t, H) = \left|\hat{I}_t^{t+H}(f) - \int_t^{t+H} f(\tau)\, d\tau\right| \doteq \gamma(t)H^{p+1} \tag{9.30}$$

with a number $\gamma(t)$, which depends on the left boundary t and on the problem. With the data ε and H of the current integration step, we can estimate $\gamma(t)$ by

$$\gamma(t) \doteq \varepsilon H^{-(p+1)} \; . \tag{9.31}$$

Suppose that $\tilde{H}$ is the stepsize for which we would have achieved the desired precision

$$\text{tol} = \varepsilon(t, \tilde{H}) \doteq \gamma(t)\tilde{H}^{(p+1)} . \tag{9.32}$$

By employing (9.31), we can compute an a posteriori approximation $\tilde{H}$ of ε and H, because

$$\tilde{H} \doteq \sqrt[p+1]{\frac{\text{tol}}{\varepsilon}} H . \tag{9.33}$$

We also call $\tilde{H}$ the *optimal stepsize* in the sense of (9.32). Should $\tilde{H}$ be much smaller than the stepsize H that we actually used, then this indicates that H was too large, and that we have possibly jumped over a critical region (e.g. a small peak). In this case we should repeat the integration step with $\tilde{H}$ as basic stepsize. Otherwise we can use $\tilde{H}$ as the *recommended stepsize* for the next integration step, because for sufficiently smooth integrands f and small basic stepsizes H, the number $\gamma(t)$ will change only little over the integration interval $[t, t + H]$, i.e.

$$\gamma(t) \doteq \gamma(t + H) \quad \text{for} \quad H \to 0 . \tag{9.34}$$

This implies

$$\varepsilon(t + H, \tilde{H}) \doteq \gamma(t + H)\tilde{H}^{p+1} \doteq \gamma(t)\tilde{H}^{p+1} \doteq \text{tol} ,$$

so that we may assume that $\tilde{H}$ is also the optimal stepsize for the next step. The algorithm of course has to verify the Assumption (9.27) as well as the Assumption (9.34), and possibly correct the stepsize.

So far, we have only considered a fixed order p, and we have determined an optimal stepsize $\tilde{H}$ for this order. The Romberg quadrature, as an extrapolation method, produces an entire series of approximations T_{ik} of various orders $p = 2k$ for the column index k, which could also vary, where the approximation error satisfies:

$$\varepsilon_{ik} \doteq |\bar{\tau}_{2k}| \, \gamma_{ik} H^{2k+1} \quad \text{for} \quad f \in C^{2k}[t, t + H] .$$

In the course of the investigation in the previous section, we worked out the error estimator $\varepsilon_{k,k-1}$ for the sub-diagonal approximation $T_{k,k-1}$ of the order $p = 2k - 2$. If we now replace the unknown error $\bar{\varepsilon} = \varepsilon_{k,k-1}$ in (9.33) by $\bar{\varepsilon}_{k,k-1}$, then we obtain the suggested stepsize

$$\tilde{H}_k := \sqrt[2k-1]{\frac{\rho \, \text{tol}}{\bar{\varepsilon}_{k,k-1}}} H ,$$

where we again have introduced the safety factor $\rho < 1$, in order to match a possible variation of $\gamma(t)$ in the interval $[t, t + h]$, compare Assumption (9.34).

For each column k, the above rule now suggests some $\tilde{H}_k$ for the next stepsize. Still missing is a criterion to choose the best from the pairs

$$(k, \tilde{H}_k) = (\text{column, stepsize suggestion}) \quad \text{for } k = 1, \ldots, k_{\max} = q/2 .$$

Remember, it was our goal to minimize cost; we could also say to get as far as possible with the least possible effort. We achieve this by minimizing the *work per unit step*

$$W_k := \frac{A_k}{\tilde{H}_k} ,$$

where the A_k are the work numbers which correspond to the sub-partitioning sequence $\mathcal{F}$. The column $\tilde{k}$ with

$$W_{\tilde{k}} = \min_{k=1,\ldots,k_{\max}} W_k$$

is in this sense "optimal", as well as the order $\tilde{q} = 2\tilde{k}$. By taking into account the data of the current integration step, we have thus found a suitable order $\tilde{q}$ and basic stepsize $\tilde{H} = \tilde{H}_{\tilde{k}}$.

By gathering all considerations of the previous sections, we arrive at the following algorithm for one step of the adaptive Romberg quadrature.

Algorithm 9.30 *One step of the adaptive Romberg quadrature.* As input, the procedure *step* gets the beginning t of the interval under consideration, the suggested column k and the stepsize H. Besides the possible success notice *done*, the output is: the corresponding values $\tilde{t}$, $\tilde{k}$ and $\tilde{H}$ for the next step, as well as the approximation I for the integral $I_t^{\tilde{t}}$ over the interval $[t, \tilde{t}]$.

> **function** $[done, I, \tilde{t}, \tilde{k}, \tilde{H}] = step(t, k, H)$
>> $done :=$ **false**;
>> $i = 1$;
>> **while not** *done* **and** $i < i_{\max}$ **do**
>>> Compute the approximations $T_{11}, \ldots, T_{kk}$ of I_t^{t+H};
>>> **while** $k < k_{\max}$ **and** $\bar{\varepsilon}_{k,k-1} >$ tol **do**
>>>> $k := k + 1$;
>>>> Compute T_{kk};
>>> **end**
>>> Compute $\tilde{H}_1, \ldots, \tilde{H}_k$ and $W_1, \ldots, W_k$;
>>> Choose $\tilde{k} \le k$ with minimal work $W_{\tilde{k}}$;
>>> $\tilde{H} := H_{\tilde{k}}$;
>>> **if** $k < k_{\max}$ **then**
>>>> **if** $H > \tilde{H}$ **then**
>>>>> $H := \tilde{H}$; (repeat the step for safety)

$$
\begin{aligned}
&\textbf{else}\\
&\qquad \tilde{t} := t + H;\\
&\qquad I = T_{kk};\\
&\qquad \textit{done} := \textbf{true}; \quad \text{(done)}\\
&\quad \textbf{end}\\
&\textbf{end}\\
&\quad i := i + 1;\\
&\textbf{end}
\end{aligned}
$$

When programming this adaptive algorithm, one encounters the unfortunate experience that in the present form it does not (yet) work as one might have hoped. It is precisely the example of the needle function which causes the trouble. The stepsizes nicely contract with decreasing orders towards the center of the needle (as expected). However, after crossing the tip of the needle, the orders still remain low and the stepsizes small. We shall briefly analyze this situation together with two further difficulties of the present algorithm.

Disadvantages of the Algorithm:

1) *Trapping of order*, as explained in the above. Once a low order $q = 2k$ is reached, and the condition $\bar{\varepsilon}_{k,k-1} \leq \text{tol}$ always satisfied, then the algorithm does not test a higher order, even though this might be advantageous. The order remains low and the stepsizes small, as we observed in the case of the integration of the needle.

2) The algorithm notices only rather late, namely only after crossing $k_{\max}$, that a suggested stepsize H was too large and that it does not pass the accuracy criterion ($\bar{\varepsilon}_{k,k-1} \leq \text{tol}$) for any column k.

3) If our assumptions are not satisfied, then the error estimator does not work. It may therefore happen that the algorithm recognizes an incorrect solution as correct and supplies it as an output. This case is referred to as *pseudo convergence*.

In the two last mentioned problems, it would be desirable to recognize early, whether the approximations behave "reasonably", i.e. entirely within our theoretical assumptions. One thus needs a so-called *convergence monitor*. The main difficulty in the construction of such a monitor is that one would have to make algorithmic decisions on the basis of information which is not (yet) available. Because of this, we try to obtain a model which hopefully describes the situation, at least in the statistical average over a large number of problems. We may then compare the actually obtained values with this model. Here we only want to discuss briefly one such possibility, which is based on Shannon's information theory. For details we refer to [19]. In this model, the quadrature algorithm is interpreted as an *encoding device*. It

Figure 9.9: Quadrature algorithm as encoding device

converts the information, which is obtained by evaluating the function, into information about the integral (see the schematic Figure 9.9). The amount of information on the input side, the *input entropy* $E_{ik}^{(in)}$, is measured by the number of f-evaluations which are required for the computation of T_{ik}. This assumes that no redundant f-evaluations are considered, i.e. that all digits of f are independent of each other. Since the values $T_{i-k+1,1}, \ldots, T_{i,1}$ are needed as input for the computation of T_{ik}, we obtain

$$E_{ik}^{(in)} = \alpha(A_i - A_{i-k} + 1)$$

with a constant $\alpha > 0$. The amount of information on the output side, the *output entropy* $E_{ik}^{(out)}$, can be characterized by the number of correct binary digits of the approximation T_{ik}. This leads to

$$E_{ik}^{(out)} = \log_2\left(\frac{1}{\varepsilon_{ik}}\right).$$

We now assume that our information channel works with a constant *noise factor* $0 < \beta \leq 1$,

$$E_{ik}^{(out)} = \beta E_{ik}^{(in)}, \tag{9.35}$$

i.e., that input and output entropy are proportional to each other. (If $\beta = 1$, then the channel is *noise free*; no information gets lost.) In our case this means that

$$\mathrm{ld}(\varepsilon_{ik}^{-1}) = c(A_i - A_{i-k} + 1) \tag{9.36}$$

with $c := \alpha\beta$. In order to determine the proportionality factor c, we need a pair of input and output entropies. In the above we required that for a given column k, the sub-diagonal error $\varepsilon_{k,k-1}$ is equal to the required precision tol, hence $\varepsilon_{k,k-1} = $ tol. By inserting this relation into (9.36), we conclude that

$$-\mathrm{ld}\,\varepsilon_{k,k-1} = -\mathrm{ld}\,\mathrm{tol} = c(A_k - A_1 + 1).$$

Having thus determined c, we can then determine for *all* i, j, which errors ε_{ij} are to be expected by our model. If we denote these errors, which the information theoretic model implies, by $\alpha_{ij}^{(k)}$ (where k is the row, from which we have obtained the proportionality factor), then it follows that

$$\operatorname{ld} \alpha_{ij}^{(k)} = -c(A_i - A_{i-j} + 1) = \frac{A_i - A_{i-j} + 1}{A_k - A_1 + 1} \operatorname{ld} \operatorname{tol} .$$

In a quite elementary manner, we have thus in fact constructed a statistical comparison model, with which we can test the convergence behavior of our algorithm for a concrete problem, namely by comparing the estimated errors $\bar{\varepsilon}_{i,i-1}$ with the values $\alpha_{i,i-1}^{(k)}$ of the convergence model. On the one hand, we thus obtain the desired convergence monitor, on the other hand, we can also estimate, how higher orders would behave. We shall omit further details. They have been worked out for a large class of extrapolation methods in [19]. For the adaptive Romberg quadrature they are implemented in the program TRAPEX (see [20]).

Obtained global precision If we ignore the safety factor ρ, then the above algorithm approximates the integral $I = I(f)$ with a *global precision*

$$|I - \hat{I}| \leq I_{\mathrm{scal}} \cdot m \cdot \operatorname{tol},$$

where m is the number of basic steps, which were obtained in the adaptive quadrature (a-posteriori error estimate). The chosen strategy obviously leads to a *uniform distribution of the local discretization errors*. This principle is also important for considerably more general adaptive discretization methods (compare Section 9.7). If one wants to prescribe a global discretization error which is independent of m,

$$|I - \hat{I}| \leq I_{\mathrm{scal}} \cdot E ,$$

then, following a suggestion by DE BOOR [13], in the derivation of the order and stepsize control, the precision tol is to be replaced by

$$\operatorname{tol} \longrightarrow \frac{H}{b - a} E .$$

This leads to smaller changes of the order and stepsize control, but also to additional difficulties and a less robust algorithm.

Example 9.31 We again return to the example of the needle impulse, whose treatment with the classical Romberg quadrature we have documented in Section 9.4.3, Table 9.5: We needed 4097 f-calls for an achieved precision of approximately 10^{-9}. In the adaptive Romberg quadrature, for a required precision of tol $= 10^{-9}$, we only need 321 f-evaluations (for 27 basic steps) with an achieved precision of $\varepsilon = 1.4 \cdot 10^{-9}$. The automatic subdivision into basic steps by the program TRAPEX is given in Figure 9.10.

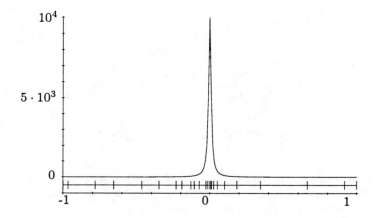

Figure 9.10: Automatic subdivision into basic steps by the program TRAPEX

9.6 Hard Integration Problems

Of course, even the adaptive Romberg quadrature cannot solve all problems of numerical quadrature. In this chapter we shall discuss some difficulties.

Discontinuous Integrands A common problem in numerical quadrature are discontinuities of the integrand f or its derivatives (see Figure 9.11). Such integrands occur e.g., when a physical-technical system is described by different models in different regions, which do not quite fit at the interface positions. If the jumps are known, then one should subdivide the integration interval at these positions and solve the arising sub-problems separately. Otherwise the quadrature program reacts quite differently. Without any further preparation, a *non-adaptive* quadrature program yields incorrect results or does not converge. The jumps cannot be localized. An *adaptive* quadrature program, such as the adaptive Romberg method, freezes at the jumps. Thus the jumps can be localized and treated separately.

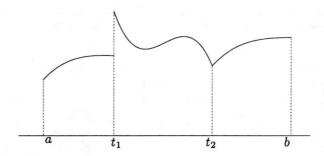

Figure 9.11: Jump of f at t_1, jump of f' at t_2

Needle impulses We have considered this problem repeatedly in the above. It has to be noted, however that in principle, every quadrature program will fail, if only the peaks are small enough (compare Exercise 9.8). On the other hand, such integrands are pretty common: just think of the spectrum of a star, whose entire radiation is to be computed. If the positions of the peaks are known, then one should subdivide the interval in a suitable way, and again compute the sub-integrals separately. Otherwise, there only remains the hope that the adaptive quadrature program does not "overlook" them.

Highly Oscillatory Integrands We have already noted in Section 9.1 that highly oscillatory integrands are ill-conditioned from the relative error viewpoint. As an example, we have plotted the function

$$f(t) = \cos(te^{4t^2})$$

for $t \in [-1, 1]$ in Figure 9.12. The numerical quadrature is powerless against such integrands. They have to be prepared by analytical averaging over sub-intervals (pre-clarification of the structure of the inflection points of the integrand).

Weakly Singular Integrands A function f, which is integrable over the interval $[a, b]$, is called *weakly singular*, if one of its derivatives $f^{(k)}$ in $[a, b]$ does not exist. As an example, take the functions $f(t) = t^\alpha g(t)$, where $g \in C^\infty[0, T]$ is an arbitrarily smooth function and $\alpha > -1$.

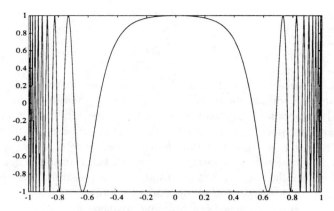

Figure 9.12: Highly oscillatory integrand $f(t) = \cos(te^{4t^2})$

Example 9.32 As an example, we consider the integral

$$\int_{t=0}^{\pi} \underbrace{\sqrt{t}\cos t}_{f(t)}\ dt\ .$$

The derivative $f'(t) = (\cos t)/(2\sqrt{t}) - \sqrt{t}\sin t$ has a pole at 0.

In the case of weakly singular integrands, adaptive quadrature programs usually tend to contract stepsize and order, and they therefore tend to be extremely slow. Non-adaptive quadrature algorithms however do not get slow, but usually false. The singularities can often be removed via a substitution.

Example 9.33 In the above example, we obtain after the substitution $s = \sqrt{t}$:

$$\int_{t=0}^{\pi} \sqrt{t}\ \cos t\ dt = 2\int_{s=0}^{\sqrt{\pi}} s^2\ \cos s^2\ ds\ .$$

This however becomes inefficient, if the substitution leads to functions which are difficult to evaluate (e.g. t^α instead of $\sqrt{t}$ for a $0 < \alpha < 1$). A second possibility consists of the recursive computation of the integral under consideration (Miller-trick), which we shall not consider here (see Exercise 9.10).

Parameter Dependent Integrands Often the integrand f depends on an additional parameter $\lambda \in \mathbf{R}$:

$$f(t, \lambda), \quad \lambda \in \mathbf{R}\ \text{parameter}\ .$$

We thus have to solve an entire family of problems

$$I(\lambda) := \int_a^b f(t, \lambda) \, dt \, .$$

The most important class of examples for such parameter dependent integrals is the multi-dimensional quadrature. Usually the integrand is differentiable with respect to λ, and so is therefore the integral $I(\lambda)$. Of course, one hopes that the approximation $\hat{I}(\lambda)$ inherits this property. Unfortunately, however, it turns out that just our best methods, the adaptive quadrature methods, do not have this property — in contrast to the simple non-adaptive quadrature formulas. There are essentially three possibilities to rescue the adaptive approach for parameter dependent problems.

The first possibility consists of carrying out the quadrature for one parameter value, storing away the employed orders and stepsizes, and use them again for all other parameter values. This is also called *freezing of orders and stepsizes*. This can only be successful, if the integrand qualitatively does not change too much in dependence of the parameter.

If, however, a peak varies with the parameter, and if this dependence is known, then one can employ *parameter dependent grids*. One transforms the integral in dependence of λ in such a way that the integrand stays the same qualitatively (the movement of the peak is e.g. counter-balanced) or, in dependence of λ, one shifts the adaptive partitioning of the integration interval.

The last possibility requires a lot of insight into the respective problem. We choose a fixed *grid adapted to the respective problem* and integrate over this grid with a *fixed quadrature formula* (Newton-Cotes or Gauss-Christoffel). In order to do this, the qualitative properties of the integrand need to be largely known, of course.

Discrete Integrands In many applications, the integrand is not given as a function f, but only in the form of finitely many discrete points

$$(t_i, f_i), \quad i = 0, \ldots, N,$$

(e.g. nuclear spin spectrum, digitalized measurement data). The simplest and best way to deal with this situation consists of forming the trapezoidal sum over these points. The trapezoidal sum has the advantage that errors in measurement data often get averaged out in the computation of the integral with an equidistant grid. If the measurement errors δf_i have the expectation 0, i.e. $\sum_{i=0}^N \delta f_i = 0$, then this is also the case for the induced error of the trapezoidal sum. This property holds only for methods where all weights are equal, and it is not true any more for methods of higher order. In the next section, we shall consider an effective method for the solution of such problems.

9.7 Adaptive Multigrid Quadrature

In the present section, we consider a second approach to the adaptive quadrature, which rests on ideas that were originally developed for the solution of considerably more complicated problems in partial differential equations (see [3]). This so-called *multi-grid approach*, or, more generally *multilevel approach*, is based on the *boundary value approach*. In the adaptive Romberg quadrature, which is based on the initial value approach, we traversed the interval in an arbitrarily chosen direction. According to the problem, we then subdivided it into sub-intervals, and then integrated over these with *local fine grids* (of the Romberg quadrature). In contrast to this, the multi-grid quadrature starts with the entire basic interval or with a coarse initial partitioning Δ^0, and step by step generates a sequence of finer *global* subdivisions Δ^i of the interval and more precise approximations $I(\Delta^i)$ of the integral. Here the grids are only refined at places, where it is necessary for the required precision, i.e. the qualitative behavior of the integrand becomes visible in the refinement of the grids. The knots condense where "a lot happens". In order to achieve this, one requires two things: a *local error estimator* and *local refinement rules*.

The local error estimator is typically realized by a comparison of methods of lower and higher order, as we have seen in Section 9.5.3 in the sub-diagonal error criterion. Here the theory of the respective approximation method enters. In the definition of refinement rules, aspects of the *data structures* play the decisive role. Thus, in fact part of the complexity of the mathematical problem is transferred to the computer science side (in the form of more complex data structures).

9.7.1 Local error estimation and refinement rules

As an example of a multi-grid quadrature, we here present a particular method where the *trapezoidal rule* (locally linear) is used as the method of lower order, and where *Simpson's rule* (locally quadratic) is used as the method of higher order. As a refinement method, we shall restrict ourselves to the *local bisection* of an interval.

We start with a sub-interval $[t_l, t_r] \subset [a, b]$ (l: left, r: right). Since we need three knots for Simpson's rule, we add the center $t_m := (t_l + t_r)/2$ and describe the interval by the triple $J := (t_l, t_m, t_r)$. The length of the interval is denoted by $h = h(J) := t_r - t_l$. A *grid* Δ is a family $\Delta = \{J_i\}$ of such intervals, which together form a partition of the original interval $[a, b]$.

By $T(J)$ and $S(J)$, we denote the results of the trapezoidal rule, as applied to the sub-intervals $[t_l, t_m]$ and $[t_m, t_r]$, and Simpson's rule with respect to the knots t_l, t_m and t_r. The formulas are given in Figure 9.13. Observe that Simpson's rule is obtained from the Romberg quadrature as $S(J) = T_{22}(J)$ (see Exercise 9.6). For sufficiently smooth functions f, $T(J)$ and $S(J)$ are

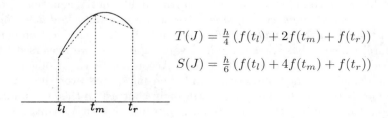

$$T(J) = \tfrac{h}{4}\left(f(t_l) + 2f(t_m) + f(t_r)\right)$$

$$S(J) = \tfrac{h}{6}\left(f(t_l) + 4f(t_m) + f(t_r)\right)$$

Figure 9.13: Trapezoidal and Simpson's rule for an interval $J := (t_l, t_m, t_r)$

approximations of the integral $\int_{t_l}^{t_r} f(t)\, dt$ of order $O(h^3)$ respectively $O(h^5)$. The error of Simpson's approximation therefore satisfies

$$\varepsilon(J) := \left|\int_{t_l}^{t_r} f(t)\, dt - S(J)\right| = O(h^5)\ .$$

By summation over all sub-intervals $J \in \Delta$, we obtain the approximation of the entire integral $\int_a^b f(t)\, dt$:

$$T(\Delta) = \sum_{J \in \Delta} T(J) \quad\text{and}\quad S(\Delta) = \sum_{J \in \Delta} S(J)\ .$$

As in the Romberg quadrature, we assume (at first not checked) that the method of higher order, the Simpson's rule, is locally better, i.e.

$$\left|T(J) - \int_{t_l}^{t_r} f(t)\, dt\right| \gg \left|S(J) - \int_{t_l}^{t_r} f(t)\, dt\right|\ . \tag{9.37}$$

Under this assumption, the sub-diagonal estimator of the local approximation error is

$$\bar{\varepsilon}(J) := |T(J) - S(J)| = [\varepsilon(J)]\ ,$$

and we can use the Simpson result as a better approximation.

In the construction of local refinement rules, we essentially follow an abstract suggestion by BABUŠKA and RHEINBOLDT [3], which they made in the more general context of boundary value problems for partial differential equations. The sub-intervals, which are obtained, when bisecting an interval $J := (t_l, t_m, t_r)$, are denoted by J_l and J_r, where

$$J_l := \left(t_l, \frac{t_l + t_m}{2}, t_m \right) \quad \text{and} \quad J_r := \left(t_m, \frac{t_r + t_m}{2}, t_r \right) .$$

When refining twice, we thus obtain the *binary tree*, which is displayed in Figure 9.14. If J is obtained by refinement, then we denote the starting

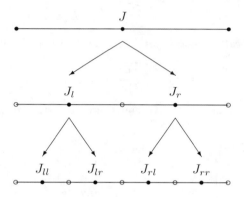

Figure 9.14: Twofold refinement of the interval $J := (t_l, t_m, t_r)$

interval of the last step by J^-, i.e. $J_r^- = J_l^- = J$.

The principle, according to which we want to proceed when listing the refinement rules, is the *equidistribution of the local discretization error* (compare Section 9.5.3). This means that the grid Δ is to be refined, such that the estimated local approximation errors of the refined grid Δ^+ are approximately equal, i.e.

$$\bar{\varepsilon}(J) \approx \text{const} \quad \text{for all} \quad J \in \Delta^+ . \tag{9.38}$$

For the estimated error of the trapezoidal rule, we make the theoretical assumption (see (9.37))

$$\bar{\varepsilon}(J) \doteq Ch^\gamma \quad \text{where} \quad h = h(J) \tag{9.39}$$

with a local order γ and a local constant C which depends on the problem.

Remark 9.34 The trapezoidal rule actually has the order $\gamma = 3$. Hidden in the constant, however, is the second derivative of the integrand, so that an order $\gamma \leq 3$ more realistically characterizes the method, if we assume that C is locally constant. In the following considerations, the order γ cancels out, so that this does not cause any trouble.

We can thus define a second error estimator $\varepsilon^+(J)$, which yields information about the error $\varepsilon(J_l)$ of the next step, in the case that we partition the interval J. Assumption (9.39) implies

$$\bar{\varepsilon}(J^-) \doteq C(2h)^\gamma = 2^\gamma Ch^\gamma \doteq 2^\gamma \bar{\varepsilon}(J) \quad \text{with} \quad h = h(J) ,$$

thus $2^\gamma \doteq \bar{\varepsilon}(J^-)/\bar{\varepsilon}(J)$, and therefore

$$\varepsilon(J_l) \doteq Ch^\gamma 2^{-\gamma} \doteq \bar{\varepsilon}(J)\bar{\varepsilon}(J)/\bar{\varepsilon}(J^-) .$$

Thus, through *local extrapolation* (see Figure 9.15), we have obtained an error estimator

$$\varepsilon^+(J) := \frac{\bar{\varepsilon}(J)^2}{\bar{\varepsilon}(J^-)} = [\varepsilon(J_l)]$$

for the unknown error $\varepsilon(J_l)$. We can therefore estimate in advance, what

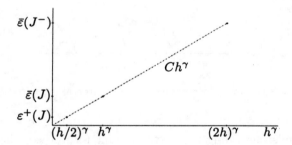

Figure 9.15: Local extrapolation for the error estimator $\varepsilon^+(J)$

effect a refinement of an interval $J \in \Delta$ would have. We only have to fix a *threshold value* for the local errors, above which we refine an interval. In order to do this, we take the maximal local error, which we would obtain from a *global refinement*, i.e. refinement of *all* intervals $J \in \Delta$, and define

$$\kappa(\Delta) := \max_{J \in \Delta} \varepsilon^+(J) . \tag{9.40}$$

In order to illustrate the situation, we plot the estimated errors $\bar{\varepsilon}(J)$ and $\varepsilon^+(J)$ in a histogram (see Figure 9.16). Already before the refinement, the

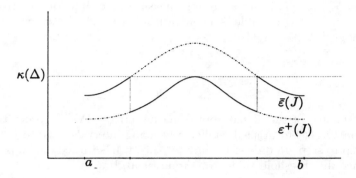

Figure 9.16: Estimated error distributions before and after global and local refinement

error at the right and left boundary is below the maximal local error $\kappa(\Delta)$, which can possibly be achieved by a complete refinement. If we follow the principle of equidistribution of the local error, then we do not have to refine any more near the right and left boundary. Refinement does only pay off in the middle region. We thus arrive at the following *refinement rule*: Refine only intervals $J \in \Delta$ for which

$$\bar{\varepsilon}(J) \geq \kappa(\Delta) .$$

This yields the error distribution which is displayed in Figure 9.16. It is obviously one step closer to the desired equidistribution of the approximation errors.

Remark 9.35 By local refinement, part of the interval J becomes two parts of the arising sub-intervals J_l and J_r:

$$\bar{\varepsilon}(J) \rightarrow \bar{\varepsilon}(J_l) + \bar{\varepsilon}(J_r) \doteq 2\varepsilon^+(J) \doteq 2^{1-\gamma}\bar{\varepsilon}(J)$$

In order that the partitioning in fact yields an improvement, the order γ has to satisfy the condition $\gamma > 1$ locally.

9.7.2 Global error estimation and details of the algorithm

A difficulty of the multi-grid quadrature is the estimation of the *global approximation error*

$$\varepsilon(\Delta) := \left| \int_a^b f(t)\, dt - S(\Delta) \right| .$$

The sum $\sum_{J \in \Delta} \bar{\varepsilon}(J)$ is not a suitable measure, since integration errors may average out. Better suitable is a comparison with the approximation of the previous grid Δ^-. If

$$\varepsilon(\Delta) \ll \varepsilon(\Delta^-) \,, \tag{9.41}$$

then

$$\bar{\varepsilon}(\Delta) := |S(\Delta^-) - S(\Delta)| = [\varepsilon(\Delta)]$$

is an estimator of the global approximation error $\varepsilon(\Delta)$. In order that the condition (9.41) be satisfied, sufficiently many intervals have to be refined from step to step. In order to guarantee this, it has turned out to be useful, to replace the threshold value $\kappa(\Delta)$ from (9.40) by

$$\tilde{\kappa}(\Delta) := \min \left(\max_{J \in \Delta} \varepsilon^+(J), \frac{1}{2} \max_{J \in \Delta} \bar{\varepsilon}(J) \right) \,.$$

The complete algorithm of the adaptive multi-grid quadrature for the computation of $\int_a^b f(t)\,dt$ with a relative precision tol now looks as follows:

Algorithm 9.36 *Simple multi-grid quadrature*

        Choose an initial grid, e.g. $\Delta := \{(a, (a+b)/2, b)\}$;
        for $i = 0$ **to** $i_{\max}$ **do**
            Compute $T(J)$, $S(J)$ and $\bar{\varepsilon}(\Delta)$ for all $J \in \Delta$;
            Compute $\bar{\varepsilon}(\Delta)$;
            if $\bar{\varepsilon}(\Delta) \leq \mathrm{tol}\,|S(J)|$ **then**
                break; (done, solution $S(\Delta)$)
            else
                Compute $\varepsilon^+(J)$ and $\bar{\varepsilon}(J)$ for all $J \in \Delta$;
                Compute $\tilde{\kappa}(\Delta)$;
                Replace all $J \in \Delta$ with $\bar{\varepsilon}(J) \geq \tilde{\kappa}(\Delta)$ by J_l and J_r;
            end
        end

The multi-grid approach obviously leads to a considerably simpler adaptive quadrature algorithm than the adaptive Romberg quadrature. The only difficulty consists in the storage of the grid sequence. However, this difficulty can be mastered fairly easily by employing a structured programming language (such as C or Pascal). In the one-dimensional quadrature, we can store the sequence as a binary tree (as indicated in Figure 9.14). In problems in more than one spatial dimension, the question of data structures often conceals a much higher complexity – consider only the refinement of meshes of tetrahedrons in three spatial dimensions.

Our current presentation of adaptive multi-grid algorithms also overcomes difficulties regarding special integrands, which we discussed in the previous section (Section 9.6). Note the case of discontinuous or weakly singular integrands, where the knots collect automatically at the critical places, without the integrator "freezing" at these places, as would be the case with the initial value approach of Section 9.5.3. The refinement strategy still works locally for these places, because it was derived for general local orders $\gamma > 1$.

Example 9.37 *Needle impulse.* We have repeatedly used this Example (9.19) for illustration purposes (for classical and adaptive Romberg quadrature). The result for the tolerance tol $= 10^{-3}$ is presented in Figure 9.17 in the case that the initial grid Δ^0 already contains the tip of the needle. The final grid Δ^9 has 61 knots, thus requiring 121 f-evaluations. The estimated total error amounts to $\bar{\varepsilon}(\Delta^9) = 2.4 \cdot 10^{-4}$, with an actual error of $\varepsilon(\Delta^9) = 2.1 \cdot 10^{-4}$. When shifting the interval asymmetrically, i.e. when the tip of the needle is not represented within the initial grid, then this does not deteriorate the result.

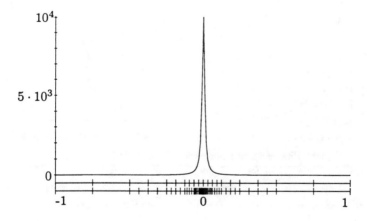

Figure 9.17: Adapted grid for the needle impulse $f(t) = 1/(10^{-4} + x^2)$ of the fifth and ninth step for the tolerance 10^{-3}.

The program can also be adapted to *discrete integrands* (it was originally developed just for this case in [83] as the so-called SUMMATOR). Here one only has to consider the case that there is no value available at a bisecting point. As always, we do the next best, and this time in the literal sense, by taking the nearest given point, which is next to the bisectional point, and thus modify the bisection slightly. Once the required precision is achieved, then for discrete integrands, and for reasons which we discussed in Section 9.6, we take the *trapezoidal sum* as the best approximation.

Example 9.38 *Summation of the harmonic series.* The sum

$$S = \sum_{j=1}^{n} \frac{1}{j} \quad \text{for} \quad n = 10^7$$

is to be computed, thus a sum of 10^7 terms. For a required precision of tol $= 10^{-2}$ resp. tol $= 10^{-4}$, the program SUMMATOR only needs 47 resp. 129 terms! In order to illustrate this, the automatically chosen grids are presented in Figure 9.18. (Observe the logarithmic scale.)

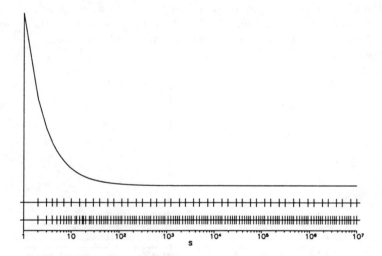

Figure 9.18: Summation of the harmonic series with the program SUMMATOR.

We finally return again to the parameter dependent case. Also in the adaptive multi-grid approach, it causes difficulties, which are similar to the ones of the other approaches, i.e., it requires additional considerations, like the ones which were described in Section 9.6. Overall, however, even for considerably more general boundary value problems (e.g. for partial differential equations), the adaptive multi-grid concept, turned out to be simple, fast and reliable.

9.8   Exercises

Exercise 9.1 Let

$$\lambda_{in} = \frac{1}{n} \int_0^n \prod_{\substack{j=0 \\ j \neq i}}^n \frac{s-j}{i-j} \, ds$$

be the constants of the Newton-Cotes formulas. Show that

$$\lambda_{n-i,n} = \lambda_{in} \quad \text{and} \quad \sum_{i=0}^n \lambda_{in} = 1 \, .$$

Exercise 9.2 Compute an approximation of the integral

$$\int_0^2 x^2 e^{3x} \, dx$$

by five-fold application of Simpson's rule and using equidistant knots.

Exercise 9.3 The n-th Newton-Cotes formula is constructed, such that it yields the exact integral value for polynomials of degree $\leq n$. Show that for *even* n, even polynomials of degree $n+1$ are integrated exactly. Hint: Employ the remainder formula of the polynomial interpolation, and use the symmetries with respect to $(a+b)/2$.

Exercise 9.4 (VAN VELDHUIZEN) Compute the period

$$P = 2 \int_{-1}^1 \frac{f(t)}{\sqrt{1-t^2}} dt \, ,$$

of the radial movement of a satellite in an orbit in the equatorial plane (apogeum height 492 km) under the influence of the flattening of the Earth. Here

a) $f(t) = \dfrac{1}{\sqrt{2g(r(t))}}$, $r(t) = 1 + (1+t)\dfrac{p_2 - 1}{2}$,

b) $g(x) = 2\omega^2(1 - p_1/x)$,

c) $\omega^2 = \dfrac{1}{4}(1-\varepsilon) + \dfrac{k}{6}$, $p_1 = \dfrac{k}{6\omega^2 p_2}$,

with the constants $\varepsilon = 0.5$ (elliptic eccentricity of the satellite orbit), $p_2 = 2.991\,924\,505\,9286$ and $k = 1.4 \cdot 10^{-3}$ (constant, which describes the influence of the Earth flattening). Write a program, which computes the integral

$$I_n := \frac{\pi}{n+1} \sum_{i=0}^n f(\tau_{in}) \, , \quad \tau_{in} := \cos\left(\frac{2i+1}{n+1} \cdot \frac{\pi}{2}\right) \, , \quad n = 3, 4, \ldots 7$$

by using the Gauss-Chebyshev quadrature. For verification purposes: $P = 2 \cdot 4.4395\ 41318\ 6376$.

Exercise 9.5 Derive the formula

$$T_{ik} = T_{i,k-1} + \frac{T_{i,k-1} - T_{i-1,k-1}}{\left(\frac{n_i}{n_{i-k+1}}\right)^2 - 1}$$

for the extrapolation table from the one of the Aitken-Neville algorithm.

Exercise 9.6 Every element T_{ik} in the extrapolation table of the extrapolated trapezoidal rule can be considered as a quadrature formula. Show that when using the Romberg sequence and polynomial extrapolation, the following results hold:

a) T_{22} is equal to the value, which is obtained by applying the Simpson rule; T_{33} corresponds to the Milne rule.

b) T_{ik}, $i > k$ is obtained by 2^{i-k}-fold application of the quadrature formula which belongs to T_{kk} to suitably chosen sub-intervals.

c) For every T_{ik}, the weights of the corresponding quadrature formula are positive.

Hint: By using b), show that the weights $\lambda_{i,n}$ of the quadrature formula, which corresponds to T_{kk}, satisfies

$$\max_i \lambda_{i,n} \leq 4^k \cdot \min_i \lambda_{i,n} .$$

Exercise 9.7 Program the Romberg algorithm by only using *one single* vector of length n (here each value of the table needs to be intermediately stored).

Exercise 9.8 Experiment with an adaptive Romberg quadrature program, test it with the "needle function"

$$I(n) := \int_{-1}^{1} \frac{2^{-n}}{4^{-n} + t^2}\ dt, \quad \text{for } n = 1, 2, \ldots$$

and determine the n for which your program yields the value zero for a given precision of $eps = 10^{-3}$.

Exercise 9.9 Consider the computation of the integrals

$$I_n = \int_{1}^{2} (\ln x)^n dx, \quad n = 1, 2, \ldots$$

a) Show that the I_n satisfy the recurrence relation

$$I_n = 2(\ln 2)^n - nI_{n-1}, \quad n \geq 2 \qquad (R)$$

b) Note that $I_1 = 0.3863\ldots$ and $I_7 = 0.0124\ldots$. Investigate the increase of the input error in the computation of

    1) I_7 from I_1 by means of (R) (forward recursion),

    2) I_1 from I_7 by means of (R) (backward recursion).

Here assume a four digit computation. Rounding errors can be neglected.

c) Use (R) as a backward recursion for the computation of I_n from I_{n+k} with starting value

$$I_{n+k} = 0 .$$

How is k to be chosen, in order to compute I_7 accurately up to 8 digits by this method?

Exercise 9.10 Consider integrals of the following form:

$$I_n(\alpha) := \int_0^1 t^{2n+\alpha} \sin(\pi t)dt \quad \text{where } \alpha > -1 \text{ and } n = 0, 1, 2, \ldots$$

a) For I_n, derive the following inhomogeneous two-term recurrence relation:

$$I_n(\alpha) = \frac{1}{\pi} - \frac{(2n+\alpha)(2n+\alpha-1)}{\pi^2} I_{n-1}(\alpha) .$$

b) Show:

$$\lim_{n \to \infty} I_n(\alpha) = 0 \quad \text{and} \quad 0 \leq I_{n+1}(\alpha) \leq I_n(\alpha) \quad \text{for } n \geq 1 .$$

c) Give an informal algorithm for the computation of $I_0(\alpha)$ (compare Chapter 6.2–3). Write a program to compute $I_0(\alpha)$ for a given relative precision.

Exercise 9.11 A definite integral over $[-1, +1]$ is to be computed. Based on the idea of the Gauss-Christoffel quadrature, derive a quadrature formula

$$\int_{-1}^{+1} f(t)dt \approx \mu_0 f(-1) + \mu_n f(1) + \sum_{i=1}^{n-1} \mu_i f(t_i)$$

of as high order as possible with fixed variable knots -1 and $+1$ respectively (Gauss-Lobatto quadrature).

References

[1] M. Abramowitz and I. A. Stegun (Ed.). *Pocketbook of Mathematical Functions*. Verlag Harri Deutsch, Thun, Frankfurt/Main, 1984.

[2] W. E. Arnoldi. The Principle of Minimized Iterations in the Solution of the Matrix Eigenvalue Problem. *Quart. Appl. Math.*, 9:17–29, 1951.

[3] I. Babuška and W. C. Rheinboldt. Error Estimates for Adaptive Finite Element Computations. *SIAM J. Numer. Anal.*, 15:736–754, 1978.

[4] Å. Bjørck. Iterative Refinement of Linear Least Squares Solutions I. *BIT*, 7:257–278, 1967.

[5] H. G. Bock. *Randwertproblemmethoden zur Parameteridentifizierung in Systemen nichtlinearer Differentialgleichungen*. PhD thesis, Universität zu Bonn, 1985.

[6] F. A. Bornemann. *An Adaptive Multilevel Approach to Parabolic Equations in two Dimensions*. PhD thesis, Freie Universität Berlin, 1991.

[7] R. P. Brent. *Algorithms for minimization without derivatives*. Prentice Hall, Eaglewood Cliffs, N. J., 1973.

[8] E. Brieskorn. *Lineare Algebra und analytische Geometrie, Band I und II*. Vieweg, Braunschweig, Wiesbaden, 1983.

[9] R. Bulirsch. Bemerkungen zur Romberg-Integration. *Numer. Math.*, 6:6–16, 1964.

[10] P. Businger and G. H. Golub. Linear least squares solutions by Householder transformations. *Numer. Math.*, 7:269–276, 1965.

[11] J. Cullum and R. Willoughby. *Lanczos Algorithms for Large Symmetric Eigenvalue Computations, Vol I, II*. Birkhäuser, Boston, 1985.

[12] W. Dahmen and A. Kunoth. Mathematische Methoden in der geometrischen Datenverarbeitung (CAGD). Vorlesungsausarbeitung Freie Universität Berlin, 1990.

[13] C. de Boor. An Algorithm for Numerical Quadrature. In J. Rice, editor, *Mathematical Software*. Academic Press, London, 1971.

[14] C. de Boor. *A Practical Guide to Splines*. Springer-Verlag, Berlin, Heidelberg, New York, 1978.

[15] P. Deuflhard. *Newton-Techniques for Highly Nonlinear Problems — Theory and Algorithms*. In preparation.

[16] P. Deuflhard. On Algorithms for the Summation of Certain Special Functions. *Computing*, 17:37–48, 1976.

[17] P. Deuflhard. A Summation Technique for Minimal Solutions of Linear Homogeneous Difference Equations. *Computing*, 18:1–13, 1977.

[18] P. Deuflhard. A Stepsize Control for Continuation Methods and its Special Application to Multiple Shooting Techniques. *Numer. Math.*, 33:115–146, 1979.

[19] P. Deuflhard. Order and Stepsize Control in Extrapolation Methods. *Numer. Math.*, 41:399–422, 1983.

[20] P. Deuflhard and H. J. Bauer. A Note on Romberg Quadrature. Preprint 169, Universität Heidelberg, 1982.

[21] P. Deuflhard, B. Fiedler, and P. Kunkel. Efficient numerical pathfollowing beyond critical points. *SIAM J. Numer. Anal.*, 18:949–987, 1987.

[22] P. Deuflhard, P. Leinen, and H. Yserentant. Concept of an Adaptive Hierarchical Finite Element Code. *Impact of Computing in Science and Engineering*, 1(3):3–35, 1989.

[23] P. Deuflhard and F. A. Potra. A Refined Gauss-Newton-Mysovskii Theorem. Preprint SC 91-4, Konrad-Zuse-Zentrum, Berlin, 1991.

[24] P. Deuflhard and F. A. Potra. Asymptotic Mesh Independence for Newton-Galerkin Methods via a Refined Mysovskii Theorem. *SIAM J. Numer. Anal.*, 29(5):1395–1412, 1992.

[25] P. Deuflhard and W. Sautter. On Rank-Deficient Pseudoinverses. *Lin. Alg. Appl.*, 29:91–111, 1980.

[26] J. J. Dongarra, C. B. Moler, J. R. Bunch, and G. W. Stewart. *LINPACK Users' Guide*. SIAM, Philadelphia, 1979.

[27] T. Ericsson and Å. Ruhe. The Spectral Transformation Lanczos Method for the Numerical Solution of Large Sparse Generalized Symmetric Eigenvalue Problems. *Math. Comp.*, 35:1251–1268, 1980.

[28] G. Farin. *Curves and Surfaces for Computer Aided Geometric Design: A Practical Guide.* Academic Press, New York, 1988.

[29] R. Fletcher. Conjugate Gradient methods. In *Proc. Dundee Biennial Conference on NUmerical Analysis.* Springer Verlag, New York, 1975.

[30] G. Forsythe. *Pitfalls in Computation or Why a Math Book Isn't Enough.*

[31] J. G. F. Francis. The QR-Transformation. A Unitary Analogue to the LR Transformation — Part 1 and 2. *Comp. J.*, 4:265–271 and 332–344, 1961/62.

[32] K. Gatermann and A. Hohmann. Symbolic Exploitation of Symmetry in Numerical Pathfollowing. *Impact of Computing in Science and Engineering*, 3(4):330–365, 1991.

[33] Carl Friedrich Gauß. *Theoria Motus Corporum Coelestium.* 1809.

[34] W. Gautschi. Computational aspects of three-term recurrence relations. *SIAM Rev.*, 9:24–82, 1967.

[35] W. M. Gentleman. Least Squares Computations by Givens Transformations Without Square Roots. *J. Inst. Math. Appl.*, 12:189–197, 1973.

[36] K. Georg. On tracing an implicitly defined curve by quasi-Newton steps and calculating bifurcation by local perturbations. *SIAM J. Sci. Stat. Comput.*, 2(1):35–50, 1981.

[37] A. George and J. W. Liu. *Computer Solution of Large Sparse Positive Definite Systems.* Prentice Hall, Eaglewood Cliffs, N. J., 1981.

[38] G. Goertzel. An Algorithm for the Evaluation of Finite Trigonometric Series. *Amer. Math. Monthly*, 65:34–35, 1958.

[39] G. H. Golub and C. F. van Loan. *Matrix Computations.* The Johns Hopkins University Press, 2 edition, 1989.

[40] G. H. Golub and J. H. Welsch. Calculation of Gauss Quadrature Rules. *Math. Comp.*, 23:221–230, 1969.

[41] I. S. Gradshteyn and I. W. Ryzhik. *Table of Integral Series and Produkts.* Academic Press, New York, San Francisco, London, 1965.

[42] W. Hackbusch. *Multi-Grid Methods and Applications.* Springer Verlag, Berlin, Heidelberg, New York, Tokyo, 1985.

[43] L. A. Hageman and D. M. Young. *Applied Iterative Methods.* Academic Press, Orlando, San Diego, New York et. al., 1981.

[44] E. Hairer, S. P. Nørsett, and G. Wanner. *Solving Ordinary Differential Equations I, Nonstiff Problems.* Springer Verlag, Berlin, Heidelberg, New York, Tokyo, 1987.

[45] C. A. Hall and W. W. Meyer. Optimal error bounds for cubic spline interpolation. *J. Approx. Theory,* 16:105–122, 1976.

[46] S. Hammarling. A Note on Modifications to the Givens Plane Rotations. *J. Inst. Math. Appl.,* 13:215–218, 1974.

[47] M. R. Hestenes and E. Stiefel. Methods of Conjugate Gradients for Solving Linear Systems. *J. Res. Nat. Bur. Stand,* 49:409–436, 1952.

[48] N. J. Higham. How accurate is Gaussian elimination? In *Numerical Analysis, Proc. 13th Biennial Conf., Dundee / UK 1989,* pages 137–154. Pitman Res. Notes Math. Ser. 228, 1990.

[49] A. S. Householder. *The Theory of Matrices in Numerical Analysis.* Blaisdell, New York, 1964.

[50] T. Kato. *Perturbation Theory for Linear Operators.* Springer Verlag, Berlin, Heidelberg, New York, Tokyo, 1984.

[51] K. Knopp. *Theorie und Anwendung der unendlichen Reihen.* Springer Verlag, Berlin, Heidelberg, New York.

[52] V. N. Kublanovskaya. On Some Algorithms for the Solution of the Complete Eigenvalue Problem. *USSR Comp. Math. Phys.,* 3:637–657, 1961.

[53] C. Lanczos. An iteration method for the solution of the eigenvalue problem of linear differential and integral operators. *J. Res. Nat. Bur. Stand,* 45:255–282, 1950.

[54] T. A. Manteuffel. The Tchebychev Iteration for Nonsymmetric Linear Systems. *Numer. Math.,* 28:307–327, 1977.

[55] J. A. Meijerink and H. A. van der Vorst. An iterative solution method for linear systems of which the coefficient matrix is a symmetric M-matrix. *Math. Comp.,* 31:148–162, 1977.

[56] J. Meixner and W. Schäffke. *Mathieusche Funktionen und Sphäroidfunktionen.* Springer Verlag, Berlin, Göttingen, Heidelberg, 1954.

[57] J. C. P. Miller. *Bessel Functions, Part II (Math. Tables X).* Cambridge University Press, 1952.

[58] M. Z. Nashed. *Generalized Inverses and Applications*. Academic Press, New York, 1976.

[59] A. F. Nikiforov and V. B. Uvarov. *Special Functions of Mathematical Physics*. Birkhäuser, Basel, Boston, 1988.

[60] H. Poincaré. *Les Méthodes Nouvelles de la Mécanique Céleste*. Gauthier-Villars, Paris, 1892.

[61] Ch. Pöppe, C. Pelliciari, and K. Bachmann. Computer Analysis of Feulgen Hydrolysis Kinetics. *Histochemistry*, 60:53–60, 1979.

[62] W. Prager and W. Oettli. Compatibility of approximate solutions of linear equations with given error bounds for coefficients and right hand sides. *Numer. Math.*, 6:405–409, 1964.

[63] I. Prigogine and R. Lefever. Symmetry breaking instabilities in dissipative systems II. *J. Chem. Phys.*, 48:1695–1701, 1968.

[64] C. Reinsch. A Note on Trigonometric Interpolation. unpublished manuscript.

[65] J. L. Rigal and J. Gaches. On the compatibility of a given solution with the data of a linear system. *J. Assoc. Comput. Mach.*, 14:543–548, 1967.

[66] W. Romberg. Vereinfachte Numerische Integration. *Det Kongelige Norske Videnskabers Selskabs Forhandlinger*, Bind 28(7), 1955.

[67] R. Sauer and I. Szabó. *Mathematische Hilfsmittel des Ingenieurs*. Springer Verlag, Berlin, Heidelberg, New York.

[68] W. Sautter. *Fehlerfortpflanzung und Rundungsfehler bei der verallgemeinerten Inversion von Matrizen*. PhD thesis, TU München, Fakultät für Allgemeine Wissenschaften, 1971.

[69] R. D. Skeel. Scaling for Numerical Stability in Gaussian Elimination. *J. Assoc. Comput. Mach.*, 26(3):494–526, 1979.

[70] R. D. Skeel. Iterative Refinement Implies Numerical Stability for Gaussian Elimination. *Math. Comp.*, 35(151):817–832, 1980.

[71] P. Sonneveld. A fast Lanczos-type solver for nonsymmetric linear systems. *SIAM J. Sci. Stat. Comput.*, 10:36–52, 1989.

[72] G. W. Stewart. *Introduction to Matrix Computations*. Academic Press, New York, San Francisco, London, 1973.

[73] J. Stoer. Solution of large systems of linear equations by conjugate gradient type methods. In A. Bachem, M. Grötschel, and B. Korte, editors, *Mathematical Programming, the State of the Art*. Springer Verlag, Berlin, Heidelberg, New York, 1983.

[74] G. Szegö. *Orthogonal Polynomials, 4th ed.* American Mathematical Societa, Providence, 1975.

[75] J. Traub and H. Wozniakowski. *General Theory of Optimal Algorithms*. Academic Press, Orlando, San Diego, San Francisco et al., 1980.

[76] L. N. Trefethen and R. S. Schreiber. Average-case stability of gaussian elimination. *SIAM J. Matrix Anal. Appl.*, 11(3):335–360, 1990.

[77] J. W. Tukey and J. W. Cooley. An Algorithm for the Machine Calculation of Complex Fourier Series. *Math. Comp*, 19:197–301, 1965.

[78] J. Varga. *Matrix Iterative Analysis*. Prentice Hall, Eaglewood Cliffs, N. J., 1962.

[79] J. H. Wilkinson. *The Algebraic Eigenvalue Problem*. Oxford University Press, 1965.

[80] J. H. Wilkinson. *Rundungsfehler*. Springer-Verlag, Berlin, Heidelberg, New York, 1969.

[81] J. H. Wilkinson and C. Reinsch. *Handbook for Automatic Computation, Volume II, Linear Algebra*. Springer Verlag, New York, Heidelberg, Berlin, 1971.

[82] G. Wittum. Mehrgitterverfahren. *Spektrum der Wissenschaft*, April 1990, 78–90.

[83] M. Wulkow. Numerical Treatment of Countable Systems of Ordinary Differential Equations. Technical Report TR 90–8, Konrad-Zuse-Zentrum, Berlin, 1990.

[84] J. Xu. Theory of Multilevel Methods. Thesis. AM 48, Pennstate, 1989.

[85] H. Yserentant. On the Multi-Level Splitting of Finite Element Spaces. *Numer. Math.*, 49:379–412, 1986.

Notation

$\mathbf{R}$	real numbers
$\mathbf{C}$	complex numbers
$\mathbf{N}$	natural numbers, $\mathbf{N} = \{0, 1, 2, \ldots\}$
$\mathrm{Mat}_{m,n}(\mathbf{K})$	(m, n)-matrices with coefficients in $\mathbf{K} = \mathbf{R}, \mathbf{C}$
$\mathrm{Mat}_n(\mathbf{K})$	(n, n)-matrices with coefficients in $\mathbf{K} = \mathbf{R}, \mathbf{C}$
$\langle \cdot, \cdot \rangle$	Euclidean scalar product, $\langle x, y \rangle = \sum_{i=1}^{n} x_i y_i$ for $x, y \in \mathbf{R}^n$
$(\cdot, \cdot)$	scalar product
$\|\cdot\|_p$	p-norm in $\mathbf{R}^n$, $\|x\|_p = \left(\sum_{i=1}^{n} |x_i|^p\right)^{1/p}$ for $x \in \mathbf{R}^n$
$B_\rho(x)$	open ball $\{y \mid \|x - y\| < \rho\}$ with radius ρ, and centered at x
$\bar{B}_\rho(x)$	closed ball $\{y \mid \|x - y\| \leq \rho\}$ with radius ρ, and centered at x
$C^n(U)$	n-times continuously differentiable functions $f : U \to \mathbf{R}$ on U.
$C[a, b]$	continuous functions on the interval $[a, b]$
$C^n[a, b]$	n-times continuously differentiable functions $f : U \to \mathbf{R}$ on the interval $[a, b]$
$\mathrm{GL}(n)$	invertible (n, n)-matrices $\{A \in \mathrm{Mat}_n(\mathbf{R}) \mid \det(A) \neq 0\}$
$\mathbf{O}(n)$	orthogonal (n, n)-matrices $\{A \in \mathrm{GL}(n) \mid AA^T = A^T A = I\}$
A^+	pseudoinverse of a matrix $A \in \mathrm{Mat}_{m,n}(\mathbf{R})$, see page 81
$N(A)$, ker A	nullspace of a linear mapping, $N(A) = \{x \mid Ax = 0\}$
$R(A)$, im A	range of a linear mapping, $R(A) = \{Ax \mid x \in \mathbf{R}^n\}$ for $A \in \mathrm{Mat}_{m,n}(\mathbf{R})$
$\mathbf{P}_n$	polynomials of degree less than or equal to n with coefficients in $\mathbf{R}$

$\mathbf{P}_n^d$	generalized polynomials of degree less than or equal to n with coefficients in $\mathbf{R}^d$
O, o	Landau symbol, see page 28
δ_{ij}	Kronecker symbol, $\delta_{ij} = 1$, for $i = j$ and 0 otherwise
$\doteq$	equal in first order approximation, see page 28
δ	input error, see page 27 ff.
κ	condition number, see page 28 ff.
$\kappa(A)$	condition number of a matrix $A \in \mathrm{Mat}_{m,n}(\mathbf{R})$, see page 32 ff.
σ	stability indicator, see page 40 ff.
L_i	Lagrange polynomials, see page 184 ff.
T_k	Chebyshev polynomials, see page 198 ff.
$P(f \mid t_0, \ldots, t_n)$	interpolation polynomial, see page 184 ff.
$[t_0, \ldots, t_n]f$	n-th divided difference, see page 191 ff.
B_i^n	Bernstein polynomials, see page 211 ff.
N_{ik}	B-spline, see page 228 ff.
$[\varepsilon]$	estimator for ε, see page 315 ff.
$\ln$	natural logarithm
$\Re z, \Im z$	real and imaginary part of the complex number z

Index